Sustainable Civil Infrastructures

About this Series

Sustainable Infrastructure impacts our well-being and day-to-day lives. The infrastructures we are building today will shape our lives tomorrow. The complex and diverse nature of the impacts due to weather extremes on transportation and civil infrastructures can be seen in our roadways, bridges, and buildings. Extreme summer temperatures, droughts, flash floods, and rising numbers of freeze-thaw cycles pose challenges for civil infrastructure and can endanger public safety. We constantly hear how civil infrastructures need constant attention, preservation, and upgrading. Such improvements and developments would obviously benefit from our desired book series that provide sustainable engineering materials and designs. The economic impact is huge and much research has been conducted worldwide. The future holds many opportunities, not only for researchers in a given country, but also for the worldwide field engineers who apply and implement these technologies. We believe that no approach can succeed if it does not unite the efforts of various engineering disciplines from all over the world under one umbrella to offer a beacon of modern solutions to the global infrastructure. Experts from the various engineering disciplines around the globe will participate in this series, including: Geotechnical, Geological, Geoscience, Petroleum, Structural, Transportation, Bridge, Infrastructure, Energy, Architectural, Chemical and Materials, and other related Engineering disciplines.

More information about this series at http://www.springer.com/series/15140

Wissem Frikha · Serge Varaksin
Antonio Viana da Fonseca
Editors

Soil Testing, Soil Stability and Ground Improvement

Proceedings of the 1st GeoMEast International Congress and Exhibition, Egypt 2017 on Sustainable Civil Infrastructures

Springer

2017 GEOMEAST

SSIGE
SOIL-STRUCTURE INTERACTION GROUP IN EGYPT

Editors
Wissem Frikha
National Engineering School of Tunis (ENIT)
Tunis
Tunisia

Serge Varaksin
APAGEO
Paris
France

Antonio Viana da Fonseca
Faculdade de Engenharia
Universidade de Porto
Porto
Portugal

ISSN 2366-3405 ISSN 2366-3413 (electronic)
Sustainable Civil Infrastructures
ISBN 978-3-319-61901-9 ISBN 978-3-319-61902-6 (eBook)
DOI 10.1007/978-3-319-61902-6

Library of Congress Control Number: 2017946470

Printed on acid-free paper

This Springer imprint is published by Springer Nature
The registered company is Springer International Publishing AG
The registered company address is: Gewerbestrasse 11, 6330 Cham, Switzerland

Preface

Toward building sustainable and longer civil infrastructures, the engineering community around the globe continues undertaking research and development to improve existing design, modeling, and analytical capability. Such initiatives are also the core mission of the Soil-Structure Interaction Group in Egypt (SSIGE) to contribute to the ongoing research toward sustainable infrastructure. This conference series "GeoMEast International Congress and Exhibition" is one of these initiatives.

Ancient peoples built their structures to withstand the test of time. If we think in the same way, our current projects will be a heritage for future generations. In this context, an urgent need has quickly motivated the SSIGE and its friends around the globe to start a new congress series that can bring together researchers and practitioners to pursue "Sustainable Civil Infrastructures." The GeoMEast 2017 is a unique forum in the Middle East and Africa that transfers from the innovation in research into the practical wisdom to serve directly the practitioners of the industry.

More than eight hundred abstracts were received for the first edition of this conference series "GeoMEast 2017" in response to the Call for Papers. The abstracts were reviewed by the Organizing and Scientific Committees. All papers were reviewed following the same procedure and at the same technical standards of practice of the TRB, ASCE, ICE, ISSMGE, IGS, IAEG, DFI, ISAP, ISCP, ITA, ISHMII, PDCA, IUGS, ICC, and other professional organizations who have supported the technical program of the GeoMEast 2017. All papers received a minimum of two full reviews coordinated by various track chairs and supervised by the volumes editors through the Editorial Manager of the SUCI "Sustainable Civil Infrastructure" book series. As a result, 15 volumes have been formed of the final +320 accepted papers. The authors of the accepted papers have addressed all the comments of the reviewers to the satisfaction of the track chairs, the volumes editors, and the proceedings editor. It is hoped that readers of this proceedings of the GeoMEast 2017 will be stimulated and inspired by the wide range of papers written by a distinguished group of national and international authors.

Publication of this quality of technical papers would not have been possible without the dedication and professionalism of the anonymous papers reviewers. The names of these reviewers appear in the acknowledgment that follows. For any additional reviewers whose names were inadvertently missed, we offer our sincere apologies.

We are thankful to Dr. Hany Farouk Shehata, Dr. Nabil Khelifi, Dr. Khalid M. ElZahaby, Dr. Mohamed F. Shehata, and to all the distinguished volumes editors of the proceedings of the GeoMEast 2017. Appreciation is extended to the authors and track chairs for their significant contributions. Thanks are also extended to Springer for their coordination and enthusiastic support to this conference. The editors acknowledge the assistance of Ms. Janet Sterritt-Brunner, Mr. Arulmurugan Venkatasalam in the final production of the 15 edited volumes "Proceedings of GeoMEast 2017".

Contents

Study of Slope Stability (Open Pit Mining, Algeria)

Mohamed Fredj[1,2(✉)], Abdallah Hafsaoui[1,3], Youcef Khedri[1,4], Riadh Boukarm[2], Radouane Nakache[3], Abderrazak Saadoun[1,2], and Kamel Menacer[1,3]

[1] Laboratory: Mineral Resources and Planning, University of Annaba, Annaba, Algeria
[2] Mining and Geology Department, Faculty of Technology, University of Bejaia, Béjaïa, Algeria
fredj_khiero@yahoo.com
[3] Mining Department, Faculty of Earth Science, University of Annaba, BP 12, Annaba, Algeria
[4] Mechanical Department, Faculty of Technology, University of Annaba, Annaba, Algeria

Abstract. The objective of our work is the study of the influence of the dynamic load due to the use of the explosive in the stability of the slope of an open pit mine, for this purpose we used the method of limit equilibrium to calculate the value of the safety factor through the geotechnical software SLIDE. On the one hand the calculation was carried out under static loading and on the other hand the dynamic loading is taken into account, which allowed us to conclude that the dynamic load influences the safety coefficient. Finally, a proposal to stabilize the slope at through a reprofiling of the bench is presented.

1 Introduction

As the mining method is turning from open pit mining to sag by mining, the vertical height of open-pit mine slope continues to increase, the slope deformation and failure mode has close relationship with regional geological structure characteristics and the rock mass structure feature, the stability is influenced largely by rock mass, joints and fissures, and blasting vibration (Qiao and Li 2004). Particularly, open-pit mine's productive blasting vibration and rainfall has an important effect on the mine slope stability. Productive blasting vibration has an indirect dangerous damage on the high and steep slope. It is mainly caused by blasting seismic wave after the blasting, especially for the joint fissures development. The high and steep rock slope containing a fault or fracture zone is more likely as this (Li et al. 2006). The rock instabilities occur when a number of factors come together, and for one reason or another, the state of precarious stability prevailing until then passes an unstable situation resulting in rocks characterized by displacement their types (sliding, flowing, falling), (Chalhoub 2006).

The stability calculation can be performed in two distinct circumstances; before or after the out break of the movement. In the first case, the slope is apparently stable; the purpose of stability calculation is to define a sliding surface that would have the

W. Frikha et al. (eds.), *Soil Testing, Soil Stability and Ground Improvement*, Sustainable Civil Infrastructures, DOI 10.1007/978-3-319-61902-6_1

best chance of appearing. In other words, the stability calculation allows both to assess the safety margin is on this side of the break, set the site the most threatened areas of instability and examine the influence some work (earthworks, buildings, blasting, earthquakes). On the safety margin which has been defined for the virgin slope. This calculation step, therefore, seems very important because it allows choosing the necessary parameters for the work to ensure the stability of the whole (book and website). Unlike the first case, and when the movement has already appeared on the slope, the stability calculation is performed to assess the safety margin between the current state of the site of the equilibrium state. In this second case, the parameter values necessary to introduce in the calculation are normally given by the investigations already carried out on site, are real values such as: The geometry of the surface of the slide, the Geotechnical characteristics of the massive and the sliding surface, etc. In this case, the stability calculation also of great interest because it can identify the causes of the emergence of the movement and define the work of consolidating, devices necessary to limit the risk (Faure 2000).

According to Schroeder (2010), the different forms of rock slope instabilities depend on factors "internal" own the massive relate to nature, to the morphology of rock masses and the characteristics of discontinuities that effect, and factors "external" can be natural or related to human activities.

2 Geological Setting

Jebel Onk is located in the North East of Algeria, in the eastern end of the mounts of Nemenmcha; last membered of the Saharan Atlas. It is the natural geographic boundary between the high plateaus Constantine and the Saharan area (Cieslinski et al. 1985, 1987); (Prian and Cortiel 1993). The study of the site area HAS year Approximately 250 hectares and belongs to the Sami mining basin Metlaoui than mine phosphate (SW Tunisia) (Mokadem et al. 2014). It contains approximately half the phosphate reserves of Algeria (Estimated at 2 trillion tons of reserves) (Amiour Dass et al. 2013), (Fig. 1).

According to the report made by EREM Cieslinski et al. 1985 to 1987, the deposit Kef-Essnoun is located south of the massif of Jebel Onk 4 km from the town of Bir El-Ater, the administrative headquarters of the region (Daira) is 2 km to the west of Djemi-Djema deposit between Jebel Moufe and Jebel Tarfaye. The coordinates of this deposit are x = y = 951 500 and 168.0 to 170.0.

The pit is excavated as benches with slope angles of 75° to 85°, 30 m in height, and 10 m in width, (Fig. 2). The depth of the base of mine is 70 m. The thickness of the phosphate layer is about 50 m; the barren covering Consists of a series of Y Persian dolomite limestone with flint, locally Overcome by lutetium limestone, Miocene sands, and quaternary alluvium.

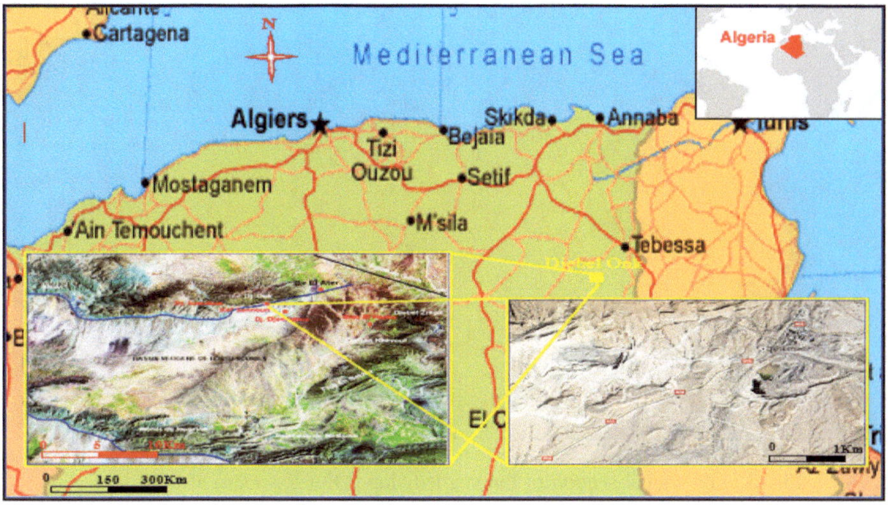

Fig. 1. Location map of the study area.

Fig. 2. Quarry of Kef Essnoun before sliding.

3 Study of Instability

Following the great landslide, which took place on September 8, 2007 in the North East side of the quarry, a large mass of rock broke off from the solid, almost completely filling the pit (estimated volume of 6,000,000 m^3), (Fig. 3).

Fig. 3. Geological section of the studied slope (Kef-Essnoun) after sliding.

According to the geological section and the position of the sliding surface at the lamination joint (marl interface phosphate), the slip is classified as a plan year. The expert reports on this shift has resulted in the fact that the probable causes of this hazard are attributed to some geological factors and operating (Benaissa 2003).

3.1 Geological Factors

- Presence of marl interface phosphates;
- The presence of a near-vertical recovery of the layer of marl, which allows the mass detached, is following a true inclined plane;
- The highly tectonized character of the massif.

3.2 Operating Factors

- Very straightened Front, almost vertically from the figure to a height of 70 m;
- Reduced dimensions of the berm;
- Presence at the foot of the front layer of marl, which played a key role in the sliding of the whole;
- The cumulative effect of shooting with explosives on the platform.

After this shift, operation has been shifted to the west side and southwest.

However, cracking was observed in February 2013, about 250 m upstream of the platform, suggesting a beginning of a new potential slip, (Fig. 4). The observed cracks are located along the outcrops of marl located stratigraphically below the level of the exploited phosphates. The most likely cause of this cracking is related much more to the vibrations generated by the rock blasting, fire with explosives that a landslide primer.

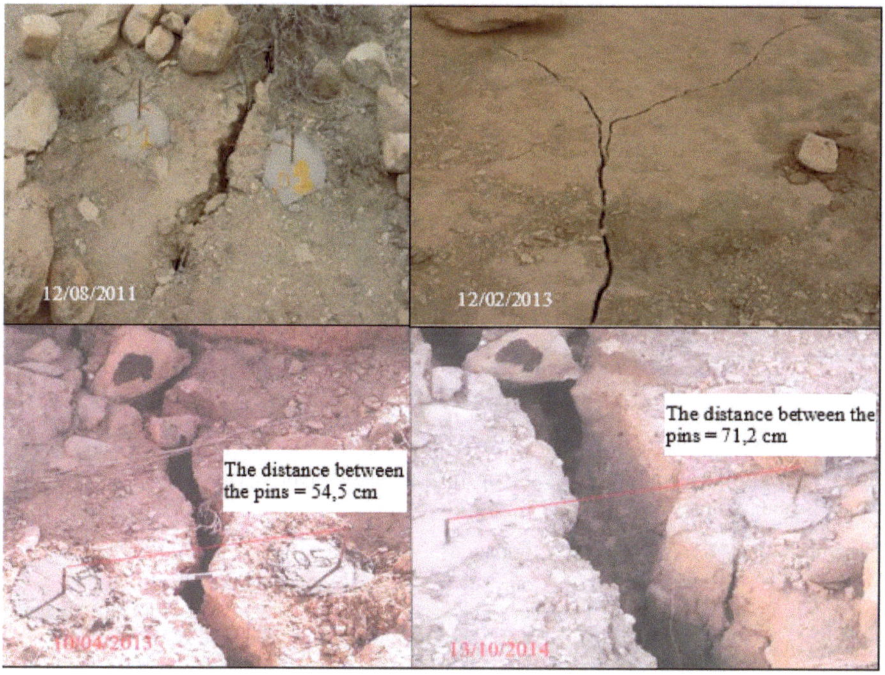

Fig. 4. Changes in cracks.

4 Methodology

In this article, we will examine the current stability of the northern flank of Kef Essnoun, (Fig. 5), and provide a potential risk of instability represented by a safety factor determined by the limit equilibrium methods (SLIDE). In our case study, the parameters used were identified in the available data submitted by the company. They come from previous tests at a laboratory in accordance with standard AFNOR Eurocode 07 (XP P 94-010), Table 1.

To achieve the modeling according to selected software, sections were cut along the northern flank of Kef-Essnoun to have profiles, with the thicknesses of the different facies.

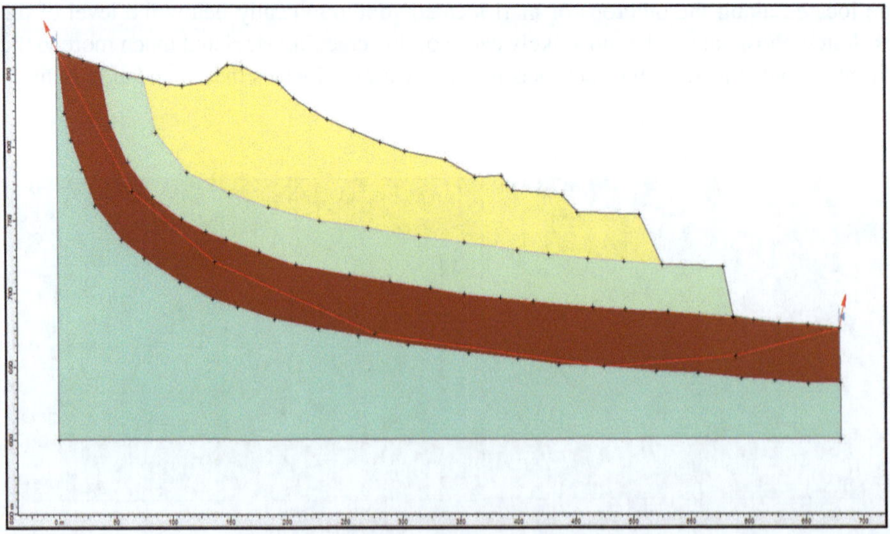

Fig. 5. Study site profile.

Table 1. The physical and mechanical properties of the different layers of along the north side of Kef-Essnoun mining.

Properties	Symbols units	Rock formations			
		Limestone Ypresian-Lutetian	Phosphate	Marl	Danio limestone-montien
Compressive strength	R_c MPa	60	25	8	40
Density unsaturated	ρ_{insat} kn/m^3	27	21	23	27
Saturated density	ρ_{sat} kn/m^3	27,46	24,81	24	27,46
Cohesion	C kn/m^2	5400	2300	160	3600
Internal friction angle	φ (°)	37	37	16	37
Dilatant	ψ (°)	7	7	0	7
Young's modulus	E MPa	27000	24000	1000	27000
Poisson's ratio	ν	0,250	0,28	0,25	0,25

5 Analysis of Stability

According to observations carried out on the site, it was deduced that the slippage affects the layer of marl. Given the lack of data regarding this facies of our study site, we opted the results of analysis of the old sliding whose geological conditions are comparable to those of current place of study to determine the most likely Geotechnical characteristics. This is to trace the value of cohesion and internal friction angle of marl, compatible for the 2007 landslide, with a critical safety factor (SF = 1).

According to the analysis results showing the variation of SF in terms of cohesion and internal friction angle of the facies of marl is found that a value of a critical safety factor (SF = 1): C = 45 kPa and $\varphi = 14°$.

The next step is to study the stability with or without the influence of vibration induced by blasting is simulated by integrating the parameter ground acceleration induced earthquake, the value obtained for horizontal and vertical acceleration estimated respectively at: Ah/g = 0.05 and Av/g = 0. 0125.

Where; Ah/g - horizontal acceleration of gravity; Av/g - vertical acceleration of gravity.

5.1 Phase Calculation

Static Load

We chose three calculation methods: Spencer, Bishop and Janbu, (Fig. 6).

SF of the values calculated according to the three methods mentioned previously shown in the following Table 2. According to the results of the safety factor, the limit equilibrium method (SLIDE), and static load (effect explosives), we find that the northern flank of Kef-Essnoun is stable, (Fig. 6).

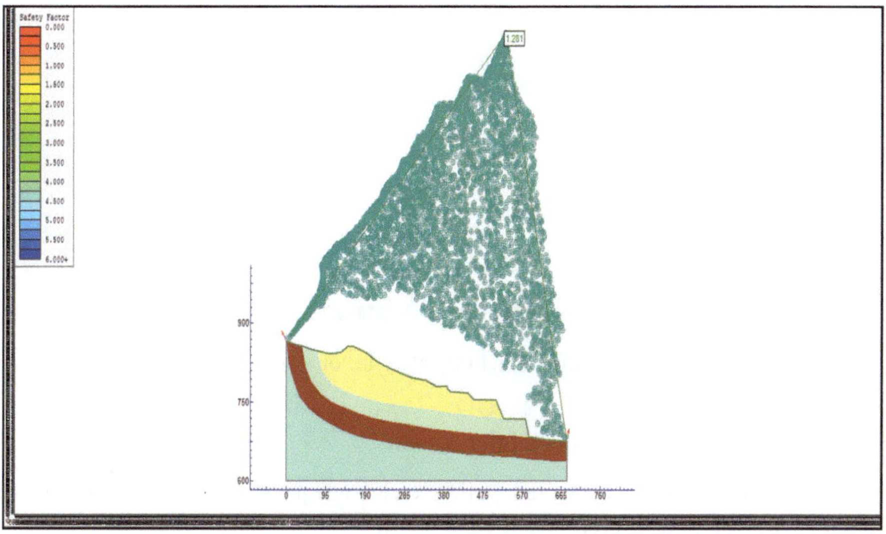

Fig. 6. The sliding surface and the safety factor value (static load).

Table 2. Value of safety factor with zero acceleration.

Methods	Bishop simplified	Janbu simplified	Spencer
Value of safety factor (SF)	1.281	1.226	1.302

Dynamic Load

SF of the values calculated according to the three methods mentioned previously shown in the following Table 3. For the same Geotechnical characteristics (specific gravity, cohesion, friction) and dynamic load (effect explosives): Ah/g = 0.05 and Av/g = 0. 0125, we find that the northern flank of Kef Essnoun is unstable and the break line is localized more exactly at the phosphate marl interface, (Fig. 7).

Table 3. Value of safety factor with acceleration.

Methods	Bishop simplified	Janbu simplified	Spencer
Value of safety factor (SF)	0.736	0.700	0.700

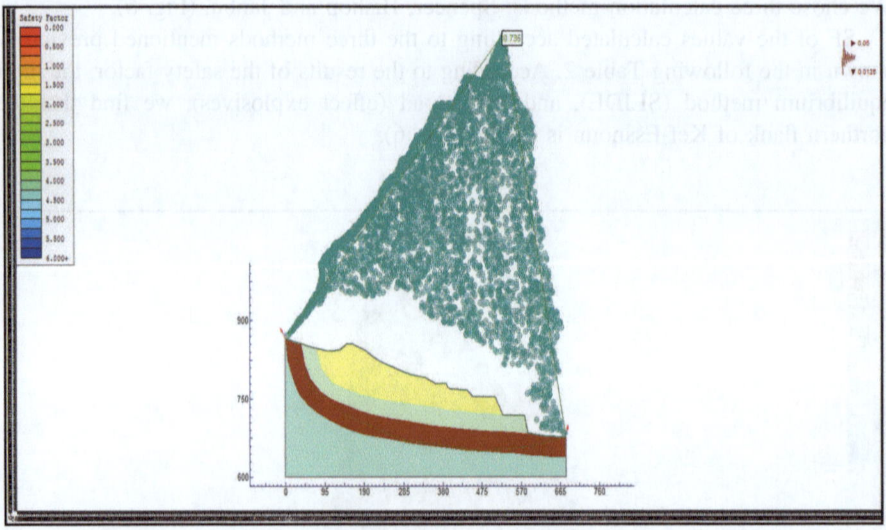

Fig. 7. The sliding surface and the safety factor value (dynamic load).

5.2 Proposal of a Method of Reinforcement

According to the analysis conducted earlier results, adequate reinforcement method to stabilize our study site is profiling. It consists of:

- The reduction of the crest to a level thus making it possible to eliminate the effect of the straightening of the layers, (Fig. 8).
- Creation of bench a height of 15 m to increase stability with the creation and widening of the platforms, (Fig. 9).

- Profiling bleachers and platforms, (Fig. 10).
- The reinforcement method chosen made it possible to increase the safety factor and thus to ensure the stability of the site, (Fig. 11).

According to the result, it is found that profiling makes it possible to increase the stability with Safty Factor of the order of 1.66.

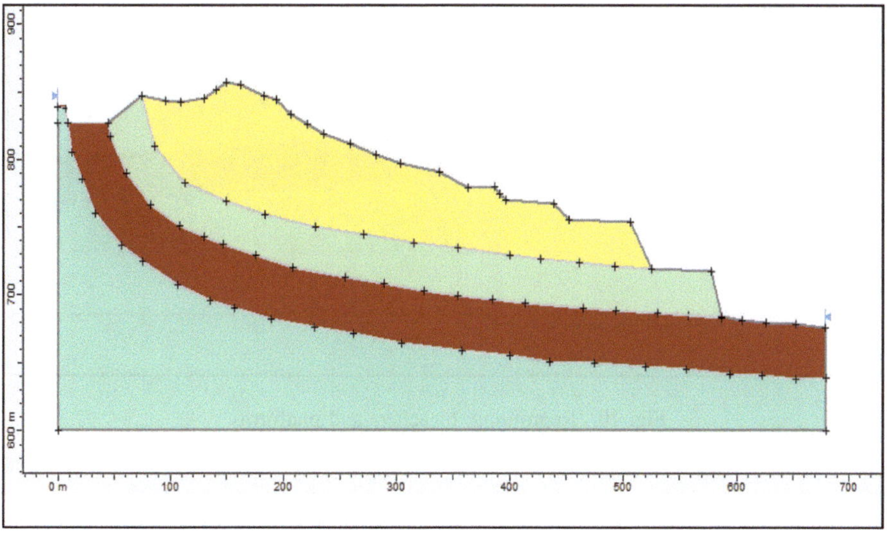

Fig. 8. Reduction of the crest.

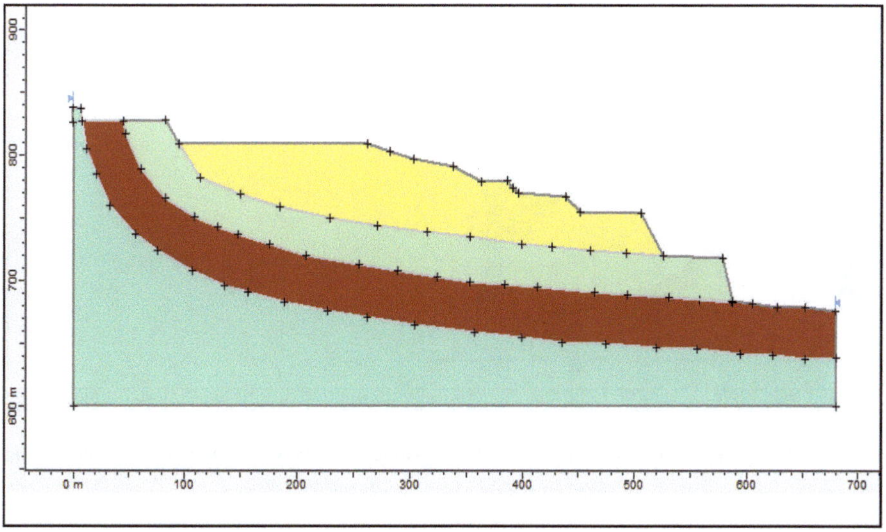

Fig. 9. Creation of stands and expanding platforms.

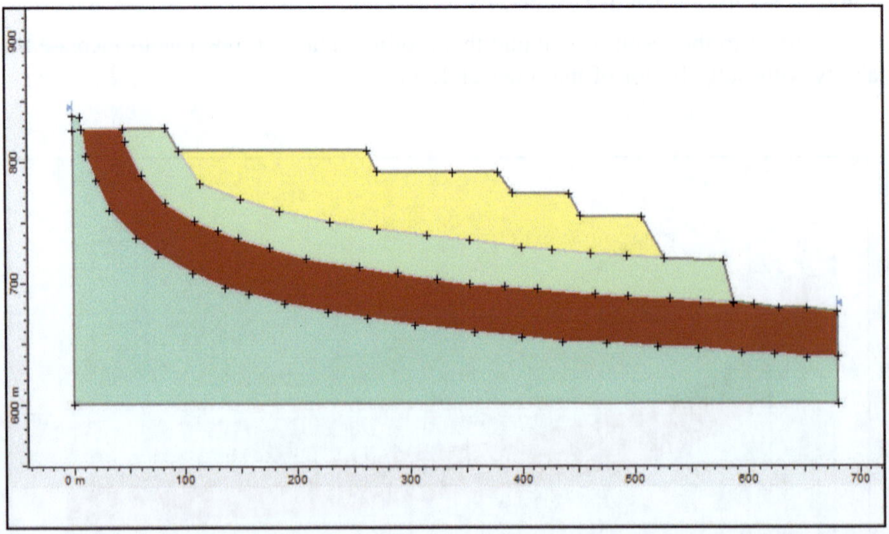

Fig. 10. Reprofiling: bleachers and platforms.

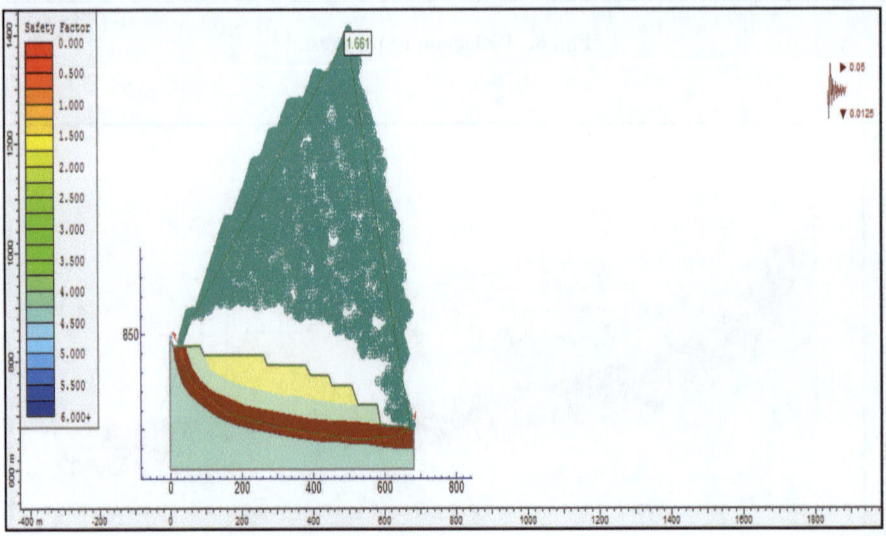

Fig. 11. The sliding surface and the safety factor value with the dynamic load after reprofiling.

6 Conclusions

The analysis of the results allows us to say that in addition to the geological and geotechnical factors that can influence the stability of a slope, the dynamic loading due to the explosive must be taken into account. The control of this parameter can be achieved by reducing the height of the bench, the widening of the working platform and the modification of the operating method of exploitation.

Acknowledgments. The author would like to thank the staff of the mining laboratory of the Annaba University (Algeria). Also, a special thanks to **A. Hafsaoui** (Annaba University - Algeria) and to **Boukarm R** (Bejaia University - Algeria) for her objective comments and corrections.

References

Dass Amiour, M., Mezghache, H., Elouadi, B.: The use of three physico-chemical methods in the study of the organic matter associated with the sedimentary phosphorites in Djebel Onk Basin, Algeria. Arab. J. Geosci. **6**, 309 (2013). doi:10.1007/s12517-011-0381-9

Benaissa, A.: Glissements de terrain: Calcul de stabilité, 2éme édition. Edition office des publications universitaires, Alger (2003)

Chalhoub, M.: Apports des méthodes d'homogénéisation numériques a la classification des massifs rocheux fracturées. Thèse de doctorat, Ecole Nationale Supérieure des Mines de Paris, spécialité géologie de l'ingénieur, pp. 39–45 (2006)

Cieslinski, S., et al.: Travaux de prospection et d'évaluation des phosphates de la région de Bir El Ater, EREM, pp. 3–10, 14–26, 39–44, 62, 63, 73–84, inédit (1985, 1987)

Faure, R.M.: L'évolution des méthodes de calcul en stabilité des pentes partie I méthodes à la rupture, Revue Française de Géotechnique (2000)

Li, W., Wang, J., Dai, L.: Safety criterion of blasting shock for the high deep slope of open-pit in mountain area. Ind. Miner. Process. **1**, 20–22 (2006)

Mokadem, N., Hamed, Y., Ben Saad, A., Gargouri, I.: Atmospheric pollution in North Africa (ecosystems-atmosphere interactions): a case study in the mining basin of El Guettar-M'Dilla (southwestern Tunisia). Arab. J. Geosci. **7**(5), 2071–2079 (2014)

Prian, G.P., Cortiel, P.: Etude de développement du gisement de phosphate de Djebel Onk (Algérie). Rapport d'expertise géologique, B.R.G.M. France, pp. 11–29, 133–149, 169–173, inédit. (1993)

Qiao, L., Li, Y.: Engineering geological model of high-steep slope damage in open cut mines. J. Univ. Sci. Technol. Beijing **26**(5), 461–464 (2004)

Schroeder, C.: Etude de stabilité des parois rocheuses. Revue de société Belge de géologie de l'ingénieur et de mécanique des roches (SBGIMR), pp. 15–18, 21–29, 34 (2010)

Prediction of Drained Settlement and Ultimate Bearing Capacity for Stone Columns Supported Foundation

Kok Shien Ng(✉)

Faculty of Civil Engineering, Universiti Teknologi MARA,
Pulau Pinang Branch, 13500 Permatang Pauh, Penang, Malaysia
ngkokshien@ppinang.uitm.edu.my

Abstract. This paper presents a validation of simplified design approach against a field study and a laboratory test. The design approach has been developed to predict drained settlement and the ultimate bearing capacity of stone columns supported foundation adopting a hyperbolic relationship to simulate the strain hardening behavior. The result of a plain footing load test is required to give two curve fitting parameters and then adapted them by introducing a reduction factor which take into account the length ratio and the area replacement ratio to predict the stress-strain characteristic of a stone column supported foundation. Good agreements are found when the predictions using the simplified approach are compared with the laboratory testing and field case study.

1 Introduction

Stone columns has been used intensively all around the world to improve the subsoil for supporting many civil infrastructure including road and railway embankment, building, factory, large storage tank, etc. that can tolerate some settlements. Design of stone columns supported foundation is often carried out in two steps. First, it has to be checked for the ultimate bearing capacity. Second, serviceability check is performed particularly on the long term drained settlement which is usually more critical.

Stone columns are used in groups and can be installed either as end bearing or floating type. Most of the stone column design is catered for end bearing columns (Balaam and Booker 1981; Priebe 1995; Han and Ye 2001; Castro and Sagaseta 2009) and very little for floating columns (Rao and Ranjan 1985; Lawton et al. 1994; Shahu and Reddy 2014; Ng and Tan 2014; Bouassida and Carter 2014). Many model tests (either laboratory or numerical) and field tests have been performed to study different aspects of stone column behavior such as settlement, consolidation rate, ultimate bearing capacity, failure modes, load sharing, group performance, etc. (Babu et al. 2013; Najjar 2013; Bouassida and Hazzar 2015). Based on limit analysis, Bouassida et al. (2009) proposed an analytical solution to predict the bearing capacity of a rigid foundation on a soil reinforced by a group of floating columns. Considering the elastic strain in the surrounding soil and plastic strain in column, Castro (2016) developed a closed-form solution to estimate the settlement of small column groups. A mechanical model has been developed by Das and Deb (2014) to study settlement as well as

W. Frikha et al. (eds.), *Soil Testing, Soil Stability and Ground Improvement*,
Sustainable Civil Infrastructures, DOI 10.1007/978-3-319-61902-6_2

bending moment and shear force of the raft foundation resting on a stone column improved ground.

Load-deformation behavior of a plain footing is a complex phenomenon where three types of deformation mode have been observed, i.e. strain hardening, strain softening and limiting capacity. With the inclusion of stone columns, the complexity is further increased due to multiple failure modes that can happen in the improved ground system such as wedge shearing, bulging, and punching (Tan and Ng 2013). Prediction of the nonlinear behavior of the stone columns reinforced foundations is difficult except by numerical approach. However, this modern computational technique requires advanced constitutive model to simulate the soil and the column behavior where the soil/column properties are sometimes difficult and expensive to obtain.

In previous study, a simplified design approach was developed to predict the long term drained settlement of a stone columns supported foundation. It was first compared against a series of finite element analysis where a good match was obtained for the load-deformation curve (Ng et al. 2014). In this study, the soundness of the design approach was further tested against a laboratory model test and a field test.

2 Simplified Design Approach

The general hyperbolic equation has been used to simulate this nonlinear stress-strain response of soils (Kondner 1963; Duncan and Chang 1970). Jeon and Kulhawy (2001) further adapted the hyperbolic equation to displacement and the equation is given below:

$$q = \frac{\Delta}{a + b\Delta} \tag{1}$$

where q is the applied bearing pressure, Δ is the footing displacement, and a and b are curve fitting parameters. The curve fitting parameters can be obtained by plotting the actual load-deformation curve in a linearized hyperbolic form as shown in Fig. 1 where a is the intercept of Δ/q versus Δ plot while b is the slope of Δ/q versus Δ plot. The parameter a is the inverse of initial tangent modulus of the ground while parameter b is related to the inverse of ultimate bearing capacity, q_{ult} of the footing system. The ultimate bearing capacity and the actual failure stress are assumed to be the same since the footing drained performance usually exhibit strain hardening behavior.

Footing supported by stone columns displays stiffer response under load as well as having a higher ultimate bearing capacity compared to a plain footing which also means smaller values for a and b parameters in the Eq. (1). The term 'plain footing' is used to refer to the un-reinforced shallow foundation. The determination of a and b parameters for improved ground is the essence of this new design approach. These two curve fitting parameters acquired from the load-deformation curve for a plain footing in drained manner is needed in the first place. Then, reduction factor, M applies to these parameters to obtain the hyperbolic curve for the reinforced stone column foundation, give as:

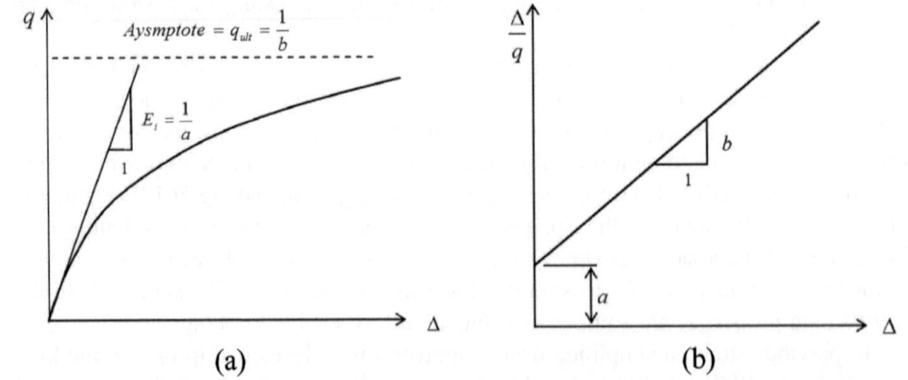

Fig. 1. (a) Hyperbolic stress-strain curve; (b) Transformation of hyperbolic relation.

$$a_1 = \frac{a}{M}$$
$$b_1 = \frac{b}{M} \tag{2}$$

which they are named as modified *a* and *b* parameters for stone columns reinforced foundation. Modified parameters are dependent on two factors: area ratio, α and column length, *L*. Area ratio is defined as the ratio of stone columns cross sectional area to the cross sectional area of the footing. Reduction factor, *M* is proposed as:

$$M = 1 + \frac{L}{L_{opt}}(2\alpha - 0.15) \quad L \leq L_{opt}$$
$$M = 1 + \frac{L}{d}(2\alpha - 0.15) \quad L \leq d \tag{3}$$

where *d* is the thickness of soft soil below footing, L_{opt} is the optimum length of stone column and L/L_{opt} and L/d are denoted as length ratio. Columns longer than a certain length which give no further improvement in settlement performance is termed optimum length, L_{opt}. The M value was first approximate from a series of parametric studies conducted during the preliminary study by means of axisymmetric finite element method using 15-node triangular element with elastic-perfectly plastic material for both columns and soils where M is taken to be equal to the basic settlement improvement factor, $n_0 = 2.575\alpha + 0.931$ (*n* = the ratio of settlement without improvement over settlement with improvement) for cases with column length longer than the optimum length However, further improvement was made to account for nonlinear behavior of soil, thus the M value becomes slightly smaller i.e. $M = 2\alpha + 0.85$ for end bearing columns or columns longer than optimum length. This value is influenced by the area replacement ratio together with the length ratio and those are the important factors in determining the stone column performance. A previous study by Tan et al. (2014) suggested that the optimum length can be computed as:

$$L_{opt}\ (m) = (2\alpha + 0.8)D + 0.5\ -\ t \tag{4}$$

where D equal to the diameter or breath, B of a footing, and t equal to the thickness of granular mat above the stone columns. The load-deformation curve for the stone column reinforced foundation can be drawn using *Eq.* (1) when the values of a_1, and b_1 are obtained. The design procedure to predict the load-deformation curve for a small column group is summarized in Fig. 2.

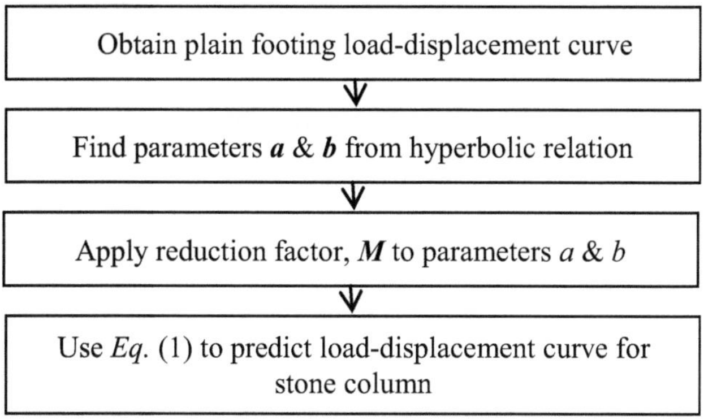

Fig. 2. Calculation procedures in hyperbolic method

The simplified design approach differs from the conventional method because it does not require the arbitrary selection of soil parameters, but is capable of simulating the yielding response of the improved ground that is strain hardening. Besides, the effect of footing embedded depth, ground water table and soil layering are mooted in this design approach since the only requirement is the stress-strain characteristic of a plain footing under similar testing condition.

3 Validation with Model Test and Field Load Test

3.1 Model Test (Wood et al. 2000)

In a model test by Muir-Wood et al., floating columns with the length, L of $1.7D$ ($>L_{opt} \approx 1.33D$) and $\alpha = 0.24$ were used to support a rigid circular footing of 0.1 m diameter. The sand columns were formed with replacement technique and placed in a square grid extending just beyond the edge of the footing. The relative density of the sand poured into the columns was around 0.5. The undrained shear strength of the surrounding soil is $C_u = 10$ kN/m^2. The displacement controlled test was conducted slowly in a drained manner until 30 mm settlement was reached.

Figure 3 shows the hyperbolic relation of the plain footing without stone columns as used in the model test. The parameter a and b are obtained as 0.01 and 0.13 respectively. Using *Eq.* (3), the reduction factor M was computed as 1.33. The

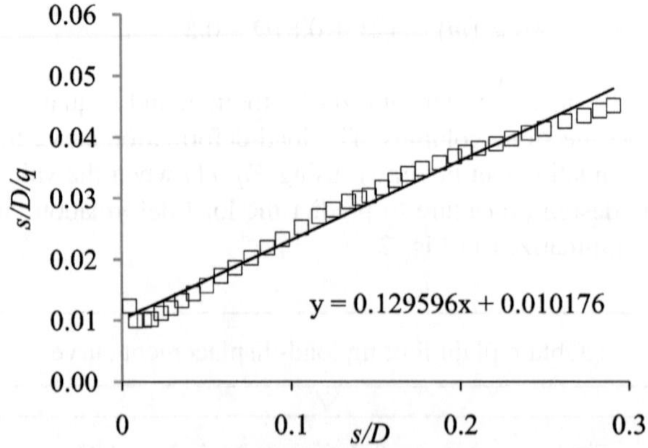

Fig. 3. Hyperbolic relation.

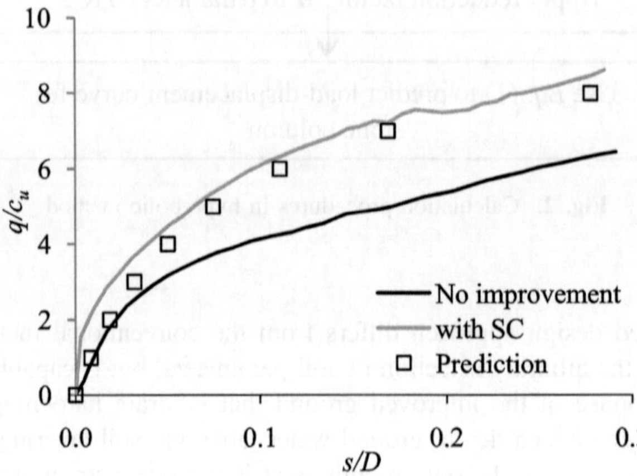

Fig. 4. Load deformation results.

prediction is well comparable to the model test as depicted in Fig. 4 despite slight overestimation of settlements, especially in the beginning of loading stage. There are probably two reasons for this overestimation. First, the footings are not only supported underneath, but with extra "buttressing" columns outside the edge of the footing which give extra constraint effect. Second, the footing was contained in a tank with 300 mm diameter, thus the boundary is only 1.5 times the footing diameter measured from the centerline and this boundary effect may have influenced the footing performance.

3.2 Field Load Test (Stuedlein et al. 2011)

A field load test (denoted as G4) on square footings of 2.7 m width supported with 5 numbers of floating stone columns. The column length is 3.05 m which is shorter than the optimum length, L_{opt} of 4.34 m. The diameter of columns is 0.76 m and the area replacement ratio, α is 0.3. The stone columns in the test are vibro-compacted. The first soil layer is 0.7 m thick hard, desiccated clay crust, followed by soft to medium stiff upper clay layer to a depth of 3.7 m, overlying sandy silt/silty sand layer with total thickness of 0.7 m. 4.5 m deep onwards, lie the fourth and deepest layer which is a slightly silty clay. Laboratory soil tests showed the upper 4.5 m was over-consolidated. A plain footing test (G3) acted as control test. Square footings in both tests were placed on the ground surface, direct contact with stone columns. The duration of load tests was approximately 500 to 600 min (8.5 to 10 h).

Using the proposed method, the reduction factor, M was calculated to be 1.32. The result of field load test was compared with the result obtained from the proposed method where a good agreement was obtained as shown in Fig. 5. The slight over prediction may probably due to the short period of test which the drained condition had

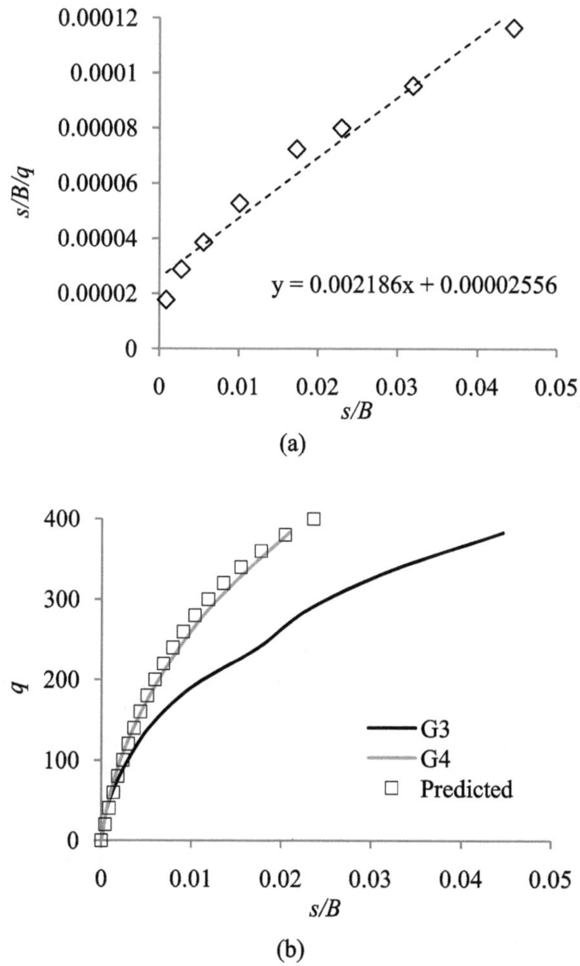

Fig. 5. Field test (a) Hyperbolic relation, and (b) Load-deformation results.

not been fully achieved. In addition, the ultimate bearing capacity for the column group was predicted to be 602 kN/m^2.

4 Conclusion

This paper presents a validation of a simplified design approach against a model test and a field load test where the results proved the method to be sufficiently accurate in estimating the drained settlement of stone columns supported foundation. The load-deformation response of the foundation can be characterized with hyperbolic function as shown in the above validation exercise. Albeit simple and easy, the method is indeed a rational method which has taken into account many intrinsic behaviors of stone columns such as the load sharing mechanism, nonlinearity and length ratio.

References

Babu, M.D., Nayak, S., Shivashankar, R.: A critical review of construction, analysis and behaviour of stone columns. Geotech. Geol. Eng. **31**(1), 1–22 (2013)

Balaam, N.P., Booker, J.R.: Analysis of rigid rafts supported by granular piles. Int. J. Numer. Anal. Methods Geomech. **5**(4), 379–403 (1981)

Bouassida, M., Jellali, B., Porbaha, A.: Limit analysis of rigid foundations on floating columns. Int. J. Geomech. **9**(3), 89–101 (2009)

Bouassida, M., Carter, J.P.: Optimization of design of column-reinforced foundations. Int. J. Geomech. **14**(6), 04014031 (2014)

Bouassida, M., Hazzar, L.: Performance of Soft Clays Reinforced by Floating Columns. In: Ground Improvement Case-Histories: Embankments with Special Reference to Consolidation and Other Physical Methods, p. 433 (2015)

Castro, J., Sagaseta, C.: Consolidation around stone columns. Influence of column deformation. Int. J. Numer. Anal. Meth. Geomech. **33**(7), 851–877 (2009)

Castro, J.: An analytical solution for the settlement of stone columns beneath rigid footings. Acta Geotech. **11**(2), 309–324 (2016)

Das, A.K., Deb, K.: Modeling of uniformly loaded circular raft resting on stone column-improved ground. Soils Found. **54**(6), 1212–1224 (2014)

Duncan, J.M., Chang, C.Y.: Nonlinear analysis of stress and strain in soils. J. Soil Mech. Found. Div. **96**(5), 1629–1653 (1970)

Han, J., Ye, S.L.: Simplified method for consolidation rate of stone column reinforced foundations. J. Geotech. Geoenviron. Eng. **127**, 597–603 (2001)

Jeon, S.S., Kulhawy, F.H.: Evaluation of axial compression behavior of micropiles. In: Foundations and Ground Improvement, pp. 460–471. ASCE (2001)

Kondner, R.L.: Hyperbolic stress-strain response: cohesive soils. J. Soil Mech. Found. Div. **89** (1), 115–143 (1963). ASCE

Lawton, E.C., Fox, N.S.: Settlement of structures supported on marginal or inadequate soils stiffened with short aggregate piers. In: Geotechnical Special Publication 40, pp. 962–974 (1994)

Najjar, S.S.: A state-of-the-art review of stone/sand-column reinforced clay systems. Geotech. Geol. Eng. **31**(2), 355–386 (2013)

Ng, K.S., Tan, S.A.: Design and analyses of floating stone columns. Soils and Foundation **54**(3), 478–487 (2014)

Ng, K.S., Tan, S.A., Sun, J.: A new approach to the estimation of settlement and ultimate bearing capacity of stone columns supported shallow foundation. In: The 8th European Conference on Numerical Methods in Geotechnical Engineering (NUMGE 2014), 18–20 June 2014, Delft (2014)

Priebe, H.J.: The Design of Vibro Replacement, pp. 31–37. Ground Engineering (1995)

Rao, B.G., Ranjan, G.: Settlement analysis of skirted granular piles. J. Geotech. Eng. **111**, 1264–1283 (1985)

Shahu, J.T., Reddy, Y.R.: Estimating long-term settlement of floating stone column groups. Can. Geotech. J. **51**, 770–781 (2014)

Stuedlein, A.W., Holtz, R.D.: Analysis of footing load tests on aggregate pier reinforced clay. J. Geotech. Geoenviron. Eng. **138**(9), 1091–1103 (2011)

Tan, S.A., Ng, K.S.: Stone columns foundation analysis with concentric ring approach. In: Proceedings of the Third International Symposium on Computational Geomechanics (ComGeo III), Krakow, Poland (2013)

Tan, S.A., Ng, K.S., Sun, J.: Column groups analyses for stone column reinforced foundation. Geotechnical Special Publication 233, pp. 597–608 (2014)

Wood, D.M., Hu, W., Nash, D.F.T.: Group effects in stone column foundations model tests. Geotechnique **50**(6), 689–698 (2000)

Considerations on the Stiffness of Sensitive Soft Soils

Lamia Touiti[1(✉)] and William Van Impe[2]

[1] LGC, Civil Engineering Laboratory - Ecole Nationale d'Ingénieurs de Tunis, Université de Tunis El Manar, Tunis, Tunisia
touiti.l@gmail.com
[2] ISSMGE, Ghent University Belgium, Gent, Belgium

Abstract. The present paper comments on the evaluation of the initial stiffness of sensitive soft soils as an outcome of several testing methods: Resonant Column (RC), Bender Element (BE) tests, Seismic cone tests and Empirical relationships (correlations). The RC, BE and common geotechnical laboratory tests were carried out on soft clays adjacent to the Tunis Lake in Tunisia. The database was completed making use of existing studies carried out on other types of Swedish soft clays from the literature related to seismic cone tests. A number of empirical correlations for determining the initial value of the shear modulus of soils in the literature was adopted in order to re-analyze the available data set. The authors suggested an adapted empirical lower bound correlation aiming at evaluating the analyzed soft soils' small strain shear modulus (G_{max}), probably even useful for less undisturbed sampling. The proposed equation aims at giving an indication of the effect of, for example, the mean effective stress (p′) and the corresponding void ratio (e) on the initial soil stiffness more clearly.

1 Introduction

The small strain shear modulus (G_{max}) of soil is an important parameter for a large variety of geotechnical design problems. Such modulus is typically associated with shear strain levels of about 10^{-4}% and below. The critical role of soil stiffness at small-strains in the design and analysis of geotechnical structures is widely accepted. For example, G_{max} is a key parameter in small-strain static and dynamic analyses (certainly covering foundation engineering problems such as those related to earthquake, wind power plants, machine foundations or rail/road traffic…) (Schnaid 2005; Stokoe et al. 2005). G_{max} can be equally important for small-strain cyclic situations (wind or wave loading).

G_{max} is reasonably well connected also to other soil properties, such as density and sample disturbance (Nash 1999). Its value depends on a number of parameters, including void ratio, confining stress, soil structure, degree of saturation, stress history and time factors. Table 1 shows a list of parameters affecting G_{max} at different levels of importance, originally listed by Vucetic and Dobry (1991) and updated by Benz (2007). As shown in this table, the most important parameters affecting the soft soils small-strain stiffness are confining pressure, void ratio, geologic age, cementation, overconsolidation ratio, plasticity index as well as the frequency of a loading pattern.

W. Frikha et al. (eds.), *Soil Testing, Soil Stability and Ground Improvement*,
Sustainable Civil Infrastructures, DOI 10.1007/978-3-319-61902-6_3

Table 1. Parameters affecting the small-strain stiffness of soils (Theenathayarl (2011), modified from the original table presented by Vucetic and Dobry (1991))

Increasing factor	Gmax
Confining pressure (σ'm)	Increases with (σ'm)
Void ratio (e)	Decreases with e
Geologic age (tg)	Increases with tg
Cementation (c)	Increases with c
OCR	Increases with OCR
Plasticity Index (PI)	Increases with PI if OCR > 1
	Stays constant if OCR = 1
Cyclic strain (γc)	–
Frequency of loading (f)	Increases with f

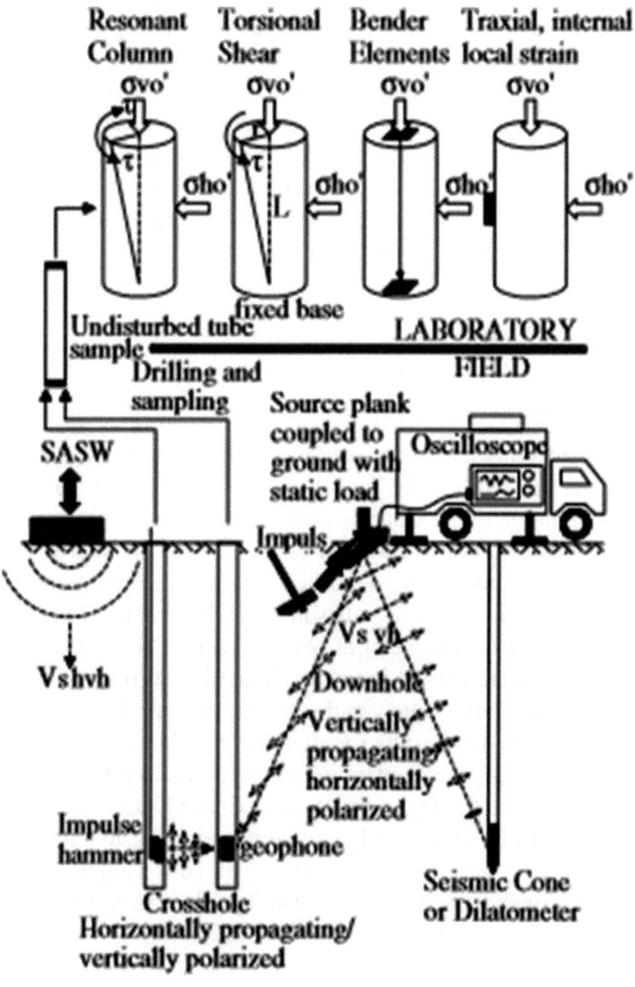

Fig. 1. Field and laboratory methods for determining shear modulus (Shneider et al. 1999).

It is common practice to qualify a reliable value of G_{max} of the soil (at induced strain levels less than 0.0001%) from the shear wave velocity:

$$G_0 = V_s^2 \cdot \rho, \text{ kPa} \tag{1}$$

Where: G0 = Initial shear modulus; ρ = mass density; Vs = shear waves velocity for a linear, elastic and isotropic medium.

Most testing techniques are consisting of a combination of standard geotechnical tests with the geophysical modulus used both in laboratory and field conditions. Such hybrid (field/lab) method is shown in Fig. 1 (Schneider et al. 1999). Field techniques, besides of the seismic cone (SCPTU) and seismic flat dilatometer, include cross hole and down hole tests in the typical geophysical version and the SASW method. Laboratory tests in turn, are described as the resonant column (RC), torsional shear and triaxial tests with local strain measurement and the measurement of the shear wave velocity based on bender element (BE). Such configurations diminish the disadvantages of each group of tests and considerably enhance the optimization of data collection (Wolski et al. 2006).

The focus of this paper is to evaluate the G_{max} using RC, BE and seismic cone test results verifying some empirical formulas, from literature and adding own test results. The analyzed data correspond to both laboratory tests conducted on Tunis soft soil (RC and bender BE tests) and collected data related to Swedish soft soils (laboratory tests and seismic cone as field tests). Having analyzed the results, the authors suggest an adapted correlation which fits well the sensitive soils.

2 Materials and Methods

2.1 Tunis Soft Soils

The soil for laboratory tests, sampled as "undisturbed" standard tube samples, was collected from the region close to the Tunis Lake in Tunisia, more specifically, from the test sites located in and near the Ghana avenue (south lake). The soil samples can be described as gray-black soft silty clay and do belong to a natural cohesive soil formation, of recent Quaternary origin, deposited in a marine environment. The deposit qualifies as a normally consolidated soft soil with illite as the dominating clay and quartz in the silt particles (Kaâniche et al. 2000; Touiti et al. 2009).

All laboratory tests were carried out at the Geotechnical Institute for Experimental Models (ISMGEO Laboratory) in Italy and at the Geotechnical laboratory of the Ghent University (Belgium). The test results summarized in Table 2 do show a clay and silt content of about 40%, an organic content in the range of 2 to 5%, a water content from 84 to 110% (slightly above the liquid limit), an undrained shear strength derived from TXCIU tests from 15 to 28 kPa and G_{max}, determined using RC and BE tests, in the range of 3 to 8 MPa. The sensitivity of the Tunis soft soils samples and the one at the Ska-Edeby site was estimated on the basis of the liquidity index (LI) and the undrained shear strength (Cu), as established by Leroueil et al. (1983):

Table 2. Identification and soil index properties of selected Tunis soft soils

Specimen name	Depth (m)	Soil type	Water content W [%]	Bulk density γh/γd [KN·m^{-3}]	Void index e [–]	Liquid limit LL [%]	Plasticity index PI [%]	Liquidity index LI [–]	Organic Matter OM [%]	$CaCO_3$ CC [%]	Cu kPa	St
RC1(SC1)	5.0–5.8	MH	98	14.8/7.5	2.7	96	41	0.85	4.5	34	28	20
RC2(SC2)	7.8	CH	92	14.2/7.4	2.1	85	49	1.10	2.84	39	28	24
BE1(SC3)	3.2–4.0	CH	111	13.8/6.5	2.6	85	58	1.13	–	–	15	22
BE2(SC3)	3.2–4.0	CH	91	14.8/7.7	2.4	80	58	0.79	–	–	20	19
BE3(SC3)	3.2–4.0	CH	84	15.1/8.2	2.3	81	58	0.67	–	–	25	18
BE4(SC4)	5.0–5.7	MH	89	14.1/7.5	2.3	79	30	1.13	2.2	30	18	23

$$Cu_r(kPa) = \frac{1}{(I_L - 0.21)^2} \tag{2}$$

$$St = Cu(I_L - 0.21)^2 \tag{3}$$

The sensitivity of Tunis soft soils ranged from 18 to 24.

2.1.1 Resonant Column Tests

The RC tests were performed using the ISMGEO laboratory Resonant Column Apparatus (4–100 Hz) which is an example of Hardin-Drnevich resonant column using "fixed-free" configuration (Fig. 2). This equipment is commonly used to study the dynamic deformation characteristics of soil.

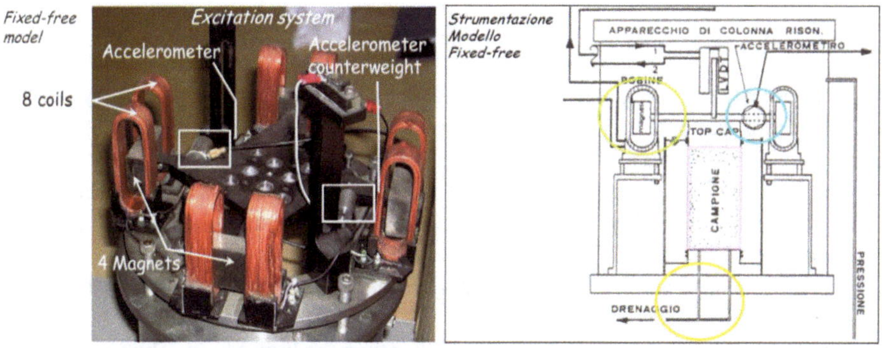

Fig. 2. Resonant column apparatus used for the tests - ISMGEO laboratory – Italy

The procedure for testing the soils with RC method followed that outlined in ASTM D 4015-07. In fact, the specimen is rigidly fixed at the base while the torsional oscillation is applied to the free end by a drive head. The frequency of vibration is gradually increased until reaching the first-mode fundamental frequency of the sample, at which measurements of the resonance frequency and amplitude of vibration are made. Knowing the geometry and the end constraints of the sample, the measured resonance frequency is then used to calculate the shear wave propagation velocity using the wave propagation equation and the theory of elasticity. The shear modulus is subsequently obtained from the derived shear wave velocity and the density of the sample.

2.1.2 Bender Element Tests

The triaxial tests were performed on 6 undisturbed Tunis soft soil specimen in three stages: saturation (back pressure method), consolidation and shearing (strain controlled mode).

The bender elements are small electro-mechanical transducers which either bend as an applied voltage is changed or generate the voltage as they bend. Bender elements were placed into soil sample during the triaxial tests (Fig. 3). The measurements of the shear wave velocity were performed at the end of each of the saturation and consolidation stage during triaxial tests (Fig. 3). In order to rule out the incorrect measurements, additional measurements of the travel time at the same stress condition but with different frequency of input signal were made.

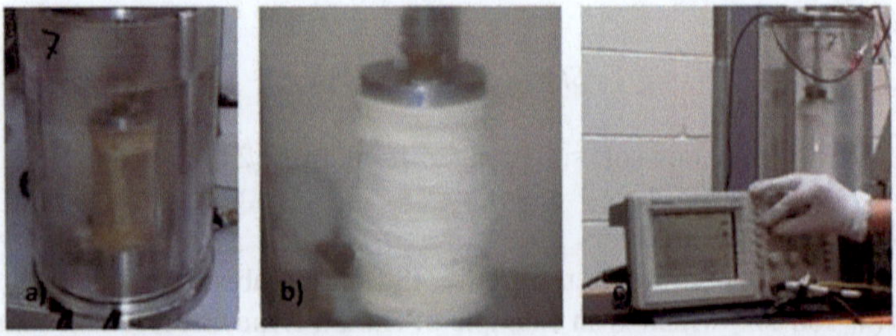

Fig. 3. (a) Tunis soft soil samples wrapped in filter paper and rubber skin. (b) Bender elements placed into Tunis soft soil sample (c) Triaxial apparatus equipped with bender elements used for the tests

2.2 Swedish Soft Soils

A geotechnical soft soil database including stiffness properties measured by seismic cone tests was elaborated. The data base was completed on the basis of several investigations and studies performed on Swedish soft soils (Larsson and Mulabdic 1991).

The analyzed Swedish test sites were selected among the well documented test fields previously used by the Swedish Geotechnical Institute (SGI): Skå-Edeby, Norrköping and Bäckebol sites. The soft soils at those sites have a thickness of 5 to 10 m predominantly marine post-glacial clay with the presence of shells and thin layers or seams of silt.

The natural water content is generally higher than the liquid limit. The sensitivity of the clays varies between 8 and 24. The undrained shear strength as determined by field vane tests reaches a minimum of 7.5 to 9 kPa at 2 m below the ground surface for the various sites and seems to increase with depth.

The geotechnical parameters of the various sites are summarized in the Table 3.

Table 3. Geotechnical parameters of selected Swedish soft soils

Site	Z (m)	W (%)	WL (%)	IP (%)	P′ (kPa)	e	Cu (kPa)	OC (%)	St	Vs m/s	OCR
Norrköping	3–10	70–118	62–84	_	33–50	1.7–3.3	9–15	_	10–20	66–98	1.2–1.6
Skå-Edeby site	3–9	64–87	55–83	30–56	17–38	1.7–2.4	8–15	1–5	6–16	–	1.1–1.2
Bäckebol site	2–8	78–100	69–86	36–51	22–37	2.0–2.7	11–15	<1%	12–24	56–85	1.2–1.9

2.2.1 Norrköping Site

According to Larson and Mulabdic (1991), the investigated soil profile consists of about 8 meters of gray varved clay (2–10 m) with thin layers or seams of silt which occur at about 5 m and do appear even more regular from 7 m depth on. The natural water content is higher than the liquid limit and varies from 120% at the upper layers to about 70% in the silty bottom layers. The soil is slightly over consolidated; the over consolidation ratio ranges from 1.2 to 1.6. The sensitivity of the clay measured with the field vane tests varies between 10 and 20. The small strain shear modulus deduced from the seismic cone tests ranges from 5 to 15 MPa.

2.2.2 Skå-Edeby Site

The test field is located on Svartsjölandet. A large number of tests has been made in that test field at Ska-Edeby (Kallstenius and Bergau 1961; Kallstenius 1963; oysterman and Lindskog 1963; Push 1970; Holtz and Linkskog 1972; Holtz and Broms 1972; Holtz and Holm 1972; BomanandBroms 1975; Wiezel 1975; Massarsch et al. 1975; Torstensson 1976; Holm and Holtz 1977; Larsson 1986; Larsson and Eskilson 1989; Larsson and Mulabdic 1991).

According to the above-mentioned studies, the soil profile (3–9 m) consists of 3 m of post-glacial clay, highly plastic and slightly organic, followed by 6 m of varved glacial clay. For both layers, the water content is higher than the liquid limit, the soil is slightly overconsolidated (OCR = 1.1 to 1.2), the undrained shear strength (field vane testing) increases with depth from 8 kPa to 14 kPa, the sensitivity varies from 6 to 16 and is common of 15 and the small strain shear modulus (seismic cone tests) varies with depth from 3 MPa to 14 MPa showing a uniform and continuous increase.

2.2.3 Bäckebol Site

The test field is located near the river GötaÄlv and concerns mainly the upper 10 m of the soil site. According to Larrsson and Mulabdic (1991), the clay consists of post glacial marine clay, with illite as the dominating clay mineral and quartz and feldspar in the silt particles, the clay content being about 60%, the organic content less than 1%, the water content reaching some 70 to 90% which is slightly above the liquid limit.

The undrained shear strength varies between 11 to 15 kPa, its sensitivity shows values ranging from 12 to 24 and the small strain shear modulus derived from the seismic cone tests varies from 5 to 11 MPa.

3 Interpretation and Discussion of RC and BE Test Results and Collected Data

3.1 Tunis Soft Soils

The results obtained in the laboratory from BE and RC tests are presented in Figs. 4 and 5. The values of the shear wave velocity and the small strain shear modulus varied from 40 to 80 m/s and 3 to 8 MPa respectively at an applied mean effective stress ranging from 10 to 50 kPa.

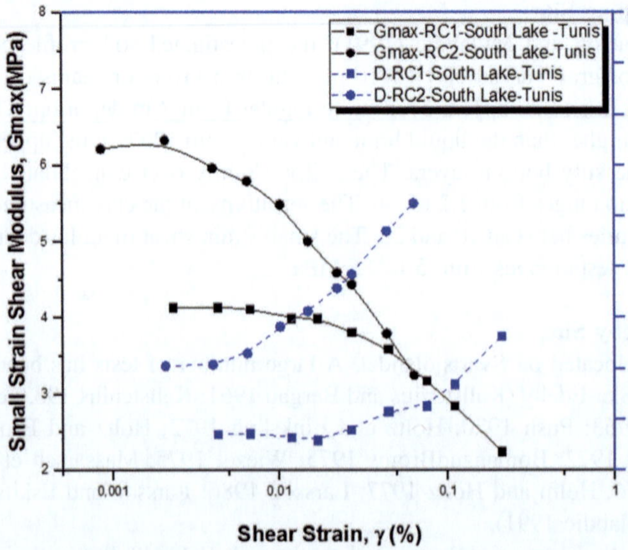

Fig. 4. Shear modulus and damping ratio for Tunis soft soil

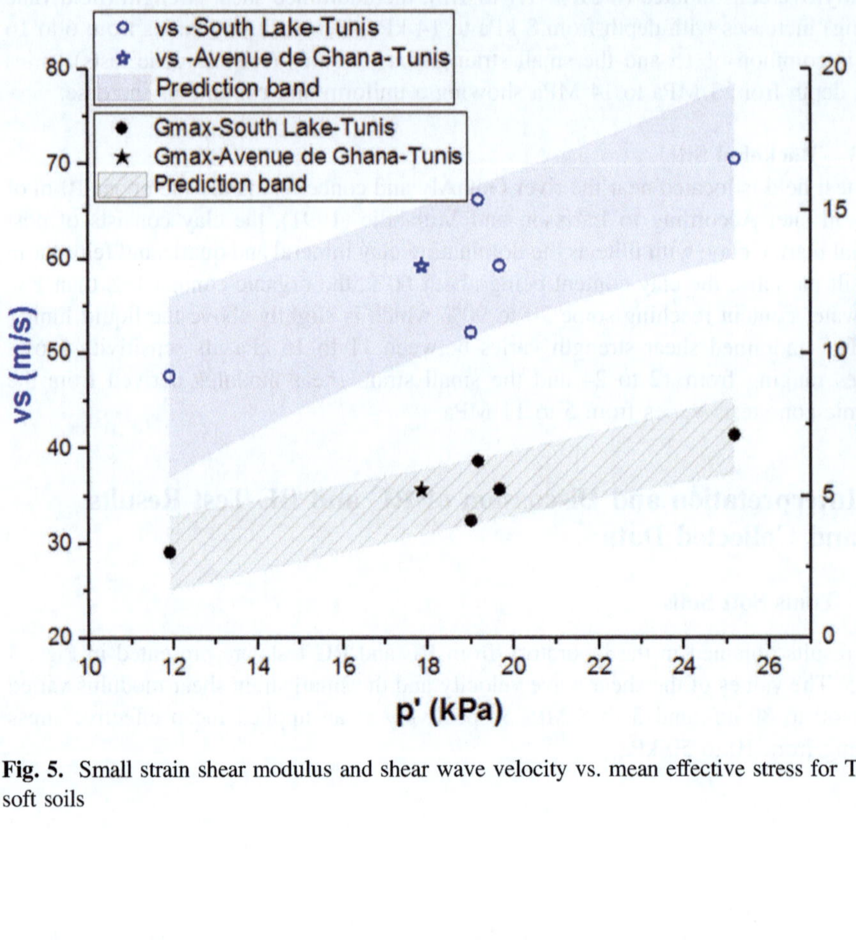

Fig. 5. Small strain shear modulus and shear wave velocity vs. mean effective stress for Tunis soft soils

The relationship between the estimated shear modulus, the mean effective stress and the void ratio for tested soft soils (Fig. 6) confirm the positive impact of the mean effective stress and the negative impact of voids on the G_{max}.

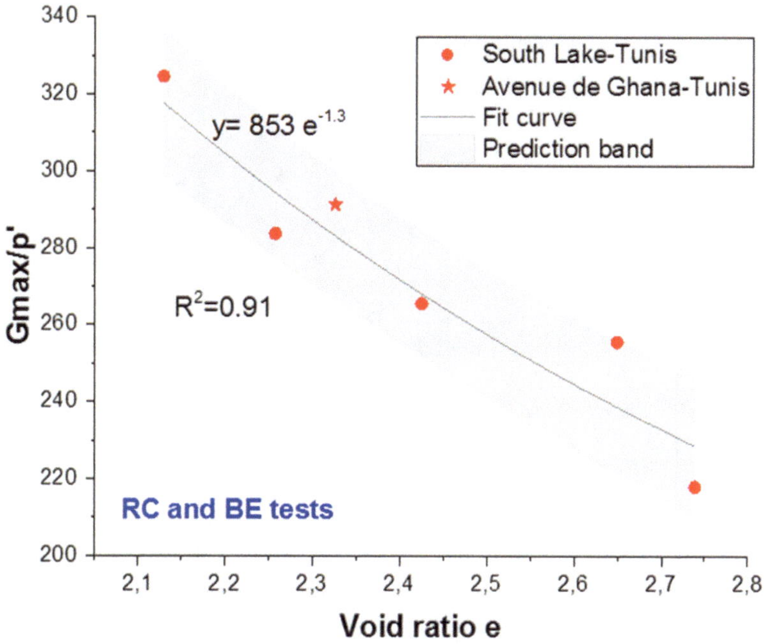

Fig. 6. Normalized Gmax/p′ deduced from authors' bender element and resonant column tests versus void ratio for Tunis soft soils

3.2 Swedish Soft Soils

The mean effective stress and the void ratio were calculated and deduced based on the collected data related to Swedish soft soils from the test fields in Norrköping, Skå-Edeby and Bäckebol sites.

The small strain shear modulus versus the mean effective stress and the void ratio for the different Swedish sites was elaborated. Swedish soft soil stiffness properties resulting from seismic cone tests were collected and re-interpreted as well and summarized in Table 4.

Table 4. The stiffness parameters of selected Swedish soft soils

Gmax (MPa)	e	P′(kPa)
3–15	1.8–3.4	17–50

3.2.1 Norrköping Site

The small strain shear modulus versus the mean effective stress and the small strain shear modulus versus the void ratio are illustrated in Figs. 7 and 8.

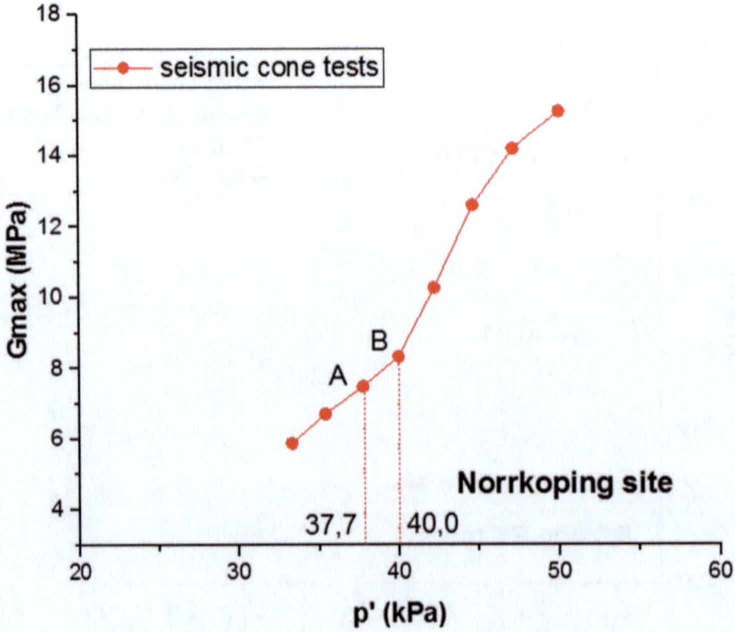

Fig. 7. Small strain shear modulus versus mean effective stress for sensitive soils in Norrköping

The singular points (A and B) at depths of 5 and 6 m (37.7 and 39.98 kPa), are indicating a change of the curving slope and a greater stiffness of the soil. This is probably due to the presence of 3 different layers: the first layer from 3 to 5 m corresponding to soft clay, the second one, from 5 to 7 m, related to the irregular varved clay with thin silt layers and the third layer appearing from 7 to 10 m and corresponding to the regular varved clay.

3.2.2 Skå-Edeby Site

The small strain shear modulus versus the mean effective stress (Fig. 9) illustrates a uniform and linear relation for each soft clay layer while the standardized small strain shear modulus versus the void ratio suggests a non-linear but continuous relationship (Fig. 10), except for 1 isolated point (C) at 6 m depth. This can possibly be explained by the presence of a thin seam with quite deviating geotechnical characteristics.

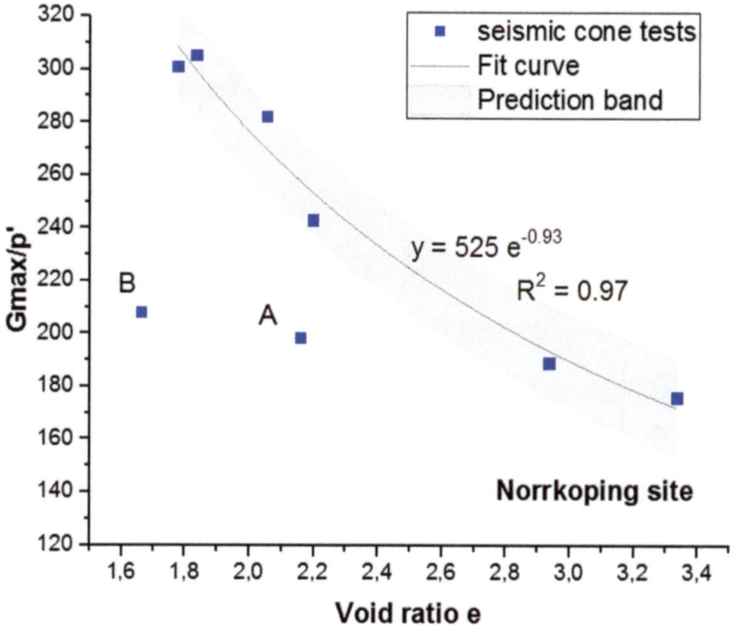

Fig. 8. Normalized small strain shear modulus versus mean effective stress for sensitive soils in Norrköping

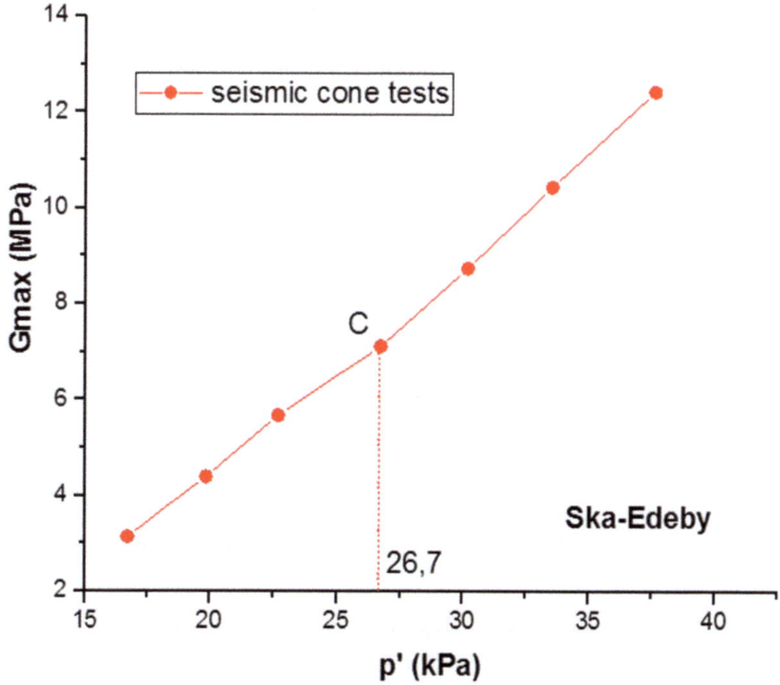

Fig. 9. Small strain shear modulus versus mean effective stress for sensitive soils in Skå-Edeby

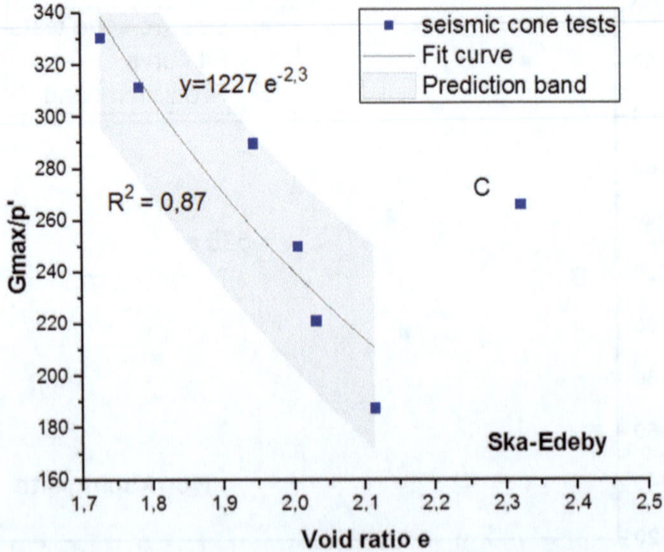

Fig. 10. Normalized small strain shear modulus versus mean effective stress for sensitive soils in Skå-Edeby

3.2.3 Bäckebol Site

Figures 11 and 12 do show a very straightforward relationship between the small strain shear modulus, the mean effective stress and the void ratio, with the exception of 2 isolated points (D and E) at 5 and 6 m, respectively. This can possibly be explained by the presence of a thin seam of other geotechnical characteristics.

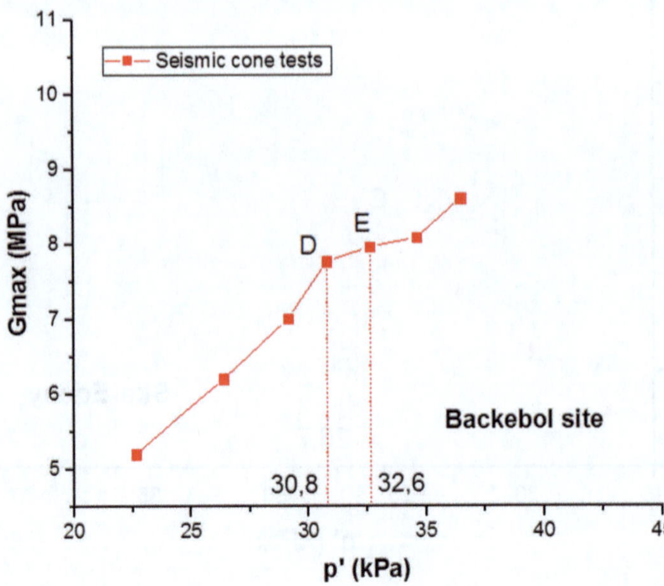

Fig. 11. Small strain shear modulus versus mean effective stress for sensitive soils in Bäckebol

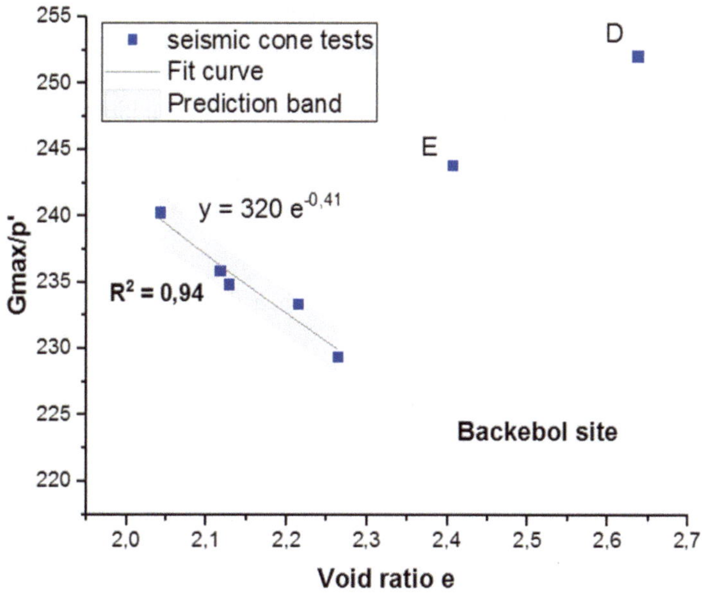

Fig. 12. Normalized small strain shear modulus versus mean effective stress for sensitive soils in Bäckebol

3.3 Empirical Correlations for Soft Soils

Many empirical correlations for the evaluation of G_{max} have been proposed in the literature. The authors would suggest a new empirical correlation for such type of soft sensitive soils, besides of the standard proposals gathered in Table 5.

Table 5. Selected empirical formulas for calculating G_{max} for soft soils

Form of the function	Authors	Kind of soil
$G0 = 625.\frac{(p' \cdot pa)^{0,5}}{(0,3+0,7e^2)}$	Hardin (1978)	Normally consolidated cohesive soils
$G0 = 625.\frac{(p' \cdot pa)^{0,5}}{(e^{1,3})}$	Jamilkowski et al. (1991)	NC cohesive soils
$G0 = 14,24.p'^{0,23}.e^{-1,26}$	Markowska-Lech et al. (2007)	Cohesive soils
$G0 = p^{0,85}.e^{-1,26}$	Gabryś and Szymańsk (2015)	NC cohesive soils

As mentioned above, G_{max} depends on a number of parameters and the void ratio is recognized as the second meaningful parameter affecting soil stiffness in combination with the effective stress level.

The collected data and the tests results indicate the possibility of estimating the small strain shear modulus in this type of soft sensitive soils from the mean effective stress and the void ratio. The authors proposed a new correlation which could be helpful in estimating the shear modulus at very small strain without the need for the shear wave velocity measurement:

$$G_{max}(MPa) = 0,53.p'.e^{-0,93} \tag{4}$$

Table 6. Laboratory fitting parameters of test soils for authors' equation

Form of the function	Coefficient			Standard error of the coefficient			Correlation coefficient	Statistics	
	a	b	c	a	b	c	R^2	Reduced Chi-Sqr	Adj. R-square
$G_{max} = a.\ p'^{b}.\ e^{c}$	0,53	1,00	−0,93	0,15	0,06	0,14	0,96	0,530	0,956

Table 6 shows the values of the coefficients as well as the basic fitting parameters for the proposed function (Eq. (4)). The proposed correlation explains at least 96% of the Gmax variation (R^2 = 0.96) (when ignoring the isolated points). It indeed very closely matches the data set. A graphical illustration of the proposed equation is illustrated in Fig. 13.

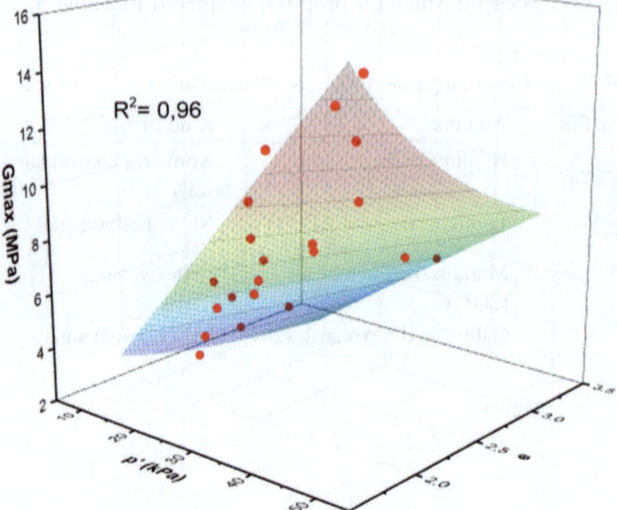

Fig. 13. Correlation of the experimental shear modulus with the mean effective stress and void ratio for sensitive soils.

To examine the reliability of the proposed empirical equation, some standard empirical equations (Table 5) were also used to verify their suitability to the data set of each site (Figs. 14, 15, 16 and 17). As can be seen from those Figures, the authors' proposed expression fits the results of the tests for Tunis, Norrköping, Skå-Edeby and Bäckebol soft soils well.

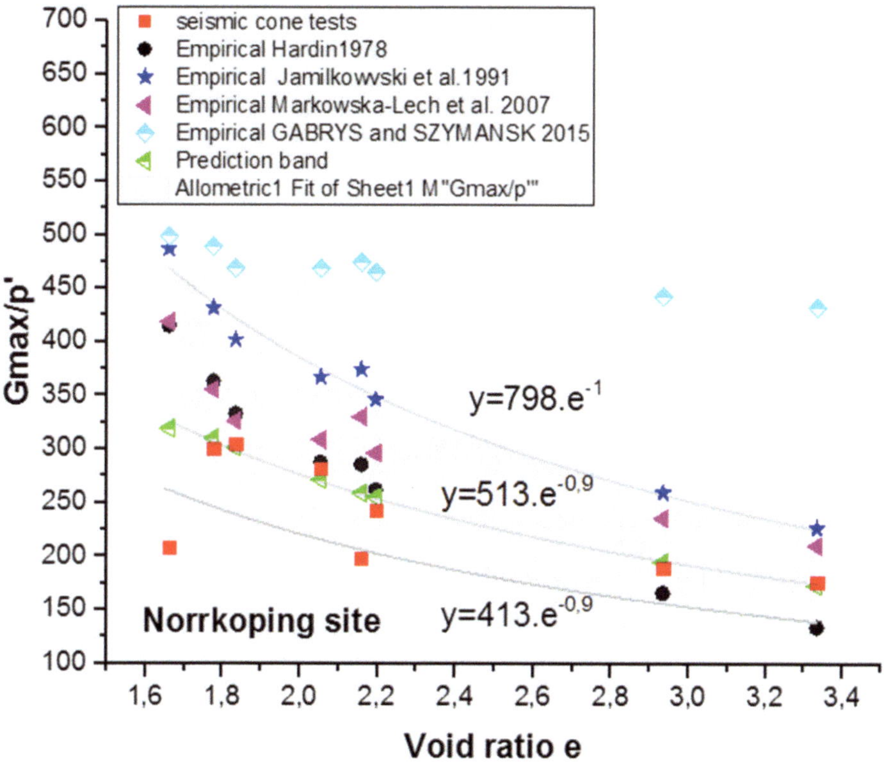

Fig. 14. Correlation of the experimental values of normalized shear modulus with the void ratio for the Norrköping sensitive soil.

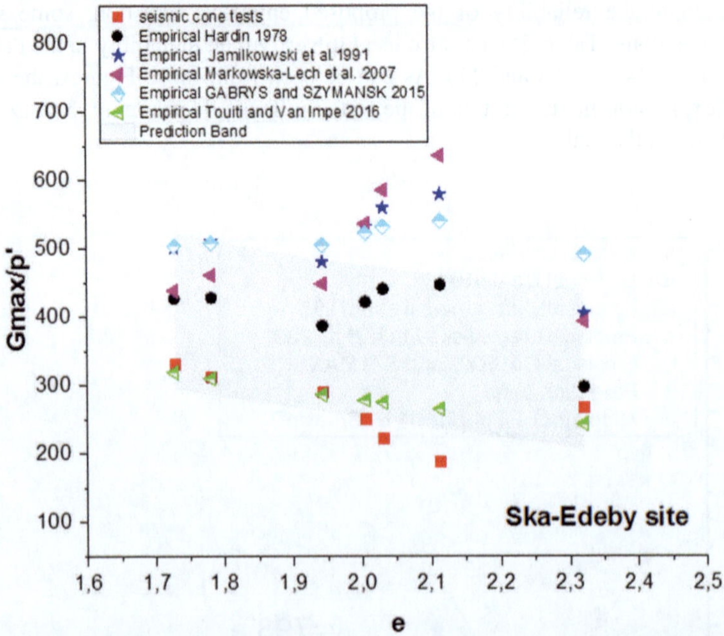

Fig. 15. The experimental and the estimated values of the normalized shear modulus with the void ratio for Skå-Edeby sensitive soil

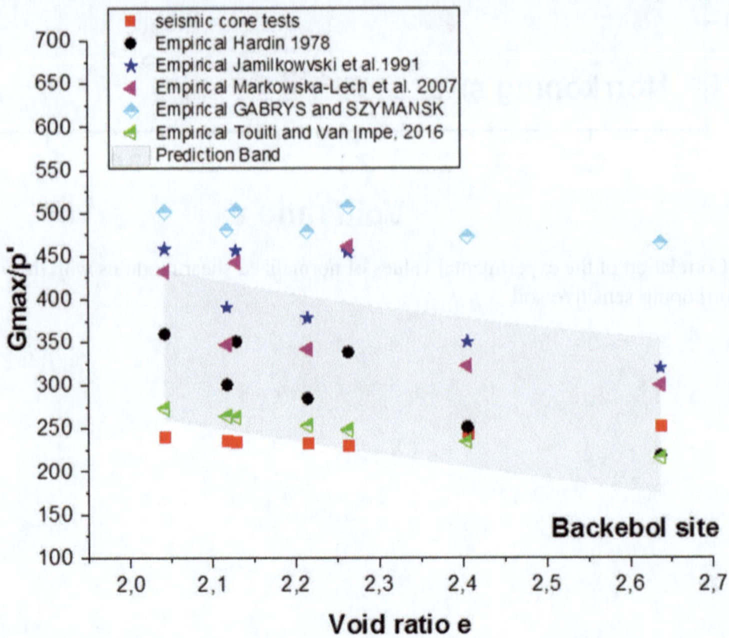

Fig. 16. The experimental and the estimated values of the normalized shear modulus with the void ratio for Bäckebol soft soil.

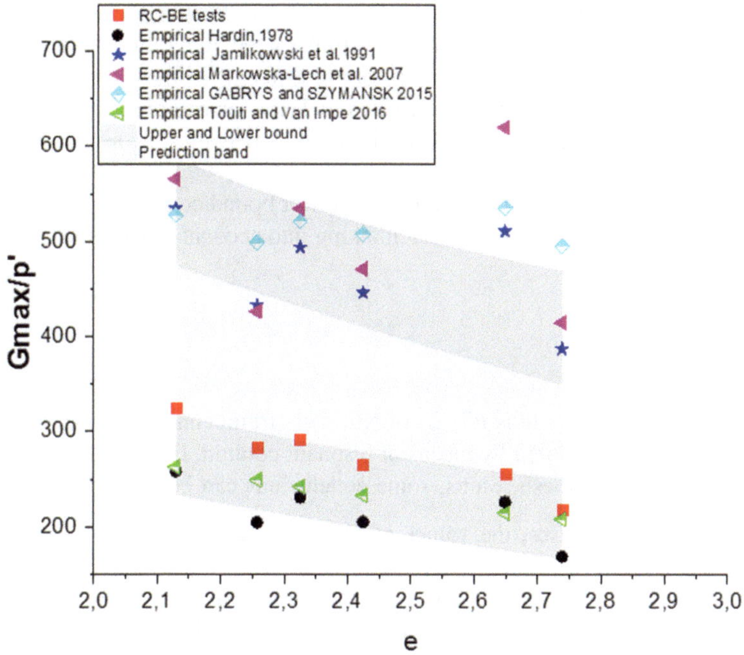

Fig. 17. The authors' experimental and the estimated values of the normalized shear modulus with the void ratio for Tunis soft soil.

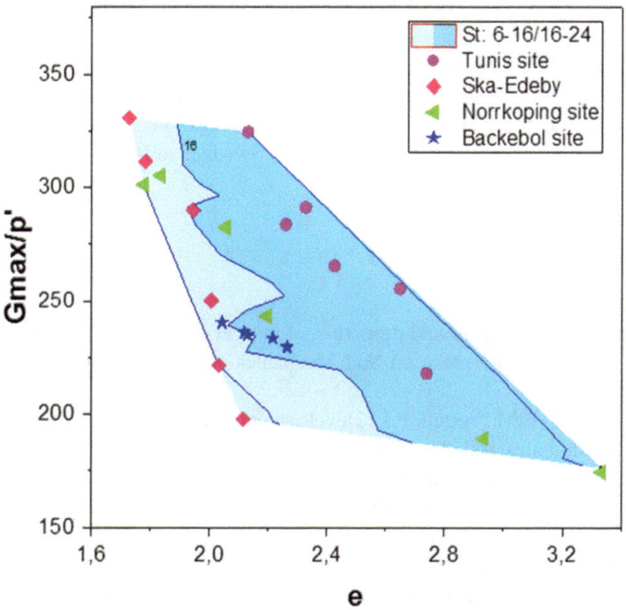

Fig. 18. The normalized small strain shear strength versus the void ratio by indicating the sensitivity range of each investigated soft soil.

The sensitivity is obviously a meaningful parameter affecting the small strain shear modulus (Fig. 18). Consequently, soft soils with the same void ratio and mean effective stress but different sensitivity values would have different stiffness.

The proposed equation fits the data related to Tunis, Norrköping and Bäckebol soft soils a lot more than those related to Skå-Edeby soft soil as it is less sensitive (Figs. 14, 15, 16 and 17).

Obviously, we are aware that the proposed lower bound correlation requires further verification by performing further research taking into account other sensitive soils and more diverse sites.

4 Conclusion

The small strain shear modulus (G_{max}) of soft soils from Tunis (Tunisia) and various sites in Sweden was measured by means of resonant column, Bender element and cone seismic tests. Based on these results, some conclusions can be drawn:

- In all the analyzed cases, the values of G_{max} do increase with an increase of the mean effective stress.
- The results clearly indicate that the G_{max} increases with a decreasing void ratio for most of the specimens, which is in agreement with results in literature.
- The authors proposed a suitable correlation pattern ($R^2 = 0.96$), enabling to cover up to 96% of the G_{max} variation.
- The soft soil sensitivity (St) is also a meaningful parameter when describing the small strain shear modulus of soft soils, however not fully and well enough understood.

Acknowledgements. The first author would like to acknowledge that all laboratory tests were conducted at the Geotechnical Institute for Experimental Models (ISMGEO Laboratory) in Italy and thank all the staff for their assistance. We would also like to thank Prof. Larsson for permitting us to use his experimental data about the Swedish soils.

References

Schnaid, F.: Geo-characterization and properties of natural soils by in situ tests. In: Proceedings of 16th International Conference on Soil Mechanics and Geotechnical Engineering, Osaka, pp. 3–45 (2005)

Stokoe II, K.H., Rathje, E.M., Axtell, P.J.: Development of an in-situ method to measure the non linear shear modulus of soil. In: Proceedings of the 16th International Conference on Soil Mechanics and Geotechnical Engineering, Osaka, pp. 751–754 (2005)

Nash, D.F.T., Lings, M.L., Pennington, D.S.: The dependence of anisotropic go shear moduli on void ratio and stress state for reconstituted Gault clay. In: Lancellotta&LoPresti, J. (eds.) Prefailure Deformation Characteristic of Geomaterial, pp. 229–238. Rotterdam (1999)

Benz, T.: Small-Strain Stiffness of Soils and its Numerical Consequences. Ph.D. thesis, University of Stuttgart (2007)

Schneider, J.A., Hoyos Jr., L., Mayne, P., Macari, E.J., Rix, G.J.: Field and laboratory measurements of dynamic shear modulus of Piedmont residual soils. ASCE Geotech. Spec. Publ. **92**, 12–25 (1999)

Wolski, W., Lipinski, M.J.: Site characterisation for geotechnical and environmental purposes. In: Proceedings of the XIIIth Danube-European Conference on Geotechnical Engineering, Ljubljana, pp. 129–150 (2006)

Kaâniche, A., Inoubli, M.H., Zargouni, F.: Développement d'un SIGG et réalisation d'un atlas géotechnique électronique. Bulletin Anglai deGéologie et d'environnement. Springer, Heidelberg (2000)

Touiti, L., Bouassida, M., Van Impe, W.: Discussion on the Tunis soft clay sensitivity. Geotechn. Geol. Eng. J. (2009). doi:10.1007/s10706-009-9263-2

Markowska-Lech, K., Lech, M., Szymanski, A.: Estimation of shear modulus from seismic tests on Pliocene clays. In: Proceedings of International Conference on Numerical Models in Geomechanics, London, pp. 153–157 (2007)

Larsson, R., Mulabdic, M.: Shear moduli in scandinavian clays. Swedish Geotechnical Institute, report n°. 40, Linkoping, pp. 40–94 (1991)

A Novel Field Device for the Measurement of Soil Collapsibility

Mehdi Mokhberi(✉) and Seyed Ayuob Rafieean

Islamic Azad University, Tehran, Iran

Abstract. Soil collapsibility or soil swelling is typically measured in the laboratory by using a single or double consolidation test. Such a test requires time and is costly: its stages include preparing the samples, transferring them to the laboratory, and conducting related tests. In most cases, geotechnical engineers have to carry out experiments during the stage of site investigation to evaluate the soil in terms of collapsibility. However, soil collapsibility might be easily estimated by a simple desert test. The device designed, developed, and used by the authors of the present study has the ability to determine the soil's collapsibility and offers its primary percentage. The experiments that were conducted on the downstream alluvium of Seevand Dam, with high collapsibility, to assess the performance of the device indicated that the device is able to appropriately evaluate the collapsibility of soil. Furthermore, comparing the results of these experiments with those of experiments using the consolidation method confirm the accuracy of the findings present study, with a high percentage. Given the probability of disturbing soils in laboratories, it might be contended that this device obtains results that are more valid than laboratory results.

Keywords: Collapsibility · Field test · Unsaturated soil · Double consolidation · Seevand Dam

1 Introduction

Collapsible soils are widespread in Iran and the world. These types of soils are usually known for their low natural humidity percentage and also low unit weight. Structures placed on collapsible soils, in the case of soil saturation, might settle unexpectedly. In the planning of most engineering structures, it is essential to pay special attention to soil collapsibility as it might destroy the foundation of a structure, destroy a dam, or produce road subsidence.

Furthermore, the existence of other technical buildings in these areas, and also given the development of cities and the necessities of the development of big cities, building residential areas, water and waste water pipes, etc. on these types of soil is of great importance (Derbyshire 2001). In the process of the development of soil mechanics science, which formally started in 1936, it was in the late 1960's and during the seventh and eighth international soil mechanics conferences that the investigation of the scientific principles of unsaturated soil was taken into consideration. In 1959, one of the first criteria of collapsibility based on dry density was offered by Clevenger (1958). After that, in 1962, Gibbs and Bara (1962) suggested using the unit weight of

W. Frikha et al. (eds.), *Soil Testing, Soil Stability and Ground Improvement*,
Sustainable Civil Infrastructures, DOI 10.1007/978-3-319-61902-6_4

dry density and soil smoothness limit as criteria for separating collapsible and non-collapsible oils. The trend of offering criteria for distinguishing collapsible and non-collapsible soils continued during the 1970's, 1980's, and 1990's, and different benchmarks were introduced by Denisov (1963), Feda (1966) and Fookes and Best (1969). In a Russian conference on soil mechanics and foundation engineering, the instability of a collapsible soil's skeleton was investigated and discussed. Jennings and Knight (1957) suggested the double consolidation test to assess the behavior of soil under conditions of saturation and loading with different levels of stress. Since the late 1980's, considerable attempts have been made to offer mathematical and computer models. In these models, the attention was on modeling the behavior of collapsible soils and the distribution process of these soils' porosity (Handy 1973; Houston et al. 1988). Since then, many studies have been carried out to estimate collapsibility, numeric modeling, failure model offering, etc. Similarly, different studies have also been conducted in Iran about the behavior of soils. For example, Habibagahi and Mokhberi (1998) investigated the effect of humidity on collapsibility estimation by modifying the relation offered by Duncan and Chang (1970) and also involving important parameters. They could offer a new model for soil failure. Tarantino et al. (2005) attempted to model the behavior of collapsible soils through laboratory and computational methods, especially the finite element method and using remolded specimens of loose soil and computer programs continue to be used. They have offered related research studies related to conducted experiments in their book.

The common methods of determining soil collapsibility require taking disturbed and non-disturbed soil samples. Such methods are not only time consuming and costly, but also change the behavior of the soil. Additionally, recreation of collapsible soils in the laboratory needs experience and special accuracy. The present research study is, in fact, an attempt to provide a small portable device to determine the collapsibility of soil. The soil collapsibility field identification device offered in the present study is used to evaluate the soil's collapsibility in non-laboratory and field contexts. The device is able to evaluate soils in terms of their potentials.

2 Materials and Method

2.1 Development of Device

The offered device consists of two sections, one for collapsibility and the other for loading. Figure 1 presents the whole of the device. In brief, the main parts of the device are as follows:

1. Stabilizer panel: This is a square rigid metal panel with sides of 30 cm and thickness of 1 cm, made from iron. There is a hole whose diameter is equal to the outer diameter of the cylinder in the limiting center of the cylinder within which the sampling cylinder is inserted.
2. Sampling cylinder: This is a hollow cylinder made from steel and is completely resistant to the materials of the soil. Its length is 13 cm.
3. Porous stone: This is placed into the sampling cylinder and above the soil sample.

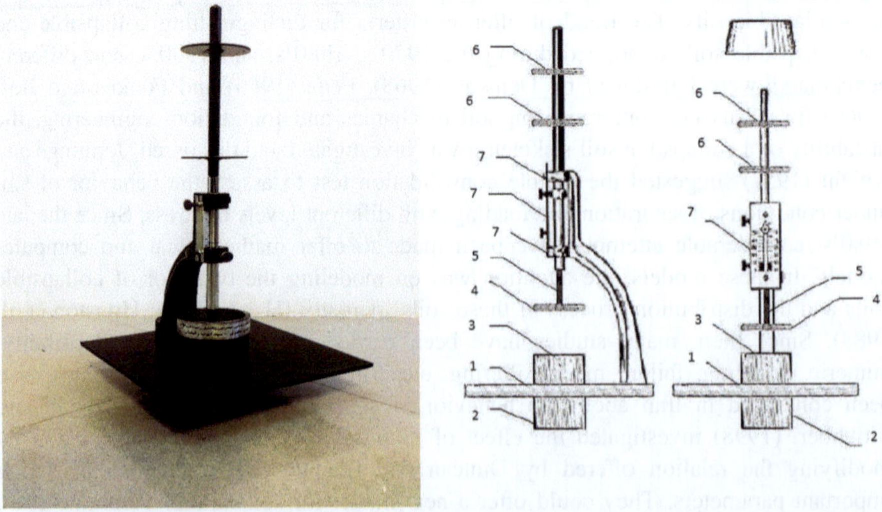

Fig. 1. Schematic design and image of the offered device to assess the soil collapsibility

4. Loading panel: This is made from steel and its diameter is equal to that of the porous stone. It is completely resistant to the soil materials and is an important point in the design of the device.
5. Moving loading bar: This is a solid metal bar of steel and is completely polished so that it can easily move as the soil settles.
6. Device Weights: The weight is imposed through the metal weights placed on the moving loading bar.
7. Sample Thickness Change Measuring Gauge: The gauge used in this device has 0.01 mm accuracy. There is a magnet in the foot of this gauge that might be connected to the cylinder limiting panel.

The information obtained from this device is used to assess the soil's collapsibility and also to calculate the indices that are about the relation between effective the stress and the strain or vacuum symptom. Additionally, the information resulting from this device might be used to increase or estimate speed, determine unsteady settlement and total settlement of a structure or collapsible soil piles.

2.2 Method

First, the device was placed and leveled in the intended area. Then, the user gradually puts pressure steadily on the two handles inserted on the cylinder stabilizer panel so that the cylinder penetrates into the soil. After that, filter paper and the porous stone were placed on the cylinder. Next, the primary load is done for 5 kPa. Then, within a 5-minute time period after the primary loading, further loading steps are gradually done at intervals of one hour in natural humidity conditions. The loading process continues until the settlement stops. The loading steps should be 10, 25, 50, 100, 200 kPa. Before each step, any change of form is observed. The amount of stress on the sample before its saturation is determined based on conditions. One hour after reaching the intended stress, any changes of the soil form in natural humidity is measured. After that, the sample is drowned by pouring water into the cylinder. Water steadily enters into the porous stone through the three holes on the panel. In this way, the whole of the sample is steadily saturated downwards in line with natural patterns. After drowning the sample, the sample form changes are read and observed at time intervals of approximately 0.1, 0.25, 0.5, 1, 2, 4, 8, 15, 30, and 60 min. It should, however, be mentioned that in soils with high penetration, collapsibility might happen quickly and makes reading the time difficult. Adding water to the sample should be in a way that it is saturated downwards and air is not imprisoned in the soil. After adding water into the cylinder and drowning the sample, the necessary time for putting on a load would continue for one day or until the primary consolidation (based on D2435 standard) is finished. The degree of sample form change is read in the above-mentioned time intervals. Finally, the information obtained from the device is used to determine the collapsibility percentage of non-saturated soils after being drowned.

2.3 Evaluation of Test Results

2.3.1 Results of Test Device by Using the Device

In this section, the results obtained from the tests using the developed device and those of an experimental consolidation device are presented. These tests were done on three points of soil in the downstream area of Seevand Dam located 100 km to the north of Shiraz, which is known to be collapsible soil. The area soil is of CL type. Table 1 presents the soil consolidation properties in this area.

Table 1. The consolidation properties of soil of the studied area

Station	e_0	e_f	$\omega_0(\%)$	C_s	C_c	G_s
P1	0.95	0.23	5	0.06	0.31	2.56
P2	0.88	0.19	6	0.05	0.3	2.56
P3	0.99	0.25	5	0.06	0.33	2.56

To find the soil collapsibility using the device, the soil changes were recorded and measured after placing the device and putting the load on the soil. Figures 2a to c and Table 2 show the soil changes before saturation, during saturation, and after saturation.

2.3.2 Laboratory Consolidation Device Results

The single consolidation tests were carried out based on standard (ASTM D 5333-03) to determine the soil's collapsibility. In this experiment, a sample of soil is placed in the oeodometer device and vertical pressure is increased until reaching the probable pressures of real earth. At this loading level, water is injected into the device, the sample is saturated, and resulting changes are recorded.

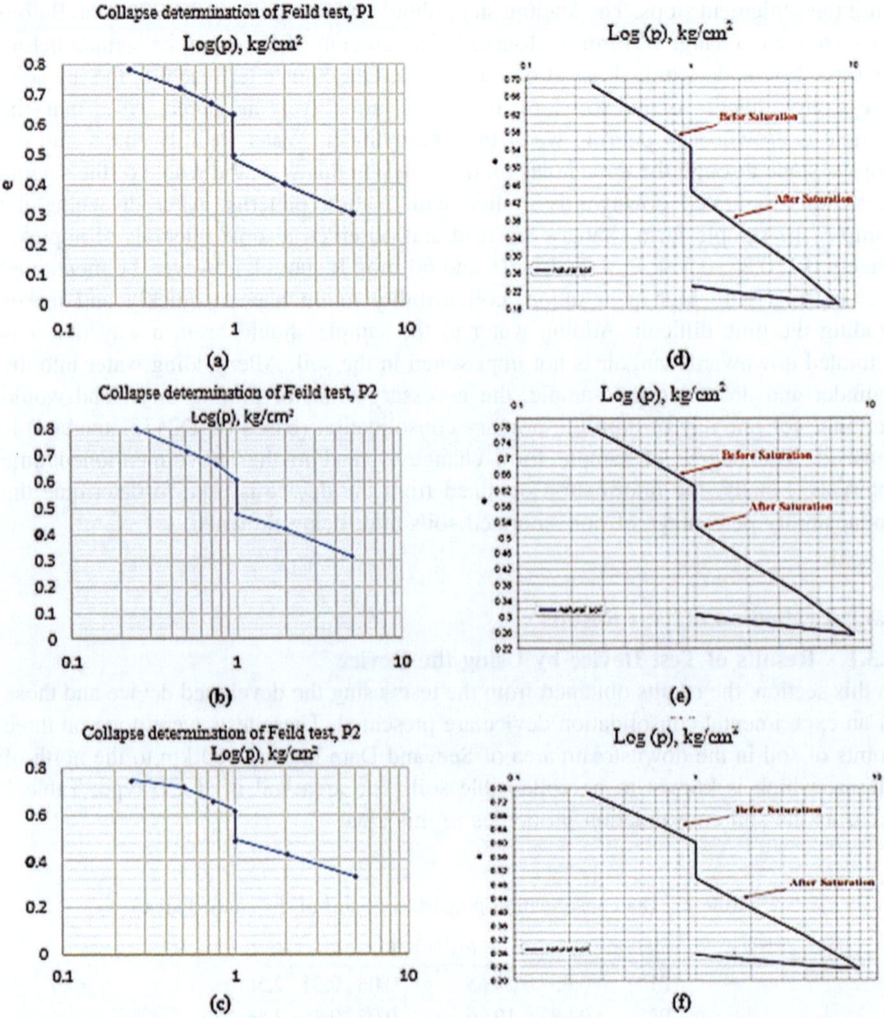

Fig. 2. Comparison of soil collapsibility from offered device (left) and single consolidation tests (right)

Table 2. Comparison of soil collapsibility from offered device and single consolidation test

Station	Collapse potential from oedometer test (%)	Collapse potential from proposed device (%)
P1	6.9	7.8
P2	6.1	7.2
P3	5.8	7.1

The consolidation graphs of three different natural soils have been presented below. It should, however, be pointed out that in the consolidation tests, stresses of 25, 50, and 100 kPa were loaded in natural humidity status and then the samples were consolidated with stresses of 100, 200, 400, and 800 kPa. After that, the loading process was done to the weight of one kilogram on one cubic centimeter. Figures 2d to f and Table 2 show the related results.

2.3.3 Comparison of Soil Collapsibility Determined by the Two Methods

The soil collapsibility was determined by using (Lutenegger and Saber 1988) relation (1).

$$I = \frac{\Delta e}{1 + e} \times 100 \qquad (1)$$

As Fig. 2 and Table 2 show, the collapsibility of the reconstructed sample is approximately 6.3 in the laboratory and 7.4 in the field, which implies the correct performance of the device.

3 Conclusions

The comparison of the results obtained from the developed device and those from an experimental consolidation device shows that the developed device has appropriately assessed the collapsibility of the area soil. Additionally, due to its description of the soil collapsibility, it has a better and more complete capability than other comparable devices. Therefore, it could be used in similar projects of the district as a suitable criterion to document the soil collapsibility. In fact, the soil collapsibility field device is a device that has removed the technical problems of existing related devices and could simulate the real and natural conditions of soil and as a result, the obtained results are completely coincident with the natural patterns of the soil. This device, in essence, increases the accuracy and output of the work considerably.

References

Clevenger, W.A.T.: Experiences with loess as foundation material. Trans. Am. Soc. Civ. Eng. **123**(1), 151–169 (1958)

Denisov, N.Y.: About the nature of high sensitivity of Quick clays. Osnov. Fudam. Mekh. Grunt. **5**, 5–8 (1963)

Derbyshire, E.: Geological hazards in loess terrain, with particular reference to the loess regions of China. Earth Sci. Rev. **54**(1), 231–260 (2001)

Duncan, J.M., Chang, C.-Y.: Nonlinear analysis of stress and strain in soils. J. Soil Mech. Found. Div. (1970)

Feda, J.: Structural stability of subsident loess soil from Praha-Device. Eng. Geol. **1**(3), 201–219 (1966)

Fookes, P.G., Best, R.: Consolidation characteristics of some kate Pleistocene periglacial metastable soils of East Kent. Q. J. Eng. Geol. Hydrogeol. **2**(2), 103–128 (1969)

Gibbs, H., Bara, J.: Predicting surface subsidence from basic soil tests. In: Field Testing of Soils. ASTM International (1962)
Habibagahi, G., Mokhberi, M.: A hyperbolic model for volume change behavior of collapsible soils. Canad. Geotech. J. **35**(2), 264–272 (1998)
Handy, R.L.: Collapsible loess in Iowa. Soil Sci. Soc. Am. J. **37**(2), 281–284 (1973)
Houston, S.L., Houston, W.N., Spadola, D.J.: Prediction of field collapse of soils due to wetting. J. Geotech. Eng. **114**(1), 40–58 (1988)
Jennings, J., Knight, K.: The additional settlement of foundations due to a collapse of structure of sandy subsoils on wetting. Proceedings (1957)
Lutenegger, A.J., Saber, R.T.: Determination of collapse potential of soils (1988)
Tarantino, A., Romero, E., Cui, Y.J.: Advanced experimental unsaturated soil mechanics. In: Proceedings of the International Symposium on Advanced Experimental Unsaturated Soil Mechanics, Trento, Italy, 27–29 June 2005. CRC Press (2005)

Moisture Variation in Expansive Subgrade Through Field Instrumentation and Geophysical Testing

Asif Ahmed[1(✉)], MD Sahadat Hossain[1], Mohammad Sadik Khan[2], Kelli Greenwood[3], and Aya Shishani[3]

[1] Civil Engineering, University of Texas at Arlington, Texas, USA
asif.ahmed0@mavs.uta.edu, hossain@uta.edu
[2] Civil and Environmental Engineering, Jackson State University, Mississippi, USA
mohammad_sadik.khan@jsums.edu
[3] Department of Civil Engineering, University of Texas at Arlington, Texas, USA
{kelli.greenwood,aya.shishani}@mavs.uta.edu

Abstract. Seasonal climatic variations in subgrade soil affect pavement responses and can reduce pavement serviceability. In addition to causing shrinkage and swelling in expansive subgrade, variations in moisture suction may alter the material properties of soil, which ultimately affect pavement performance. The current study monitored the seasonal variation of subgrade moisture content, rainfall, and pavement deformation of a section over State Highway 342, in Dallas, Texas. Moisture sensors were installed at different depths up to 4.5 m. The soil was found to be highly plastic clay (CH) in the selected site. In addition to acquiring real-time moisture data from sensors, geophysical testing was also conducted on the slope of the pavement. Electrical Resistivity Imaging (ERI) was carried out at the slope of the instrumented pavement section to observe the moisture flow at the edge of the pavement. Based on the field monitoring data, moisture variation ranged from 5% to 14%, with higher moisture contents correlating with rainfall events. While moisture sensors provided point information, resistivity imaging yielded a continuous portrayal of subsurface moisture flow. Furthermore, rainfall-associated deformation of the pavement was monitored. Based on the monitored data, it was observed that pavement deformation varied with rainfall. A total deformation of 38 mm was recorded over the monitoring period.

1 Introduction

The engineering properties of expansive soils are highly dependent on moisture content changes in the active zone. Such soils undergo volumetric deformation that gradually damages essential infrastructure such as foundation slabs, bridges, roadways, and residential homes. Present in both humid and arid/semi-arid environments, expansive soils cover nearly a quarter of the area of the United States (Nelson and Miller 1992). Annually, expansive soils alone incur more financial losses to US property owners than

W. Frikha et al. (eds.), *Soil Testing, Soil Stability and Ground Improvement*,
Sustainable Civil Infrastructures, DOI 10.1007/978-3-319-61902-6_5

earthquakes, flood, hurricanes, and tornadoes combined (Jones and Jefferson 2012). In a typical year, the associated financial losses can be as high as 15 billion dollars (Jones and Jefferson 2012). Their low stiffness, light loading, and extended presence over the country render pavements especially susceptible to deterioration caused by expansive subgrades. Continuous volumetric deformation of the problematic soils increases the pavement's roughness, resulting in reduced serviceability. Consequently, a better understanding of the causes of pavement distress and the behavior of expansive soils is necessary for acceptable pavement performance and design (Sebesta 2002).

Climatic factors, especially moisture variations, induce substantial changes in the structural characteristics of pavement systems. Fluctuations in moisture content can have major impacts on the long-term performance of pavements. Principal sources of moisture variation are rainfall, intrusion from cracks, freeze-thaw cycles, leakage, and evapotranspiration. Moisture variation affects the hydraulic conductivity, shear strength, chemical diffusivity, specific heat, and thermal conductivity of a soil Lu 2015). Furthermore, the success or failure of a pavement system is dependent on the support provided by subgrade layers. An increase in moisture content has been shown to decrease the resilient modulus, which quantifies the support that the subgrade can offer (Mehrota 2011). Moreover, variation in moisture content may cause swelling and shrinkage in highly plastic subgrade soils. Surficial cracks are generated due to this behavior and require departments of transportation to increase expenditures related to pavement maintenance (Zapata and Houston 2008). Because expansive subgrade is seldom identified as the source of pavement failure, maintenance routines typically only consist of roadway surface treatment. Consequently, surface roughness can reappear a short period of two years and six months later. For 18 out of 52 Texas Department of Transportation (TxDOT) districts, such problems are commonplace (Wanyan et al. 2010).

To consider the possible effects of moisture in the design procedures, many researchers have conducted field-based analyses to relate subgrade characteristics with environmental factors. Bayomy and Salem (2004) instrumented five different sites in Ohio. After monitoring for a period of five years (1999–2003), the researchers reported the presence of seasonal moisture variation. Manosuthkij (2008) monitored four instrumented sites in Texas and found that when the mean moisture content, or the difference between the maximum and minimum moisture content of each month, was greater than 20%, edge cracking was likely to occur. In Ohio, Heydinger (2003) obtained both seasonal variation and temporal variation due to precipitation at two instrumented sites.

In addition to installing moisture sensors that monitor moisture variation, geophysical investigation can also be significant when observing the moisture flow in subgrade. ERI is widely used in hydrogeological, environmental, and geotechnical research due to its potential to reveal the subsurface image (Aizebeokhali 2010). It is a convenient method for determining spatial variations of moisture in subsoil (Kibria and Hossain 2012). In the geotechnical and geomorphological fields, ERI is gaining popularity for the user-friendly data acquisition system inherent to the method. The technology is used in geotechnical applications to determine unknown foundation depths, investigate slope and foundation failures, detect cracks and the existence of sinkholes, record moisture variation in pavement base materials, and track soil movement with accurate results. ERI can measure both the horizontal and vertical variability of soil arrangement over several meters of a perspective area (Tabbagh et al. 2000).

The resistivity profile provides a comprehensive interpretation of the subsurface condition. The most interesting and important advantage of using ERI is that the resulting resistivity profile offers a clear picture of the moisture distribution within the test area. Accordingly, numerous studies have utilized ERI to study moisture distribution in soil. Benderitter and Schott (1999) observed the water content variation in an unsaturated soil and the effect of rainfall by observing the short-term variation of resistivity. Clarke (2006) measured subgrade resistivity over nine months to determine the active zone, edge moisture variation distance, and long-term equilibrium moisture beneath the covered area. The author also monitored the seasonal fluctuations of moisture at the edge. The study found that the most extreme changes occurred at the surface and diminished with depth. Kean et al. (1987) studied the moisture migration in the vadose zone using resistivity tomography. Finally, Zhou et al. (2001) performed spatial and temporal monitoring of soil water content using electrical resistivity tomography.

During the current study, both dielectric sensors and two-dimensional electrical resistivity imaging techniques were utilized to study the moisture variations in expansive subgrade and pavement slope. The objective of the study was to determine the variation of subgrade moisture due to both seasonal variation and precipitation events. The study was conducted in a state highway designated as SH 342 in Ellis County, Texas, USA. Sensors were installed up to 4.5 m depth beneath the driving lanes. ERI was conducted at the slope of the pavement during different seasons. A rain gauge was also installed to record the precipitation events.

2 Project Background and Site Description

The effects of moisture on soil depend on the soil properties. Based on accessibility, distance from the flooding zone, and TxDOT recommendations, a section on State Highway (SH) 342, a two-lane hot mix asphalt road, was selected for investigation. Two 3.3 m (11 ft.) wide lanes with a 3.3 m (11 ft.) wide shoulder on each side of the road were present at the site. Details of the chosen section can be seen in Fig. 1. The project site was located near Lancaster, Texas, on the border between Ellis and Dallas counties. While the pavement was fairly level, edge cracking and other pavement distresses were observed. Adjacent to the pavement, the ground was covered with grass. Additionally, dense trees could be observed on both sides approximately 6 m (20 ft.) from the pavement. To the west, a rail line ran parallel to the roadway. Further northeast, the road overpassed Bear Creek, while a residential area was located a short distance to the southeast. No water body was directly observed nearby.

Fig. 1. State Highway (SH 342) in Dallas, Texas, USA

3 Field Instrumentation

A comprehensive field instrumentation plan was implemented at SH 342 to monitor the moisture variations in the subgrade soil. Decagon 5TM moisture sensors, an ECRN-100 high resolution tipping-bucket rain gauge, an EM50 data logger, and a horizontal digitilt inclinometer probe were installed at the site. The moisture sensors and the 85 mm horizontal inclinometer casings were installed from the centerline to the edge of the northbound lane, as seen in Fig. 2.

Continuous monitoring of moisture content, rainfall, and vertical deformation were obtained from field instrumentation. Data loggers were programed to take hourly readings of moisture content, and rainfall data. Furthermore, the pavement site was visited monthly to obtain inclinometer readings. Fifteen moisture sensors and six 3 m (10 ft.)

long horizontal inclinometer castings were installed. The depth of the sensors ranged from 1.2 m (4 ft.) to 4.5 m (15 ft.) below the pavement.

(a) (b)

Fig. 2. (a) Instrumentation Plan, (b) Instrumentation at the site SH 342

4 Soil Properties

The geotechnical properties of soil samples collected from SH 342 during field instrumentation were obtained through laboratory testing. Upon investigation of the particle size distribution, the samples contained very fine subgrade soil. All samples were composed of over 85% clay. The liquid limits ranged between 50% and 64%, while plasticity ranged between 28% and 42%. As specified by the Unified Soil Classification System (USCS), sieve analysis and Atterberg limits results indicated that the soil samples were high-plastic clay (CH). Specific gravity ranged between 2.68 and 2.72, with an average of 2.70. Optimum moisture content was determined to be 22%, and dry density at the optimum moisture content was found to be 18.9 kN/m^3.

5 Moisture Variation in Subgrade by Sensor

Subgrade moisture sensor data was recorded along the center and edge of pavement at SH 342 from March 2014 to April 2016. Moisture content variation was determined through analysis of the recorded moisture data. The following discussion presents interpretable results obtained from the center borehole, which were representative of data obtained from remaining sensors at the edge.

To obtain moisture data, sensors were installed in the center borehole at depths of 1.2 m (4 ft.), 2.4 m (8 ft.), 3.6 m (12 ft.), and 4.5 m (15 ft.). Moisture variation in the center borehole is shown in Fig. 3. While rainfall was immediately detected by the shallower sensors, the deepest sensor (TM 15, located at 4.5 m [15 ft.] depth) observed no detectable moisture content changes in response to rainfall. TM 15 recorded an

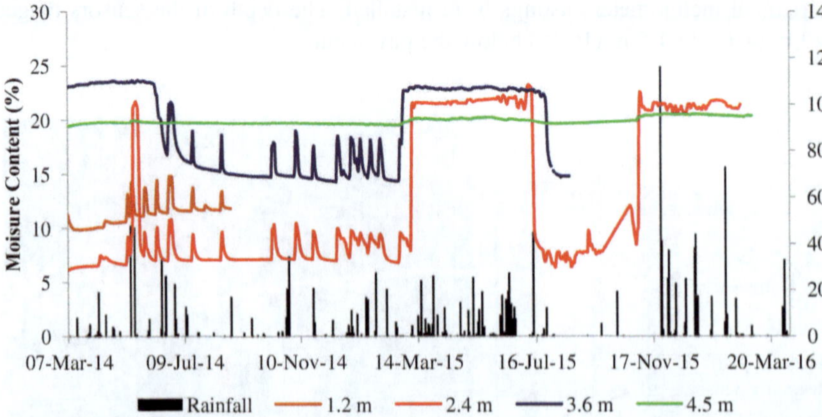

Fig. 3. Moisture variation across center borehole (BH-5)

average soil moisture content of 19%, which corresponded to the point of saturation at the depth of 4.5 m (15 ft.). Moisture contents stabilized to equilibrium moisture contents at all depths. Respectively, equilibrium moisture contents for TM 4, TM 8, and TM 12 were 11%, 6%, and 15%. Moisture readings immediately following rainfall events rose 4% to 15% in amplitude, with increases limited to the temporary saturation point at the respective depths. Shallow sensors experienced greater increases in moisture content. After draining, moisture contents returned to equilibrium. Temporary changes in moisture content are important to study in the case of expansive clay soils. Brief fluctuations in moisture can significantly affect the volumetric deformation of the soil.

The monthly variation in total rainfall and maximum moisture content recorded by the sensors is shown in Fig. 4. At shallow depths, the rises and falls in maximum moisture content mirrored the behavior of rainfall. That is, when precipitation

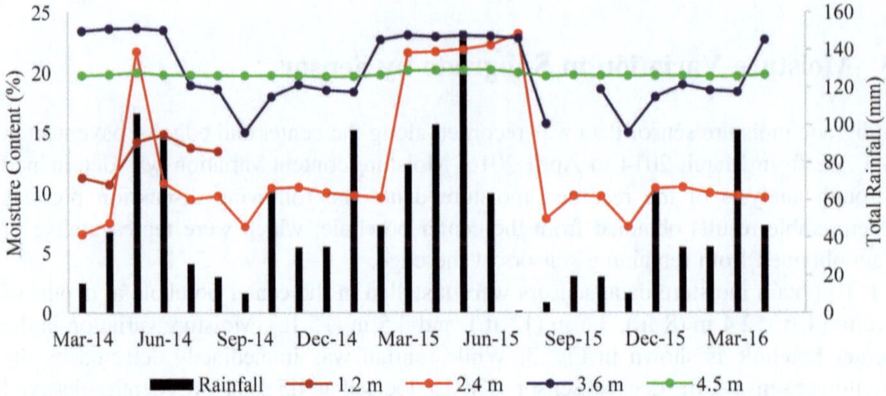

Fig. 4. Maximum moisture variation across center borehole (BH-5)

experienced a decline during the dry season, maximum moisture content also dropped. For example, the dry periods from May to September 2014 and May to August 2015 saw sensors at 2.4 m (8 ft.) experience a net decrease of up to 15% in recorded moisture content. Conversely, rising levels of rainfall during the wet season produced increasing magnitudes of moisture content. For instance, the accumulation of nearly 91 mm (3.6 in.) of rainfall between March and May 2014 returned a net increase of 15% in moisture content. Due to the expansive nature of clayey subgrade, the pavement at SH 342 could be expected to deform seasonally in response to the rises and falls in moisture content.

While subgrade moisture content varies in response to rainfall events, moisture content may also vary due to seasonal effects. Seasonal contributions to moisture variation can depend on the soil compaction, soil type, ground water table, and initial moisture content. Consequently, some locations observe seasonal variations while others do not. After five years of monitoring in Ohio, Heydinger (2003) reported that a seasonal effect on moisture variation existed at the study site. In Texas, Manosuthikij (2008) found a seasonal contribution to moisture variation at sites in Fort Worth and Houston, whereas the study site in San Antonio did not experience a seasonal trend. To ascertain if a seasonal pattern existed at SH 342, electronic resistivity imaging (ERI) was performed across the grass slope adjacent to the pavement. This method of geophysical testing provides a continuous picture of moisture content distribution. Comparison of ERI results obtained from a monthly basis could help conclude whether moisture variation at the site underwent seasonal changes.

6 Theory of ERI

ERI is an active geophysical method which measures the electric potential differences at specific locations after injecting a controlled electric current at other locations. By controlling the current injected in an entirely homogeneous half-space, a resistivity value can be calculated for the subsurface by measuring the resulting electric potential difference. Figure 5 illustrates the concept of subsurface electric current flow and the influence of subsurface heterogeneities.

The ERI method follows Ohm's law in that resulting potential differences are measured by transferring artificially generated currents to the soil. The electrical resistivity depends on several factors of the subsurface condition. Main factors influencing resistivity include the size, shape, arrangement, mineralogy, voids distribution, porosity, connectivity between the particles, degree of saturation, concentration of soluble minerals, and temperature of the subsurface soil.

For a simple soil body, the resistivity ρ (ohm-m) is defined as:

$$\rho = R(A/L) \tag{1}$$

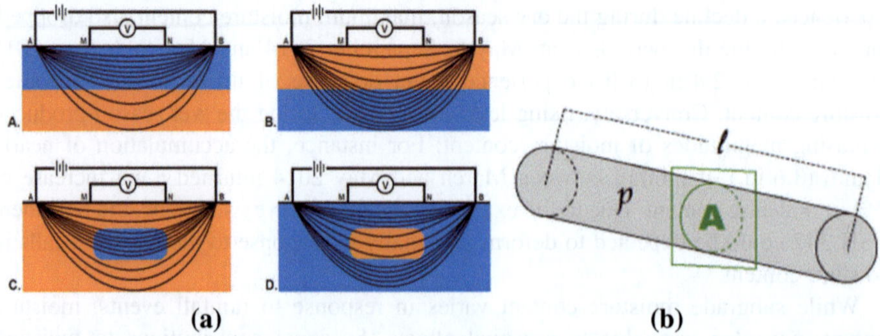

Fig. 5. (a) Variations in subsurface electric current density (Aizebeokhai 2010) (b) Relation between resistance and resistivity (Tabbagh et al. 2000)

Where, R is the electrical resistance, L is the length of the cylinder (m) and A is the cross-sectional area (m^2). The electrical resistance of a cylindrical body is defined by Ohm's law:

$$R = V/I \tag{2}$$

Where, V is the potential difference measure in Volt, and I is the current in Ampere.

7 Geophysical Testing by Resistivity

While moisture sensors returned distinct moisture content recordings, further information regarding moisture variation was desired. Thus, geophysical testing in the form of resistivity imaging was utilized. Geophysical testing provides a continuous picture of the moisture variation beneath the pavement. Since sensors were not installed beneath the grass slope adjacent to the pavement, geophysical testing allowed for determination of the moisture variation beneath the slope beside the pavement.

Instrumentation required for ERI included a super-sting R8/IP resistivity meter manufactured by Advanced Geosciences Institute (AGI) and a switch box with a 12-volt battery for power supply. RI testing was conducted with 28 electrodes placed at 0.9 m (3 ft.) intervals, resulting in a test line of 24.7 m (81 ft.). Improved resolution in the horizontal and vertical directions was obtained by employing a dipole-dipole array (Manzur et al. 2016). The schematic of the layout of testing and site photos are presented in Fig. 6. The collected data were analyzed in the Earth Imager 2D software (AGI 2004), which uses a forward modeling technique to calculate apparent resistivity values from the field data. The software uses non-linear least-squares optimization technique to yield the final output as 2D resistivity image of the subsurface.

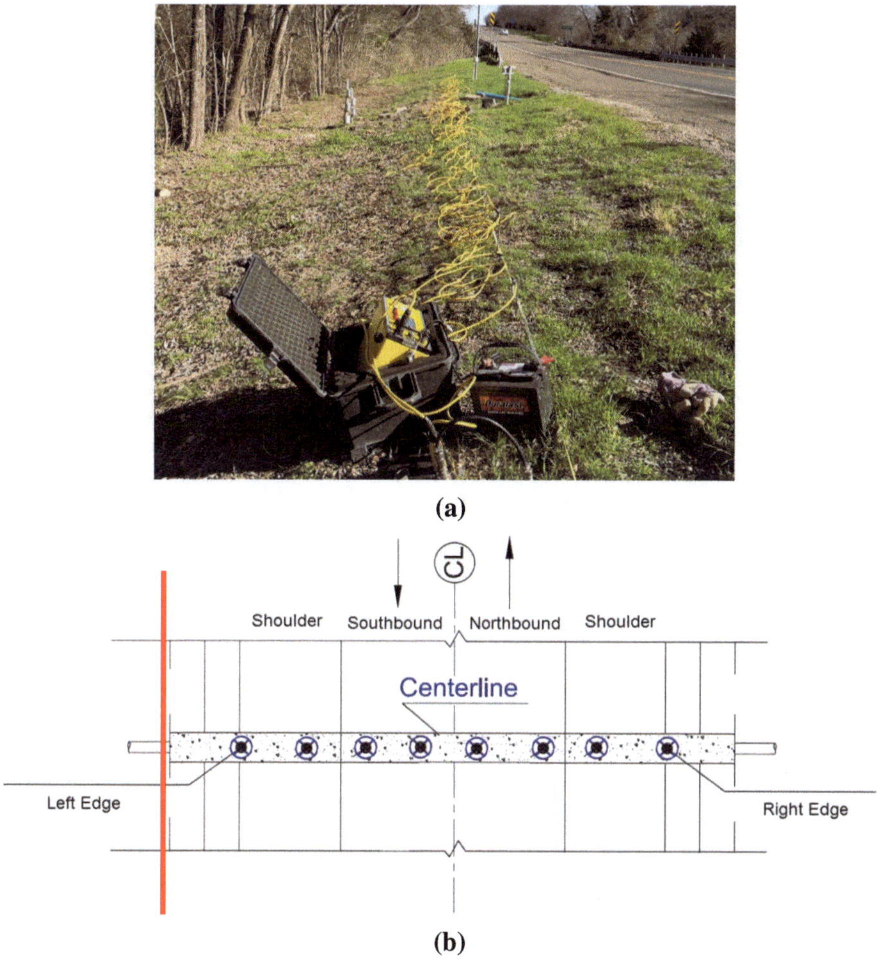

Fig. 6. (a) Field setup for resistivity (b) schematic at slope of pavement

8 Moisture Variation in Subgrade by Resistivity

ERI provided a continuous picture of the moisture content distribution at different levels and locations of the subgrade soil. An inverse relationship between resistivity and saturation was used to interpret the qualitative data presented in Fig. 7. High readings of resistivity corresponded to lower moisture contents, while low readings of resistivity translated to higher moisture contents. The scale from red to blue in Fig. 7 indicated levels of high and low resistivity values, respectively. ERI performed during the dry season (from May to October) returned high resistivity readings and produced the image depicting moisture content distribution shown in Fig. 7(a). Conversely, ERI testing during the wet season (from November to April) produced Fig. 7(b) with low resistivity values. ERI conducted immediately following a rainfall event revealed the

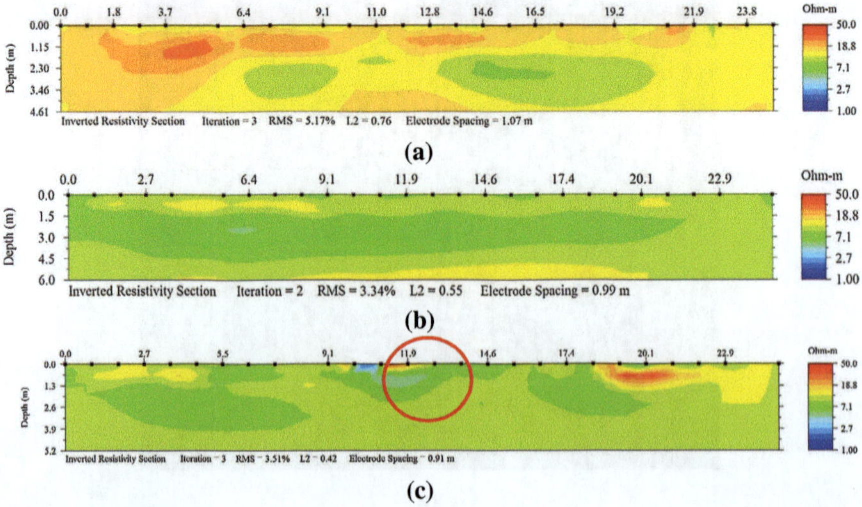

Fig. 7. Resistivity variation in (a) dry period, (b) wet period, and (c) moisture intrusion through edge after rainfall

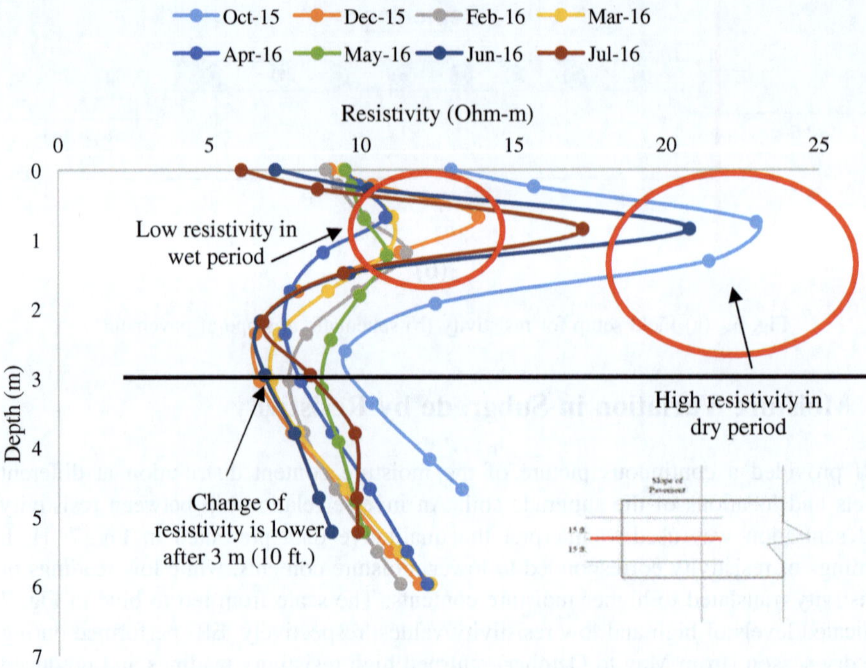

Fig. 8. Variation of resistivity across depth in central section

presence of cracks or holes in the pavement. An apparent edge crack appeared in May 2016, as seen from the region of low resistivity developing in Fig. 7(c). Low resistivity values where higher values were previously recorded can indicate the presence of cracks, potentially serving to determine which areas of pavement need the most maintenance.

Results of resistivity variation across depth produced the plot shown in Fig. 8. Resistivity values ranged between 7 to 13 ohm-m between the months of November and April, translating to high readings of moisture content. Thus, the period was identified as the wet season. In contrast, the period between May and October returned resistivity values as high as 23 ohm-m, corresponding to the dry season. Geophysical testing revealed the presence of a seasonal trend. Given the relationship between resistivity and moisture content, a seasonal effect on resistivity suggested that a seasonal effect on moisture content variation also existed. Below a depth of 3 m (10 ft.), resistivity did not undergo significant fluctuations. This finding corroborated findings obtained from other studies conducted in the same area, which identified 3–3.66 m (10–12 ft.) as the depth of the active zone (Khan et al. 2015).

9 Deformation Analysis

Both temporal and seasonal changes in subgrade moisture content induce volumetric deformation in expansive clay soils. Pavement performance can be negatively affected as a response to subgrade deformation. To monitor the pavement deformation at SH 342, monthly inclinometer results were obtained over a period of two years and are presented in Fig. 9.

The inclinometer profile showed that swelling and shrinking of the expansive subgrade caused up to 38 mm (1.5 in.) of vertical movement across the pavement. A cyclical trend in deformation was readily observed by plotting pavement deformation

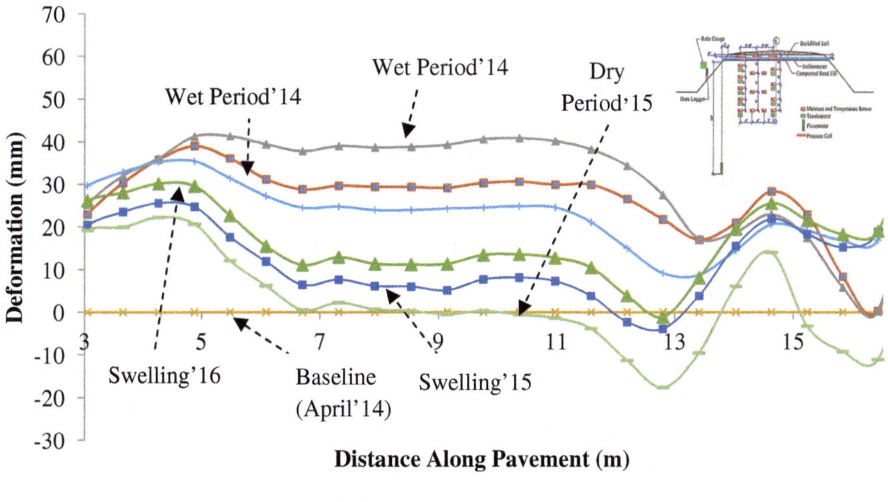

Fig. 9. Variation of deformation along distance in central section

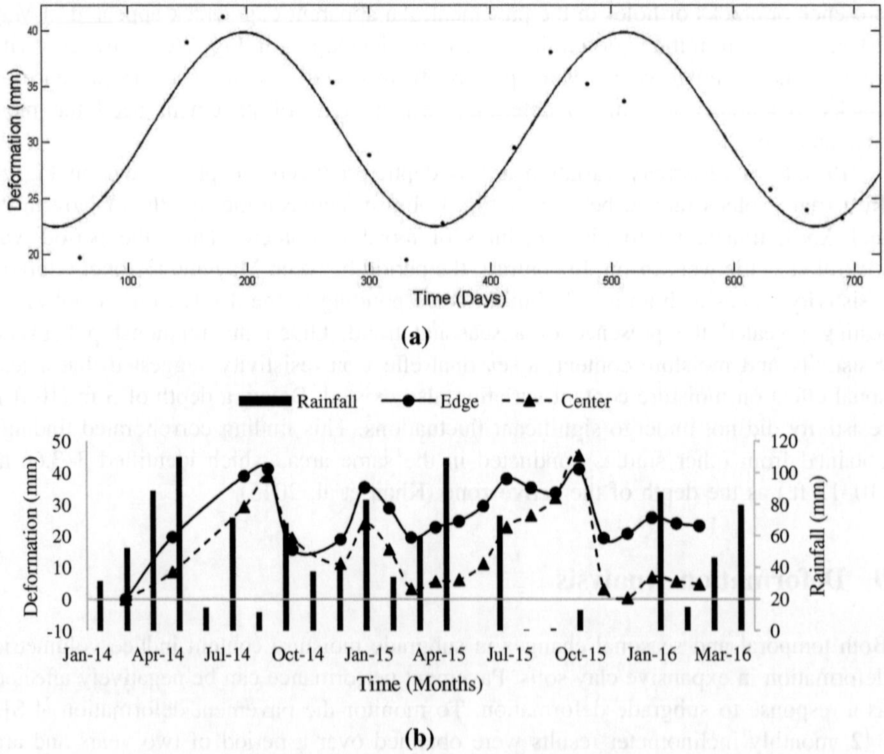

Fig. 10. (a) Seasonal trend in deformation, (b) movement in response of rainfall events

of edge over time (Fig. 10a). Seasonal effects were further examined by including precipitation in a plot of deformation over time (Fig. 10b). Changes in pavement elevation were seen to correlate with rainfall events. For example, the center of the pavement experienced 38 mm (1.5 in.) of swelling as the result of 114 mm (4.5 in.) of rainfall during the wet season from March to May 2014. After the summer of 2014, during which precipitation levels declined, the pavement dropped by almost 25 mm (1 in.). Another pattern of swelling and shrinkage was observed in October 2015, when the pavement dropped 38 mm (1.5 in.) in response to a decrease in rainfall. However, pavement elevation rose again as precipitation levels increased upon the arrival of the wet season. Pavement deformation mirrored rainfall patterns which followed a seasonal trend. Thus, pavement deformation was also concluded to adhere to a seasonal trend.

10 Conclusions

The current study was conducted to determine the variation in subgrade moisture content resulting from seasonal effects and rainfall events. A two-lane road identified as SH 342 in Dallas, Texas, USA was instrumented with moisture sensors to record moisture content and a rain gauge to record precipitation. Geophysical testing using electrical resistivity imaging was conducted to determine the presence of a seasonal effect on subgrade moisture variation. Pavement deformation was recorded using a horizontal inclinometer installed at the site. The observed results are summarized below.

- Laboratory testing indicated that subgrade soil at SH 342 was made of highly plastic clay.
- Moisture sensors installed in the pavement captured temporary net increases of 4% to 15% moisture content following rainfall events.
- Electrical resistivity imaging performed across the grass slope adjacent to the pavement captured the presence of a seasonal effect on moisture variation. The detection of water intrusion also revealed the development of cracks in the pavement.
- Subgrade shrinkage and swelling was observed to follow a seasonal trend. Wet periods (from November to April) caused a rise in pavement elevation, while dry periods (from May to October) resulted in a drop. Up to 38 mm (1.5 in.) of deformation was recorded.

The observation of moisture variation beneath pavement can be linked to deformation behavior of the pavement built on expansive subgrade. A complete deformation model in response to the moisture variation can be effective to predict the pavement performance. It will ultimately lead to better pavement preservation technique.

Acknowledgments. The project was funded by Texas Department of Transportation (TxDOT) and is greatly acknowledged.

References

Advanced Geosciences, Inc.: Instruction Manual for Earth Imager 2D Version 1.7.4. Resistivity and IP Inversion Software. Advanced Geosciences, Inc., Austin (2004)

Aizebeokhai, A.P.: 2D and 3D geoelectrical resistivity imaging: theory and field design. Sci. Res. Essays **5**(23), 3592–3605 (2010)

Bayomy, F., Salem, H.: Monitoring and modeling subgrade soil moisture for pavement design and rehabilitation in Idaho: Phase III: Data collection and analysis (Final report). University of Idaho: National Institute for Advanced Transportation Technology (2004)

Benderitter, Y., Schott, J.J.: Short time variation of the resistivity in an unsaturated soil: the relationship with rainfall. Eur. J. Environ. Eng. Geophys. **4**(1), 37–49 (1999)

Clarke, C.R.: Monitoring long-term subgrade moisture changes with electrical resistivity tomography. In: Unsaturated Soils, pp. 258–268 (2006)

Heydinger, A.G.: Evaluation of seasonal effects on subgrade soils. Transp. Res. Rec. J. Transp. Res. Board **1821**(1), 47–55 (2003)

Khan, M.S., Hossain, S., Kibria, G.: Slope stabilization using recycled plastic pins. J. Perform. Constructed Facil. **30**(3), 04015054 (2015)

Kean, W.F., Waller, M.J., Layson, H.R.: Monitoring moisture migration in the vadose zone with resistivity. Ground Water **27**(5), 561–562 (1987)

Kibria, G., Hossain, M.S.: Investigation of geotechnical parameters affecting electrical resistivity of compacted clays. J. Geotech. Geoenviron. Eng. **138**(12), 1520–1529 (2012)

Jones, L.D., Jefferson, I.: Expansive soils. In ICE Manual of Geotechnical Engineering, Volume 1: Geotechnical Engineering Principles, Problematic Soils and Site Investigation, pp. 413–441. ICE Publishing, London (2012)

Lu, Y.: Temperature effect on unsaturated hydraulic properties of two fine-grained soils and its influence on moisture movement under an airfield test facility (MS thesis). Arizona State University (2015)

Manosuthikij, T.: Studies on volume change movements in high PI clays for better design of low volume pavements (Doctoral dissertation). University of Texas at Arlington (2008)

Manzur, S.R., Hossain, M.S., Kemler, V., Khan, M.S.: Monitoring extent of moisture variations due to leachate recirculation in an ELR/bioreactor landfill using resistivity imaging. Waste Manage., 38–48 (2016)

Mehrotra, A.: Evaluating the influence of moisture variation on resilient modulus for unsaturated pavement subgrades (MS Thesis). Louisiana State University (2011)

Nelson, J., Miller, J.D.: Expansive Soils: Problems and Practice in Foundation and Pavement Engineering. Wiley, New York (1992)

Sebesta, S.: Investigation of maintenance base repairs over expansive soils: year 1 report (Publication FHWA/TX-03/0-4395-1). FHWA, Texas Department of Transportation (2002)

Tabbagh, A., Camerlynck, C., Cosenza, P.: Numerical modeling for investigating the physical meaning of the relationship between relative dielectric permittivity and water content of soils. Water Resour. Res. **36** (2000). doi:10.1029/2000WR900181. ISSN: 0043–1397

Wanyan, Y., Abdallah, I., Nazarian, S., Puppala, A.J.: Expert system for design of low-volume roads over expansive soils. Transp. Res. Rec. J. Transp. Res. Board **2154**(1), 81–90 (2010)

Zapata, C.E., Houston, W.N.: Calibration and validation of enhanced integration climatic model for pavement design (Report No. 602). Transportation Research Board, National Research Council, Washington, DC, USA (2008)

Zhou, Q.Y., Shimada, J., Sato, A.: Three-dimensional spatial and temporal monitoring of soil water content using electrical resistivity tomography. Water Resour. Res. **37**(2), 273–285 (2001)

Geotechnical Properties of Sediments by In Situ Tests

Sara Rios[1(✉)], António Viana da Fonseca[1], Nuno Cristelo[2], and Claver Pinheiro[1]

[1] CONSTRUCT-GEO, Department of Civil Engineering, Faculty of Engineering, University of Porto, Porto, Portugal
{sara.rios,viana}@fe.up.pt, clavergiovanni@gmail.com
[2] Department of Engineering, CQVR, University of Trás-os-Montes e Alto Douro, Vila Real, Portugal
ncristel@utad.pt

Abstract. River sediments are mainly composed by intermediate materials, between sand and clay, for which partial drainage conditions apply. In these cases, the interpretation of CPTU tests may be wrong since existing correlations are based on fully drained or fully undrained conditions. This paper presents results from CPTU tests performed in a river area to evaluate whether partial drainage conditions were observed. The results, presented in terms of the normalized velocity, show that great part of the analyzed profiles are in this condition. For this reason, the angle of shearing resistance was presented as a conservative estimate of the soil strength in these areas.

1 Introduction

Harbors management involves the execution of dredging operations to overcome excessive deposition of sediments in the harbor's basin and channel accesses, ensuring proper navigability conditions. Sediment disposal management depends on the nature of the dredged materials, which generally comprises fine materials such as sand (particles dimensions in the range of 0.2 to 0.06 mm), silt (between 0.06 and 0.002 mm) and clay (below 0.002 mm). For that reason, a proper in situ geotechnical characterization of such sediments is necessary before a dredging operation. While the geotechnical behavior or sands and clays is relatively well understood with very well established assumptions (e.g., usually drained response and use of effective stress parameters for sand and, undrained response and undrained shear strength for clays), intermediate materials such as silts and various mixtures with sand and clay, that occur naturally and are frequently found worldwide, violate such assumptions. Behavior of intermediate soils (including sands with fines) can be transitional between sands and clays or can exhibit unique behavior that differs from both sands and clays. As a result, a given intermediate soil may simultaneously exhibit properties like a sand, while other properties may be more similar to a clay.

In particular, the extent to which penetrometer testing occurs under drained, undrained or intermediate conditions depends on the soil consolidation characteristics relative to the penetration rate and size of the penetrometer (Finnie and Randolph 1994).

W. Frikha et al. (eds.), *Soil Testing, Soil Stability and Ground Improvement*,
Sustainable Civil Infrastructures, DOI 10.1007/978-3-319-61902-6_6

For that reason, a reliable evaluation of the mass permeability is a preliminary requirement for interpretation of piezocone (CPTU) tests (Randolph and Hope 2004; Schneider et al. 2007; Kim et al. 2008; DeJong and Randolph 2012; Mahmoodzadeh and Randolph 2014). As reported by Schnaid et al. (2004), for soils with permeability in the range of 10^{-5} to 10^{-8} m/s, the simplest accepted approach of a broad distinction between drained (gravel and sand) and undrained (clay) conditions cannot be applied to the interpretation of in situ tests without a great deal of uncertainty. Implications are that the cone resistance varies with penetration rate depending on whether the soil is contractive or dilatant as defined by the state parameter framework proposed by Jefferies and Been (Jefferies and Been 2006). As reported by DeJong and Krage (2014), if the soil has a large positive state (state > 0.20; signifying very loose conditions, i.e., contractive behaviour), in drained cone penetration, no excess pore pressure would develop, but the drained tip resistance would be relatively low since the soil is contractive. In undrained cone penetration large positive excess pore pressures would be generated, which reduces the effective stress in the soil, and hence reduces the cone tip resistance relative to the drained tip resistance ($q_{un} < q_{dr}$). If the state is highly negative (state < −0.20; dense conditions; i.e., dilatant behaviour) in drained cone penetration, this would again result in no development of excess pore pressures, but the magnitude of the drained tip resistance would be much higher than for the loose condition. In undrained cone penetration negative excess pore pressures would develop, increasing the effective stress, and making the undrained tip resistance higher than the drained tip resistance ($q_{un} > q_{dr}$).

This means that for the standard penetration rate, a partial drainage in a loose soil would result in lower estimation of the soil strength relative to a fully drained condition, and a higher estimation of the soil strength relative to a fully undrained condition. Consequently, for the soil layers identified by the Robertson (2009) unified approach as "sands", the soil resistance generally evaluated by the angle of shearing resistance will be underestimated if partial draining conditions occur due to the presence of fines. On the contrary, for the soil layers identified as silts and clays, where the undrained strength would generally be used, an overestimation of this parameter will occur in partial draining conditions.

Discussion of this topic will be made in this paper for the specific case of CPTU tests in river sediments of a harbor site.

2 Test Site and Procedures

The study area is near the city center of Vila do Conde, in north Portugal, located in the left bank of Ave river where the new shipyards of Vila do Conde were installed. This area is very close to the river mouth as indicated in Fig. 1, which means that it suffers from the tidal influence. In fact, the tides in this region of the Atlantic coast are very wide frequently achieving a difference of 4 m of water height between the low and high tide. This had an important effect not only in the tests execution but also in the water table evaluation. Since the tests were performed on shore in an area that was only dry (and accessible) on the low tide the hour to perform the tests had to be judiciously chosen and there was no time for unexpected problems. Moreover, the water table depth, very useful in the interpretation of pore pressure measurements in CPTU tests,

Fig. 1. Site area satellite view (Google maps: https://www.google.pt/maps/@41.3403005,-8.7439096,1462m/data=!3m1!1e3)

was not constant throughout the tests. It is possible that it can also vary from layer to layer depending on the soil permeability, as less permeable layers take more time to reduce the pore pressures when the tide falls.

This area was chosen because it is a preferential area of sediment deposition, not only from Ave river but also from another small stream that converges to Ave river in this specific point. Unfortunately, it is not possible to see it in Fig. 1 because part of this stream is piped. On the other hand, this area was also a deposition place of dredged sediments from Ave river in the past due to the current need of assuring river navigability.

At this site three points were selected (Fig. 2), providing that they were all accessible in the low tide. In these points cone penetrometer tests (CPTU) were performed. The test procedure included first the execution of the light dynamic probing, followed by CPTU test until a maximum tip resistance (qc) of 20 MPa was obtained since the aim was only to study the loosest layers of sediments. At the maximum depth, dissipation tests were performed but without success as the pore pressure remained constant. It can be argue that dissipation tests could have been performed at the higher excess pore pressure but since the water table was quite difficult to evaluate, it was even more difficult to know the best depth to perform dissipation tests. After removing the cone penetrometer, the water table depth was measured in the hole left by the cone, which was used in the interpretation of CPTU tests although it does not adjusts to the data as it will be shown below. A simple auger was also used to extract some undisturbed samples for future analysis that are out of the scope of this paper. Notwithstanding, it is worth mentioning that by visual inspection the soil is mainly composed by sand with fines and in some parts the sand becomes darker with more cohesion as an indication that the amount of clay/silt particles increases.

Fig. 2. Zoom of the site area with identification of the three test points

For the CPTU tests, a standard cone was used with a cone area of 10 cm^2, a penetrometer diameter of 3.57 cm and a cone area ratio (a) of 0.58. The tests were performed at the standard rate of 2 cm/s.

3 CPTU Interpretation Methodology

The tests were interpreted by the unified approach proposed by Robertson (2009) from the basic CPTU parameters (q_c, f_s and u_2). However, the main expressions used to obtain some of the parameters will be identified here since there are several correlations proposed by different authors for the same parameter.

The soil unit weight (γ) was evaluated by value obtained through expression (1) from Robertson and Cabal (2010):

$$\frac{\gamma}{\gamma_w} = 0.27 * \log(Rf) + 0.36 * \log\left(\frac{q_t}{p_a}\right) + 1.236 \quad (1)$$

where,

q_t is the corrected cone resistance as defined by Robertson (2009): $q_t = q_c + (1 - a) * u_0$
R_f is the friction ratio as defined by Robertson (2009): $R_f = f_s/q_c$
p_a is the atmospheric pressure
γ_w is the water unit weight

The constrained modulus (M) was evaluated by one of the expressions proposed by Robertson (2009) as follows,

$$M = \alpha_M (q_t - \sigma_{v0}) \tag{2}$$

where

$\alpha_M = Q_{tn}$ when $Q_{tn} \leq 14$
$\alpha_M = 14$ when $Q_{tn} > 14$

The permeability coefficient (k) was obtained by expression (3) from Robertson (2010a),

$$\begin{aligned} k &= 10^{(0.952-3.04*Ic)} \text{ if } 1.0 < \text{Ic} < 3.27 \\ k &= 10^{(0.952-3.04*Ic)} \text{ if } 3.27 < \text{Ic} < 4.0 \end{aligned} \tag{3}$$

Since it was not possible to evaluate the horizontal consolidation coefficient (c_h) from dissipation tests, it was estimated by the consolidation theory according to Eq. (4) taking into account the constrained modulus (M) and the permeability coefficient (k) obtained as described above:

$$c_h = \frac{k * M}{\gamma_w} \tag{4}$$

This enabled the calculation of the normalized velocity (V) as proposed by Randolph and Hope (2004) and DeJong et al. (2013), by expression (5) and Fig. 3.

$$V = \frac{v * d}{c_h} \tag{5}$$

where,

v is the cone penetration rate (in this case is the standard v = 2 cm/s)
d is the penetrometer diameter (is this case d = 3.57 cm)

Note that according to Fig. 3, a standard rate of 2 cm/s induces partial drainage conditions from fine sands through most silts.

In terms of strength parameters, the angle of shear resistance, can be obtained by the following expression proposed by Robertson (2010b) in function of the angle of shearing strength at critical state (ϕ'_{cv}) assumed 30º,

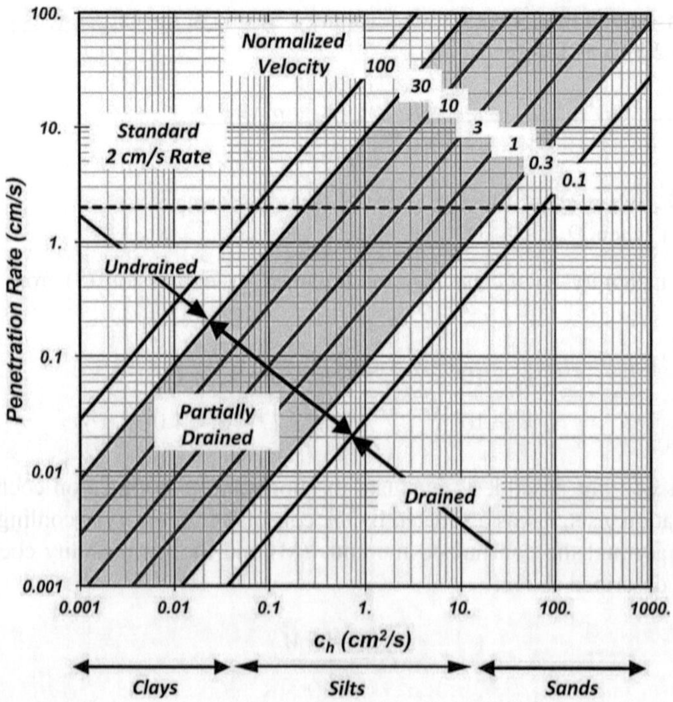

Fig. 3. Field decision chart for 10 cm^2 cone presenting relation between coefficient of consolidation, penetration velocity and normalized velocity (DeJong et al. 2013)

$$\phi' = \phi'_{cv} + 14.44\left(\log Q_{tn,cs}\right) - 22.31 \quad (6)$$

where,

$$Q_{tn,cs} = K_C Q \quad (7)$$

$$\begin{aligned} K_C &= 1.0 \quad \text{if, Ic} < 1.64 \\ K_C &= 5.581\,I_{C^3} - 0.403\,I_{C^4} - 21.63\,I_{C^2} + 33.75\,I_C - 17.88 \text{ if, Ic} > 1.64 \end{aligned} \quad (8)$$

4 Tests Results

Figures 4, 5 and 6 present the obtained results for the three points indicated in Fig. 2. The water levels present (indicated as WT) are the ones obtained after removal of the penetrometer. On point 1 the soil is mainly composed by sand with fines and a few layers of silt. There is also clay between 3 and 4 m of depth. This point is the most affected by the dredged operations on the river, as it is close to the breakwater that separates the Ave river from the shipyards bay, so it is likely that the surface sand layers came from the river. On point 2 the surface layers are more silty while sand

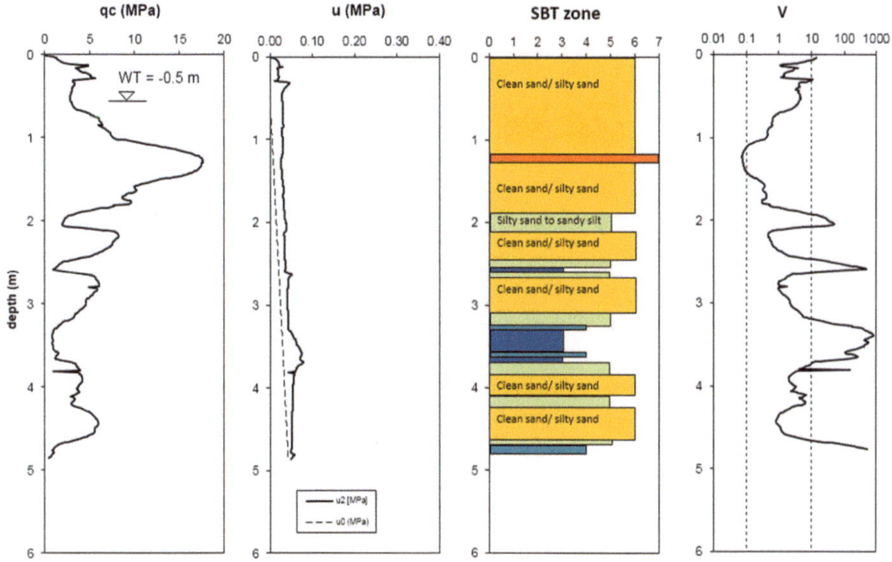

Fig. 4. CPTU test results for the first point

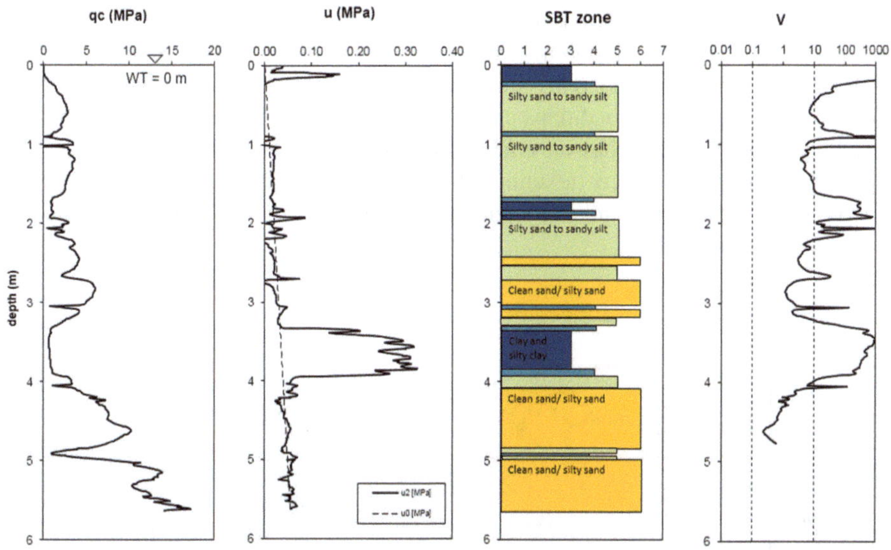

Fig. 5. CPTU test results for the second point

layers appear with depth. This point is the closest to the water in the bay and for that reason the water table is higher here. It is interesting to notice that both in points 1 and 2 the same clay layer appears between 3 and 4 m of depth. The third point, located near old wood piles (possibly used in a previous small pier), shows an intercalation of sand and silt layers.

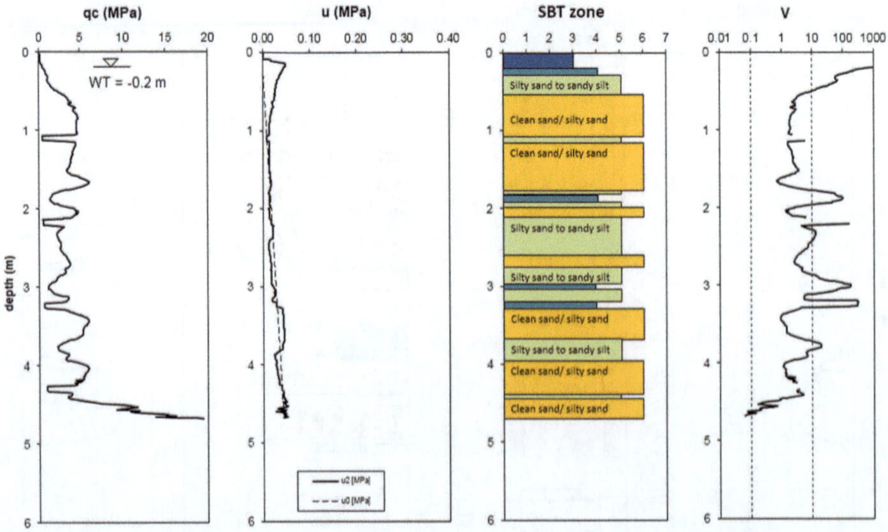

Fig. 6. CPTU test results for the third point

When the normalized velocity is computed it is observed that a significant part of the tests profiles has normalized velocities between 0.1 and 10, which are associated with partial drainage conditions according to Schnaid et al. (2004), Schneider et al. (2008), DeJong et al. (2013). This means that with exception of some layers of finer soils for which undrained conditions could be acceptable without major errors, the soil (mainly in point 1 and 3) has an intermediate behavior. Since the estimation of the angle of shearing resistance (assuming fully drained conditions) corresponds to a conservative estimation of the soil strength according to the explanation presented in

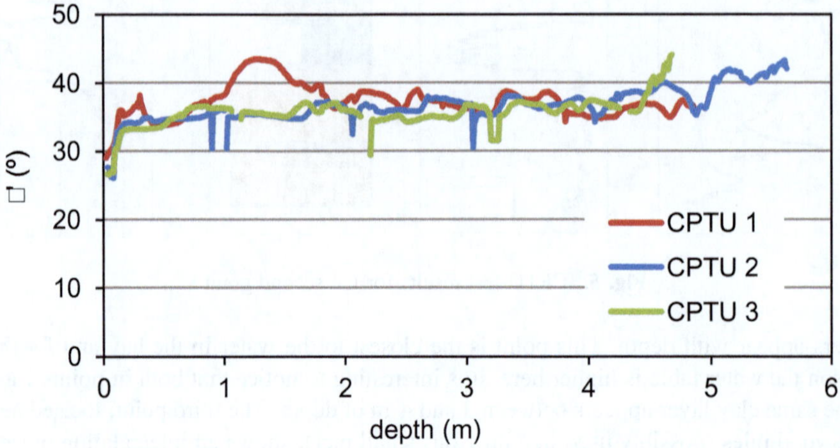

Fig. 7. Angle of shearing resistance profiles: (a) first point; (b) second point; (c) third point

the introduction section of this paper, these profiles were plotted in Fig. 7. It is clear that in these areas of intermediate soil behavior the angle of shearing resistance is between 35° and 40°, which is a reasonable value for loose well graded sandy soils (Hough 1957).

5 Conclusions

This paper presents results from CPTU tests performed in a river area influenced by the tides and dredging operations. The main purpose was to evaluate the suitability of CPTU tests to evaluate the permeability and strength properties of river sediments composed mainly by intermediate materials between clay and sand. The normalized velocity proposed by Randolph and Hope (2004) showed that this is actually the case for the soil presented in this paper, since normalized values between 1 and 10 were found in a great part of the analyzed profiles, indicating partial draining conditions. Although laboratory tests are needed to verify the soil hydraulic and mechanical properties obtained in situ, it is possible that the obtained angle of shearing resistance between 35° and 40° is a safe estimation of the effective strength of this soil.

Acknowledgments. The authors would like to acknowledge the MCTES/FCT (Portuguese Science and Technology Foundation of Portuguese Ministry of Science and Technology) for their financial support through the SFRH/BPD/85863/2012 scholarship, which is co-funded by the European Social Fund by POCH program, the project CONSTRUCT (POCI-01-0145-FEDER-007457) funded by COMPETE 2020, and CNPQ (the Brazilian council for scientific and technological development) for its financial support in 201465/2015-9 scholarship of the "Science without borders" program.

References

DeJong, J.T., Randolph, M.F.: Influence of partial consolidation during cone penetration on estimated soil behavior type and pore pressure dissipation measurements. ASCE J. Geotech. Geoenviron. Eng. **138**(7), 777–788 (2012). doi:10.1061/(ASCE)GT.1943-5606.0000646

DeJong, J.T., Jaeger, R.A., Boulanger, R.W., Randolph, M.F., Wahl, D.A.J.: Variable penetration rate cone testing for characterization of intermediate soils. In: Coutinho, R., Mayne, P. (eds.) Proceedings of the 4th International Conference on Site Characterization (ISC-4), Porto Galinhas, Brasil, pp. 25–42 (2013)

Finnie, I.M.S., Randolph, M.F.: Punch-through and liquefaction induced failure of shallow foundations on calcareous sediments. In: Proceedings of 17th International Conference on the Behavior of Offshore Structures, Massachusetts Institute of Technology, Boston, vol. 1, pp. 217–230 (1994)

Hough, B.K.: Basic Soil Engineering. Ronald Press, New York (1957)

Jefferies, M., Been, K.: Soil Liquefaction: A Critical State Approach. CRC Press, Boca Raton (2006)

Kim, K., Prezzi, M., Salgado, R., Lee, W.: Effect of penetration rate on cone penetration resistance in saturated clayey soils. ASCE J. Geotech. Geoenviron. Eng. **134**(8), 1142–1153 (2008)

Krage, DeJong: Variable penetration rate cone testing for liquefaction evaluation of sand with fines. U.S. Geological survey final technical report, Davis, California, USA (2014)

Mahmoodzadeh, H., Randolph, M.F.: Penetrometer testing: effect of partial consolidation on subsequent dissipation response. ASCE J. Geotech. Geoenviron. Eng. **138**(7), 777–788 (2014). doi:10.1061/(ASCE)GT.1943-5606.0000646

Randolph, M.F., Hope, S.: Effect of cone velocity on cone resistance and excess pore pressures. In: Proceedings of International Symposium on Engineering Practice and Performance of Soft Deposits, Yodogawa Kogisha, Osaka, pp. 147–152 (2004)

Robertson, P.K.: Estimating In-Situ soil permeability from CPT and CPTu. In: 2nd International Symposium on Cone Penetration Testing, Huntington Beach, USA, May 2010 (2010a)

Robertson, P.K.: Estimating In-Situ state parameter and friction angle in sandy soils from CPT. In: 2nd International Symposium on Cone Penetration Testing, Huntington Beach, USA, May 2010 (2010b)

Robertson, P.K.: Interpretation of cone penetration tests – a unified approach. Can. Geotech. J. **46**, 1337–1355 (2009). doi:10.1139/T09-065

Robertson, P.K., Cabal, K.L.: Estimating soil unit weight from CPT. In: 2nd International Symposium on Cone Penetration Testing, Huntington Beach, USA, May 2010 (2010)

Schnaid, F., Lehane, B.M., Fahey, M.: In situ test characterization of unusual geomaterials. In: Viana da Fonseca, A., Mayne, P. (eds.) Proceedings of the 2nd International Conference on Site Characterization (ISC-2), Porto, Portugal, vol. 1, pp. 49–75 (2004)

Schneider, J.A., Lehane, B.M., Schnaid, F.: Velocity effects on piezocone tests in normally and overconsolidated clays. Int. J. Phys. Modell. Geotech. **7**(2), 23–34 (2007)

Schneider, J.A., Randolph, M.F., Mayne, P.W., Ramsey, N.R.: Analysis of factors influencing soil classification using normalized piezocone tip resistance and pore pressure parameters. ASCE J. Geotech. Geoenviron. Eng. **134**(11), 1567–1586 (2008)

Comparison of the CO_2 Record of Different Slope Stabilization Methods

Dennis Gross[1(✉)], Susanne Kytzia[2], and Armin Roduner[1]

[1] GEOBRUGG AG, Romanshorn, Switzerland
[2] University of Applied Sciences, Rapperswil, Switzerland

Abstract. In times of increasing concern for the environment, the CO_2-footprint method is widely accepted to evaluate the use of greenhouse emitting production processes for a given product. This allows for comparisons between products and makes environmentally friendly choices of a product more transparent to decision makers.

In the field of slope stabilization, the high tensile steel wire mesh TECCO® in combination with soil or rock nailing is the state of the art in many countries. Compared to shotcrete, load transfer capacity is equal or higher. The open mesh leaves enough space for plants to grow through the system.

Recent CO_2-footprint evaluation taking into account production and transport of the material of an equivalent slope stabilization measure shows that the TECCO® system has a very low CO_2-footprint. The methods used for assessing the impact were, the level of carbon dioxide emissions from burning of fossil fuels as well as all other emissions which contribute to climate change. These other emissions have been recorded and weighted according to their specific contribution to give an overall index, "Global Warming Potential" (GWP). Compared to shotcrete, the CO_2-footprint of the mesh solution is 4–5 times lower! One reason for the good result is the high CO_2-emitting level of concrete in general. Furthermore, less material weight and also transport costs for the same stabilizing effect of the product on the slope also account for a better CO_2 balance of the mesh solution.

1 Introduction

The flexible TECCO® slope stabilization system is a proven protection system using a high-tensile steel wire mesh in combination with soil or rock nails for stabilizing slopes endangered by slides and break-outs in loose and solid rock. Special spike plates actively brace the system against the subsoil, positively influencing the deformation behavior of the protection system. This is an open, transparent system with no possibility of hydrostatic water pressure building up behind the mesh covering (Tables 1, 2 and 3).

The TECCO® system offers an alternative approach for slope stabilization compared to anchored, reinforced shotcrete. Due to the fact that the TECCO® system requires less construction materials, it can be expected that this approach accounts less for climate change than regular methods do. This was clearly confirmed in a study done by the Institute of Civil and Environmental Engineering at the University of Applied Sciences Rapperswil (HSR), Switzerland.

W. Frikha et al. (eds.), *Soil Testing, Soil Stability and Ground Improvement*,
Sustainable Civil Infrastructures, DOI 10.1007/978-3-319-61902-6_7

Table 1. Used materials in flexible facing solutions.

TECCO – system		Material	Weight (kg)
Pinning (WT 32 gross)	Nail = 4.0 m	Steel	1817.3
	Nail = 6.0 m	Steel	4088.9
	Injection for nails	Cement, w/c = 0.4	23400.0
Wire mesh	TECCO mesh	High-strength wire	1963.6
	Corrosion protection: galvanization	Zinc	169.6
	Corrosion protection: galvanization	Aluminium	8.9
	Gripping plate	Sheet steel	495.0
	Corrosion protection for gripping plate	Zinc coating	123.8
	Press claws	Steel	
	Corrosion protection	Zinc coating	
	Stranded rope (as side guys, top/bottom)	High-strength wire	100.0
	Corrosion protection for stranded rope	Zinc coating	0.5
	Spiral rope anchor = 4.0 m (top/bottom)	High-strength wire	42.4
	Corrosion protection for spiral rope anchor	Zinc coating	0.2
	Injection for spiral rope anchors	Cement	896.0
Planting	TECMAT erosion protection mat	Polypropylene	714.0
Total material weight			34 *t*

Materials used for a slope stabilization structure constructed using flexible TECCO® wire netting for a slope 100 m long.

Table 2. Used materials in hard facing solutions.

Shotcrete placement		Material	Weight (kg)
Pinning (WT 28 gross)	Nails = 4.0 m	Steel	1081.9
	Nails = 5.0 m Nails	Steel	6713.7
	Injection for Nägel	Cement, w/c = 0.4	40350.0
Shotcrete	Total shotcrete {spraying loss ~25%)		564360.0
	Clips	Steel	6324.8
	Top plate	Steel	589.2
	Corrosion protection for top plate	Zinc coating	
Drainage	Drainage pipes	PVC	64.0
Total material weight			619 *t*

Materials used for a slope stabilization structure constructed using shotcrete placement for a slope 100 m long.

Table 3. Standard processes.

Material	Description	Unit	C02 - fossil	GWP
Steel				
Reinforcing steel	63% converter steel, low-alloyed, 37% electric steel, unalloyed and low-alloy	[kg C02/kg]	0.54	1.25
Low-alloyed converter steel	This process produces primary steel	[kg C02/kg]	0.72	1.7
Cement				
Portland cement, strength class Z 42.5, at plant	90% clinker, 5% base additives, 5% plaster	[kg C02/kg]	0.764	0.829
Shotcrete				
Concrete, exacting	375 kg cement, 150 kg water, 1880 kg gravel	[kg C02/kg]	0.121	0.135
Transport				
Truck, 28 t		[kg C02/tkm]	0.174	0.223
Transoceanic freight		[kg C02/tkm]	0.0001	0.011
Coatings				
Zinc	Zinc for zinc coating	[kg C02/kg]	0.483	2.46
Aluminium	Aluminium for corrosion protection	[kg C02/kg]	1.82	11.7
Others				
Drainage pipes for the shot-crete	Polyvinyl chloride pipes for draining the slope	[kg C02/kg]	0.02	2.16
Erosion protection mat	Liner to go underneath the Tecco mesh made of polypropylene	[kg C02/kg]	0.002	2

List of the standard processes used as well as their environmental impacts.

2 The TECCO® High-Tensile Slope Stabilisation System

The use of wire meshes and wire rope nets for flexible slope stabilization has stood the test in many cases and frequently offers an alternative to solid concrete or shotcrete constructions. The open structure of meshes and nets enables the entire surface to be grassed over. Most often used for slope stabilization are wire meshes with a tensile strength of ca. 50 kN/m, respective wires from ca. 500 N/mm^2. Taking into account an economic nail spacing however, these are often unable to absorb the occurring forces and transfer them precisely to the nails.

The development of a wire mesh of high-tensile steel with a wire tensile strength of at least 1770 N/mm^2 offers an interesting possibility for efficient slope stabilization which can be dimensioned using adapted soil and rock static dimensioning models.

In principle the TECCO® slope stabilization system comprises the following elements:

- High-tensile TECCO® steel wire mesh
- TECCO® system spike plates
- Standard soil or rock nails available on the market, e.g. type GEWI, TITAN or IBO.

TECCO® high-tensile steel wire mesh used as standard for slope stabilization consists of a 3 mm thick high-tensile steel wire coated with aluminum-zinc for corrosion protection (GEOBRUGG SUPERCOATING®). The 83 mm × 143 mm diamond-shaped mesh forms are produced by a simple twisting process. The TECCO® steel wire mesh has a tensile strength of at least 150 kN/m. The three-dimensional structure of the mesh positively influences the mesh-subsoil interaction. Together with a resulting favorable friction this also offers a good basis for grassing over using hydro or spray seeding.

In form, size and bending strength, the diamond shaped system spike plates were adapted to the TECCO® mesh through numerous bending, punching and shearing tests to achieve optimal supporting behavior of the stabilization system. Close contact and where possible by lightly pressing-in the spike plates, the mesh can be optimally braced against the subsoil to be stabilized.

3 Method

3.1 Basic Principle

The basic principle for this study is the method of Life Cycle Assessment (LCA) which analyses the overall impact to the environment of a structure from the production of the raw materials to the disposal of the structure as described in ISO 12040:2006 and ISO 14044:2006. In contrast to LCA, our analysis focuses on emissions relevant for climate change, emissions characterized by their «Global Warming Potential» (according to Guinee et al. 2001). To indicate this focus we refer to our study as «CO_2 - Footprint».

3.2 Example Slope

The climate relevant emissions were estimated for two comparable slope stabilization systems and an example slope with a height of 8.5 m and a length of 100 m: One with the TECCO® system and one with a conventional shotcrete cover. The analysis of the life cycle is limited to the production of the construction materials and their transport to the construction site since no significant differences in climate relevant emissions from construction, operation, maintenance, removal and disposal can be expected (Figs. 1 and 2).

Fig. 1. TECCO® mesh and system spike plates

Fig. 2. Example of application in Queensland, Australia

3.3 Used Materials

The materials listed in the tables below were used in these two structures. These quantities correspond to a stabilization structure consisting of a slope 100 m long in each case as represented in the diagrams shown in Fig. 3.

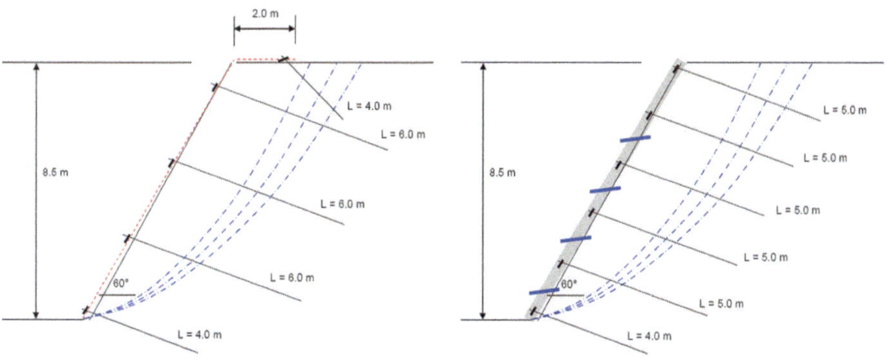

Fig. 3. Slope stabilization with flexible TECCO® mesh (left) and shotcrete (right)

3.4 System Definition

The service life analysis will be limited to the manufacture of the building materials and their transport to the construction site. Construction, operation, maintenance, removal and disposal will not be taken into consideration (see Fig. 4).

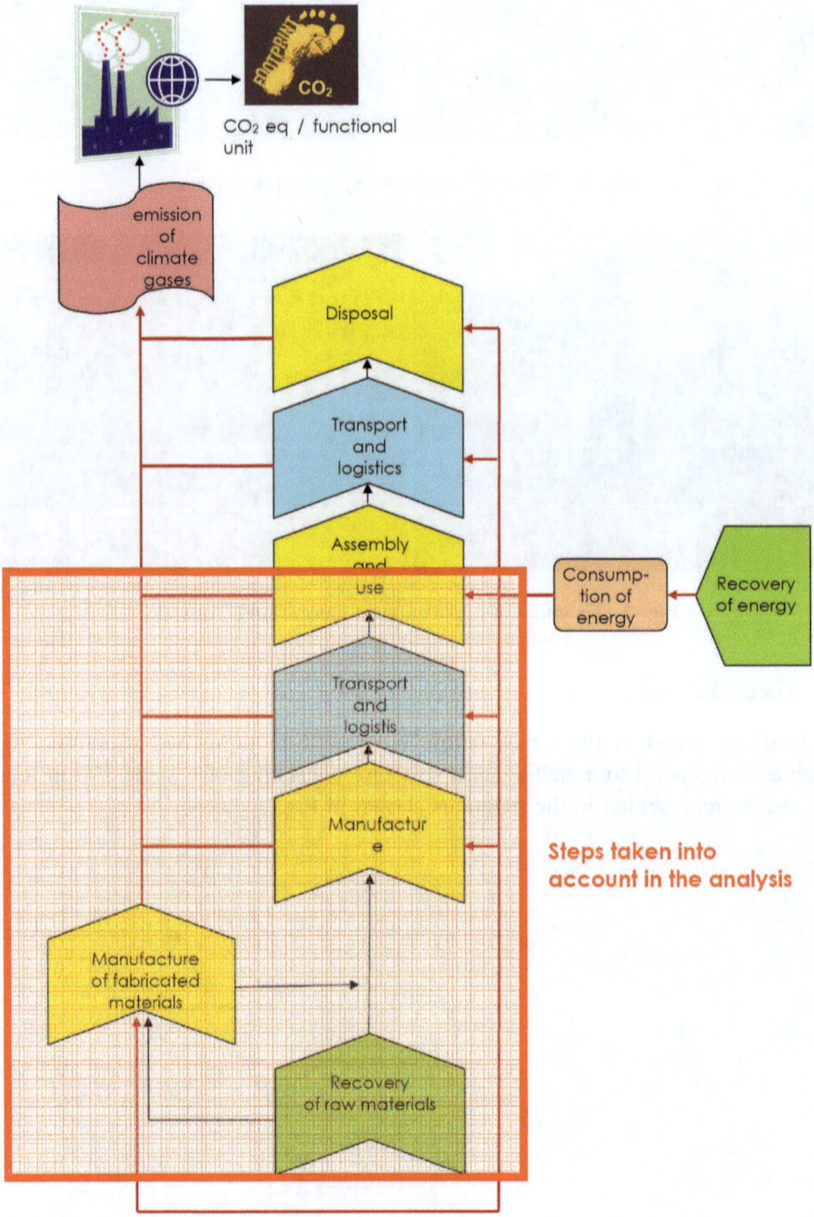

Fig. 4. Life cycle of a product and CO2 footprint (schematic diagram)

This definition is based on the following considerations:

- Studies published on life cycle assessment of structures are based on the assumption that the creation of the structure will only make a relatively small contribution towards the pollution of the environment. Kasser (1998) and Geiger and Fleischer (1997) come to the conclusion that in the case of residential buildings, less than 1% of the cumulative energy demand of a building can be attributed to construction. The construction of the structures considered here is not expected to result in (relatively) higher energy expenditure.
- It has been assumed that in general, there will be no expenditure on operation and maintenance.
- When the two structures are removed, it is mainly steel and concrete which will need to be dealt with. If these two waste materials are collected separately, then their processing and disposal will only result in very low energy consumption and thus only to a low level of emissions which might affect climate change.

In addition, the following processes involved in the manufacture of building materials were not taken into consideration:

- The galvanizing of the TECCO® mesh: here, only the manufacture of the zinc or aluminum coating has been taken into consideration. However, the energy consumption associated with galvanizing itself has not been taken into consideration.
- The refinement of the steel for the TECCO® mesh.

This simplification is necessary as these processes cannot be assessed using standard data.

However, it can be assumed that the energy consumption associated with galvanization and refinement is negligibly low compared to the energy consumption associated with the steel manufacturing process.

3.5 Data Used as a Basis for the Life Cycle Inventory Analysis

The following table provides information on the processes used and the associated environmental impacts.

The emissions from the manufacture of materials which might affect climate change (incl. the recovery of materials, the manufacture of fabricated materials and the supply of energy) as well as from transport processes have been taken from the Ecoinvent database from the Swiss Federal Institute of Technology (ETH) domain (see Ecoinvent center 2007).

Within the context of a conservative estimate of emissions (consideration of the "worst case scenario"), in case of doubt, processes which produce the highest level of environmental pollution will be chosen in each case. This includes above all:

- The selection of distances and means of transport. These have been based on an average distance of 100 km for suppliers from Switzerland, 500 km for suppliers from Germany and 700 km for suppliers from the rest of Central Europe. A (28 t) truck has been used as the means of transport for all journeys from European destinations. Transport from China has been based on a distance of 8000 km for transport by ship and 500 km for transport by truck.

- The selection of steel grades: as no information was available on the proportion of recycling and alloying, the process of manufacturing 'low-alloy oxygen steel' which creates a comparatively high level of emissions has been taken as a basis.

3.6 Method for Assessing Impact

The emissions which might affect climate change have been shown by the level of carbon dioxide emissions from the burning of fossil fuels (in kg of "CO_2 - fossil") on the one hand. On the other hand, all the emissions which contribute towards climate change have been recorded and weighted according to their specific contribution - relating to carbon dioxide as a reference variable - and added together to give an overall index. This overall index is known as the "Global Warming Potential" ("GWP" for short); the unit in which it is measured is kg CO_2 equivalent. This method of assessment has been used as part of the Life Cycle Assessment in many studies and is internationally recognized (see Guinee et al. 2001 and Frischknecht et al. 2003).

4 Results

The comparison of the total climate relevant emissions during the life cycle of the two structures clearly shows that the TECCO® system accounts significantly less for the greenhouse effect than the anchored shotcrete cover (see Fig. 5). This is the case for both

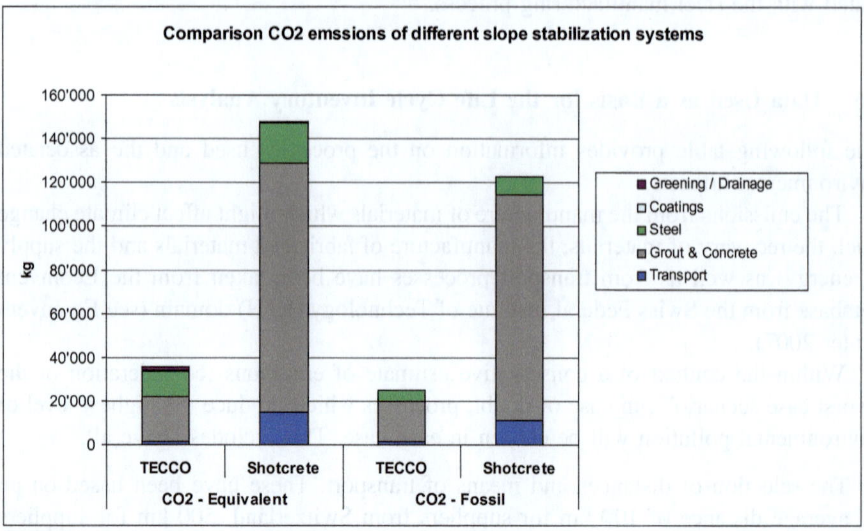

Fig. 5. Climate relevant emissions during the life cycle of two comparable slope stabilization structures built with two different methods.

«CO_2 - fossil» and the «Global Warming Potential (GWP)». The environmental impact of the TECCO® system is approx. 4–5 times smaller with both evaluation methods.

The difference can be explained with the differences in the used quantities of materials. In the structure executed with anchors and shotcrete, approx. 14'700 kg of steel, 40'300 kg of cement and 564'000 kg of concrete are required. For the same slope with the TECCO® system, approx. 8'100 kg of steel and 23'400 kg of cement are used. These differences have a big effect on the environmental impact due to both the production of the materials and also their transport to site.

For the TECCO® system it can be concluded that:

- The anchorage accounts more for the greenhouse effect than the mesh and the greening.
- The used cement has the biggest contribution to the total environmental impact, followed by the steel (in relation 3 to 1).
- The effect of the transport is relatively small with 5% of the total impact.

For the system with the anchored shotcrete cover:

- The shotcrete cover accounts more to the greenhouse effect with 70% of the total environmental impact than the anchorage with approx. 30%.
- The used cement has the biggest contribution to the total environmental impact, followed by the steel (in relation 12 to 1).
- The effect of the transport is with approx. 9% more significant than the impact of the steel.

5 Conclusion

On the base of these results, it can be concluded that the TECCO® system accounts less for the greenhouse effect than a comparable slope stabilization system executed with anchored shotcrete. The difference between the two systems is significant as the here presented study shows. The environmental impact of the slope stabilization with a TECCO® system is approx. 4–5 times smaller than with shotcrete.

Since the difference between the two systems is quite big, a detailed analysis considering all the processes during the life cycle and a more detailed examination of the steel production and the transports would not lead to fundamentally new findings. This can be stated for the assumptions of life span, maintenance and repair. The result is robust here as well.

References

Ecoinvent Center, Ecoinvent Database V2.0 (2007). http://www.ecoinvent.ch. Editor. Dubendorf, CH

Frischknecht, R., Jungbluth, N., Althaus, H.-J., Doka, G., Dones, R., Hischier, R., Hellweg, S., Humbert, S., Margni, M., Nemecek, T., Spielmann, M.: Implementation of life cycle impact assessment methods. Final report ecoinvent 2000 No. 3. Swiss Centre for Life Cycle Inventories, Dübendorf (2003)

Geiger, B., Fleischer, T.: Stoffliche und energetische Lebenszyklusanalysen von Wohngebäuden. In: Gesamtheitliche Betrachtung von Energiesystemen. VDI-1328. Düsseldorf (1997)

Guinee, J.B., Gorrée, M., Heijungs, R., Huppes, G., Kleijn, R., de Koning, A., van Oers, L., Wegener Sleeswijk, A., Suh, S., de Haes, H.A., de Bruijn, H., van Duin, R., Huijbregts, M.A. J., Lindeijer, E., Roorda, A.A.H., Weidema, B.P.: Life cycle assessment; An operational guide to the ISO standards; Characterisation and Normalisation Factors. Leiden (2001)

Kasser, U.: Gebäude gesamtenergetisch beurteilt. Sonderdruck aus: Schweizer Architekt, Nr. 13 (1998)

Bearing Capacity of Strip Foundation on Soft Soil Reinforced with Stone Columns Using Method of Slices

M. Khalifa[1(✉)], M. Etezad[2], A. Hanna[1], and M. Sabry[3]

[1] Department of Building, Civil and Environmental Engineering, Concordia University, 1455 De Maisonneuve Blvd. W. Montréal, Québec, Canada
[2] Golder Associates Ltd., 6925 Century Avenue, Suite #100, Mississauga, ON L5N 7K2, Canada
[3] Bechtel Corporation, 12011 Sunset Hills Road, Suit 110, Reston, VA 20190, USA

Abstract. Stone columns are recognized as an environmently-friendly and cost-effective ground improvement technique. Stone columns are used in soft soil to increase the bearing capacity, reduce the settlement, increase the rate of settlement and reduce the liquefaction potential of the ground.

This paper presents an analytical model utilizing the method of slices to predict the ultimate bearing capacity of the soil reinforced with a group of stone columns. The soil within the failure zone was divided into slices and the limit equilibrium technique was adopted to perform the analysis. Shear forces and passive earth pressure on the boundaries of each slice were determined. By utilizing a circular failure plane, the minimum inter-slice force coefficients were determined.

The analysis was carried out using the Morgenstern-Price method to estimate the failure surface together with the bearing capacity of the reinforced ground. The failure surface was determined by trial and error to estimate the minimum factor of safety. The ultimate bearing capacity was defined by increasing the foundation load until the factor of safety of one was obtained. Results of the present theory were compared with those available in the literature, where a good agreement between the two was noted.

1 Introduction

Stone columns technique is a ground improvement method widely used over the past decades. Stone columns are stiffer and have higher shearing resistance than the native soft cohesive soil. This results in an increase of the soil bearing capacity and reduce in the corresponding settlement (Etezad et al. (2015); Mitchell et al. (1985); Muir Wood et al. (2000); Priebe (1995)). Furthermore, stone columns have higher permeability than the native surrounding soil, which leads to reduction of the drainage path and accordingly reduces the time required to complete the consolidation settlement (McKelvey et al. (2004)).

W. Frikha et al. (eds.), *Soil Testing, Soil Stability and Ground Improvement*,
Sustainable Civil Infrastructures, DOI 10.1007/978-3-319-61902-6_8

Unit cell concept was one of the first methods used to estimate the bearing capacity of the ground reinforced with stone columns. The bulging mode of failure was used to model this process. Accordingly, the bearing capacity of a single stone column was predicted by estimating the horizontal capacity of the clay soil around the pile. In this method, the capacity of the group is the total capacities of the individual columns. Hughes and Withers (1974) used elastic-plastic theory developed by Gibson and Anderson (1961) to calculate the maximum vertical stress that can be carried by a single stone column due to bulging failure. Similar approach has also been used by Balaam and Booker (1981).

Hu (1995) performed laboratory tests on a group of end bearing and floating stone columns. Shear and punching mode of failure were reported in this study. He concluded that the collapse pattern for soil mass reinforced by stone columns changes from general shear failure mechanism to punching shear based on the stone column's length. The group interaction reported by Hu (1995) was confirmed by the numerical analysis carried out by Lee and Pande (1998); Muir Wood et al. (2000).

Priebe (1995) reported two methods to estimate the bearing capacity of a footing on the group of stone columns based on the general shear failure and the equivalent homogeneous composite. In the first approach, the weighted average values of the friction angle and the cohesion were calculated along the failure plane, and the bearing capacity was estimated using theories for homogeneous soils. In the second approach, a larger footing width was assumed, and the bearing capacity was determined using the new footing width and the shear properties of the unreinforced soil. (Lee et al. 1998) introduced the concept of composite properties of the reinforced soil. They used finite element technique to estimate the bearing capacity and settlement of the reinforced soil. Bouassida et al. (2009) presented design charts to determine the ultimate bearing capacity of a group of floating stone columns. In their study, the friction between the footing and the soil was neglected as well as the distribution of stone columns.

An analytical model was developed by Etezad et al. (2015) using limit equilibrium method and equivalent soil properties under the footing to calculate the bearing capacity of a rigid footing placed on the ground reinforced with stone columns. The method utilize the general shear failure and slip surface, which was deduced from the results of the numerical model of Hanna et al. (2013).

Many of the theories developed to predict the bearing capacity utilize the theory for homogeneous soil as a simplified assumption. In the present paper, the method of slices was adopted to estimate the bearing capacity of strip footing on a compacted inhomogeneous soil. Terashi et al. (1991) conducted a series of centrifuge tests and full-scale tests to estimate the bearing capacity of the improved ground by compacted sand piles. The experimental results agreed well with the bearing capacity calculated using a circular type of slip surface. Morgenstern-Price method was utilized in this study to calculate the bearing capacity of a footing rests on clay soil reinforced with stone columns.

2 Bearing Capacity Calculations Based on Slip Circle

A 2-D model made of soft clay reinforced with stone columns was developed. The ground was loaded with a uniform pressure to simulate the case of strip footing. The failure zone was divided into 50 slides and the forces acting on each slide are shown in Fig. 1. The limit equilibrium method by Morgenstern et al. (1965) method was used in the computer program SIDE V-6.020, developed by Rocscience (2012), to calculate the minimum factor of safety of the circular slip surface (Fig. 2). The analysis was also check with the Bishop's Simplified method for comparison (Bishop 1955). The footing was loaded to the ultimate capacity in the form of uniform pressure applied on the ground surface.

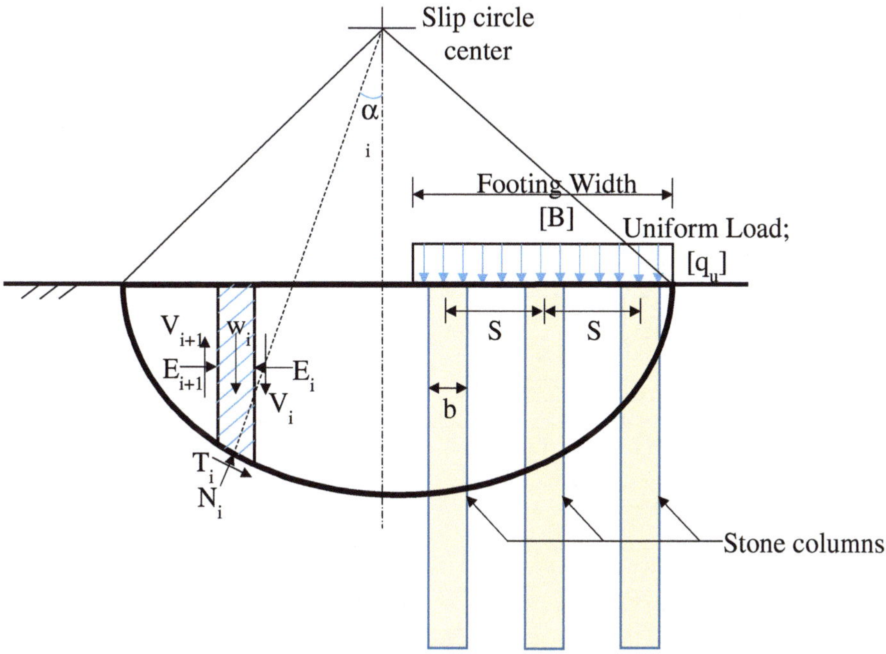

Fig. 1. Forces acting on a slide in the slip circle

2.1 Validation

The results obtained from the present analysis was compared with the laboratory and numerical results available in the literature (Hanna et al. (2013); Hu (1995); McKelvey et al. (2004)). Table 1 presents these comparison, where a good agreement can be noted. This further validate the methods of Morgenstern-Price and the Bishop's Simplified method of slices as viable techniques to estimate the bearing capacity of clay soil reinforced by a group of stone columns. However, Bishop's Simplified method generally overestimates the factor of safety as compared to Morgenstern-Price method (Turnbull et al. 1967). Nevertheless, both approaches underestimate the bearing capacity of the reinforced soil.

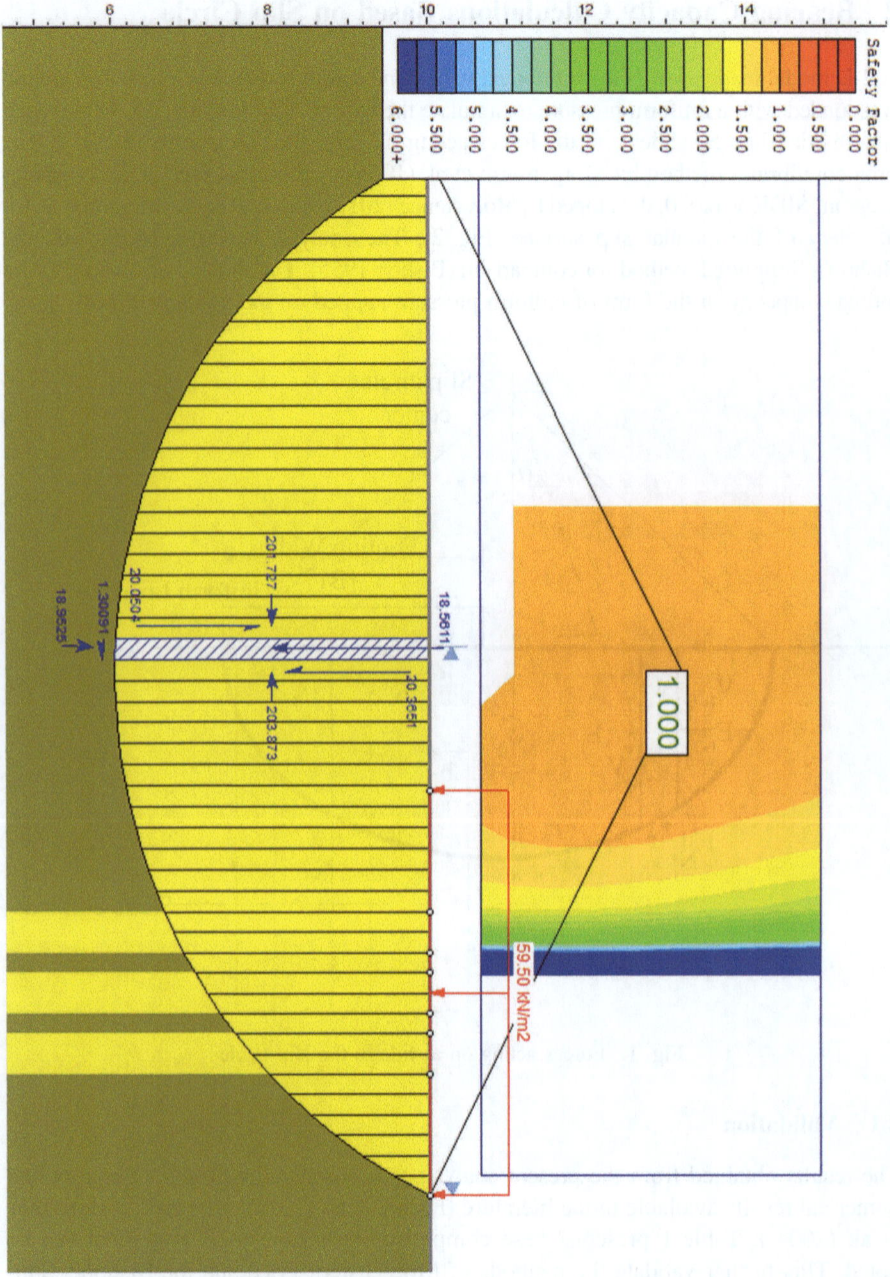

Fig. 2. An example of the bearing capacity calculation using slip circle method

Table 1. Comparison between the bearing capacity of the reinforced soil estimated by the method of slices and the available experimental and numerical results.

No.	Clay properties			Stone properties		Area replacement ratio, As%	Footing width, B [m]	Surcharge load, q [kPa]	q_u, Measured [kPa]	q_u, B. S. method [kPa]	% Difference from measured	$q_{u,}$ Morgenstern Price method [kPa]	% Difference from measured
	Cu [kPa]	ϕ c [Degree]	γc [kN/m³]	[Degree]	γs [kN/m3]								
1a	32	0	14	34	17.3	24	0.09	0	272	222	20%	218.8	18%
2a	20.5	0	9.9	34	20.3	40	0.05	0	160	165	1%	161	3%
3b	10.5	0	13.1	30	15.47	30	0.1	0	75	73	4%	71.7	3%
4b	11.5	0	13.1	30	15.47	30	0.1	0	79	80	1%	78.5	1%
5c	5	12	13	45	21	35	2.5	2.6	352	270	24%	268.2	23%
6c	5	13	13	40	19	35	2.5	2.6	280	250	12%	246.5	11%
7c	5	15	14	45	21	35	2.5	2.8	420	380	9%	380.2	10%
8c	15	15	14	45	21	35	2.5	2.8	660	644	2%	646	2%
9c	10	13	13	40	19	30	2.62	2.6	366	270	27%	267	26%
10c	15	13	13	40	19	30	2.62	2.6	458	350	24%	346.5	24%

a experimantal work of McKelvey et al. (2004)
b experimental work of Hu (1995)
c numerical work of Hanna et al. (2013)

2.2 Parametric Study

In this study, the results are presented in the form of improvement ratio (IR), which is defined as the ratio of the capacity of improved soil to the capacity of the unimproved soil. In this analysis, the effect of the stone columns spacing to diameter ratio (S/D), and the number of stone column rows (N) on the bearing capacity at a given replacement ratio (As %) was performed. Each row of the stone columns was assumed as a trench with a width (b) and the spacing between trenches was considered the same as the spacing between columns as shown in Fig. 3. The trench width was calculated using Eq. 1. In this case, the replacement ratio (As) was calculated using Eq. 2. The ratio of S/D was examined in the range of 1 to 3.0 as given in Table 2.

$$b = \frac{\pi \times D^2}{4 \times S} \tag{1}$$

$$A_s = \frac{N \times b}{B} = \frac{N \times \pi \times D^2}{4 \times S \times B} \tag{2}$$

Where,

N is the number of stone columns rows, b is the trench width, B is footing width, D is stone column diameter, and S is the spacing between stone columns.

Moreover, this method was also used to determine the effect of the undrained strength of the clay soil (Cu), and the replacement ratio (As %). In this analysis, the replacement ratio ranged from 10% to 30% which widely used in practice (Hanna et al. (2013); Hu (1995)). The range of the other parameters believed to govern the bearing capacity are presented in Table 3.

The effect of the column arrangement on the bearing capacity was also investigated. Figure 4 presents the spacing/ diameter ratio of stone columns versus replacement ratios, assuming all other parameters are constant. It can be noted that the bearing capacity increases with the increase of the replacement ratio. Also. It can be noted that the spacing between stone columns slightly influences on the improvement ratio. For low replacement ratio (<10%), the improvement ratio is almost constant, which agreed well with Castro (2014) observation. However, the improvement ratio reduces with the increase of the columns spacing ratio; for higher replacement ratio (20%, and 30%), which confirm that for small spacing between columns, the lateral support from the surrounding soil increases and accordingly, will show significant improvement.

Regarding the effect of the clay shearing resistance, it is noticeable that the improvement ratio reduces with the increase of shear strength of clay soil (Cu) for the same spacing/ diameter ratio as shown in Fig. 5. Contrary, the improvement ratio raised by the increase of the stone columns shearing resistance angle (ϕ) as shown in Fig. 6.

The influence of stone columns number under the footing has been also investigated. In this analysis, the columns number varied from N = 1 to N = 4. The columns diameter ranged from 0.3 m to 1.6 m, which covers the maximum and minimum ranges that may be used in practice. Figure 7 presents the effect of the stone columns number on the bearing capacity, and it can be noted that there is no remarkable change

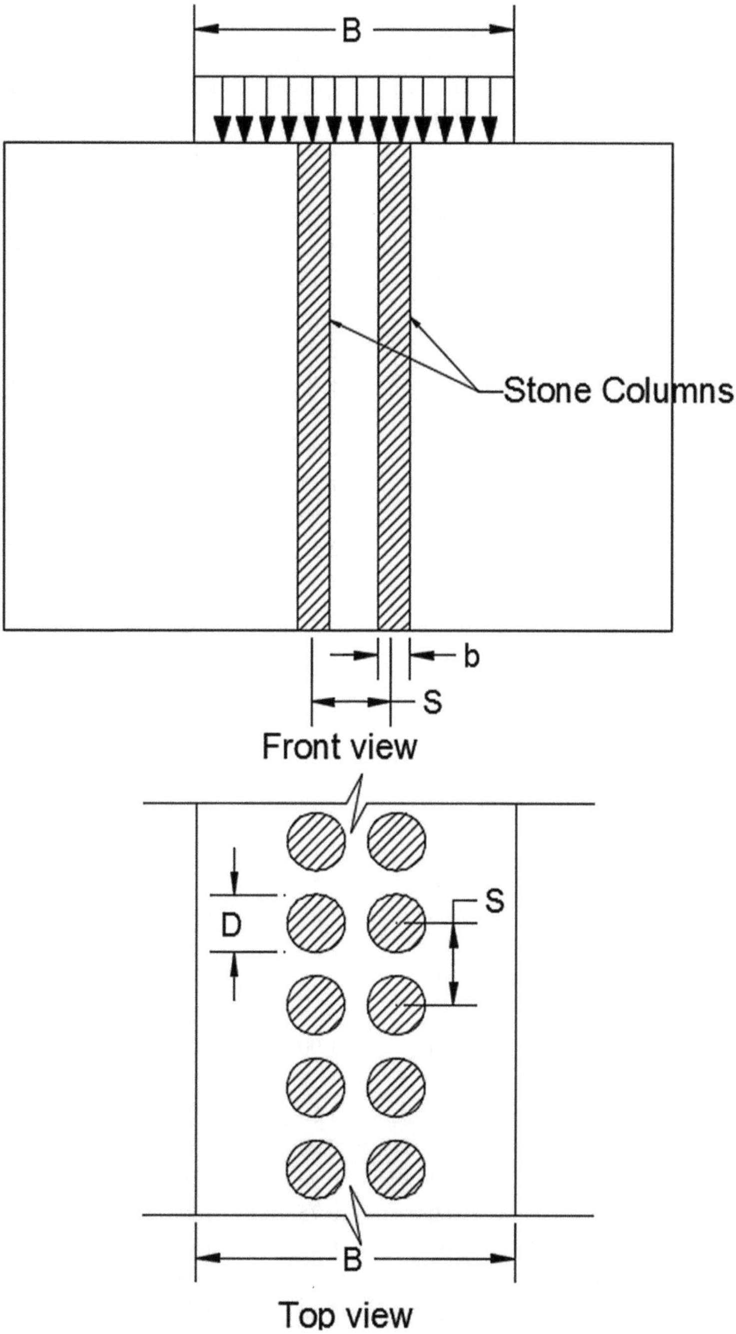

Fig. 3. Stone columns arrangement

Table 2. Range of stone columns dimensions used in parametric study

Area replacement ratio, As %	No. of columns, N	Column diameter, D [m]	Column spacing, S [m]	Spacing/Diameter ratio, S/D	Trench width b [m]
10%	1	0.70	0.70	1.00	0.50
	2	0.35	0.38	1.09	0.25
	2	0.40	0.50	1.25	0.25
	2	0.45	0.63	1.40	0.25
	2	0.49	0.75	1.53	0.25
	2	0.63	1.25	1.98	0.25
	2	0.89	2.50	2.80	0.25
20%	1	1.30	1.30	1.00	1.00
	2	0.69	0.75	1.09	0.50
	2	0.80	1.00	1.25	0.50
	2	0.89	1.25	1.40	0.50
	2	0.98	1.50	1.53	0.50
	2	1.26	2.50	1.98	0.50
	2	1.69	4.50	2.66	0.50
	3	0.46	0.50	1.09	0.33
	3	0.53	0.67	1.25	0.33
	3	0.59	0.83	1.40	0.33
	3	0.65	1.00	1.53	0.33
	3	0.84	1.67	1.98	0.33
	3	1.00	2.33	2.34	0.33
30%	1	2.10	2.10	1.00	1.50
	2	1.04	1.13	1.09	0.75
	2	1.20	1.50	1.25	0.75
	2	1.34	1.88	1.40	0.75
	2	1.47	2.25	1.53	0.75
	2	1.89	3.75	1.98	0.75
	3	0.69	0.75	1.09	0.50
	3	0.80	1.00	1.25	0.50
	3	0.89	1.25	1.40	0.50
	3	0.98	1.50	1.53	0.50
	3	1.13	2.00	1.77	0.50
	4	0.52	0.56	1.09	0.38
	4	0.60	0.75	1.25	0.38
	4	0.67	0.94	1.40	0.38
	4	0.73	1.13	1.53	0.38
	4	0.85	1.50	1.77	0.38

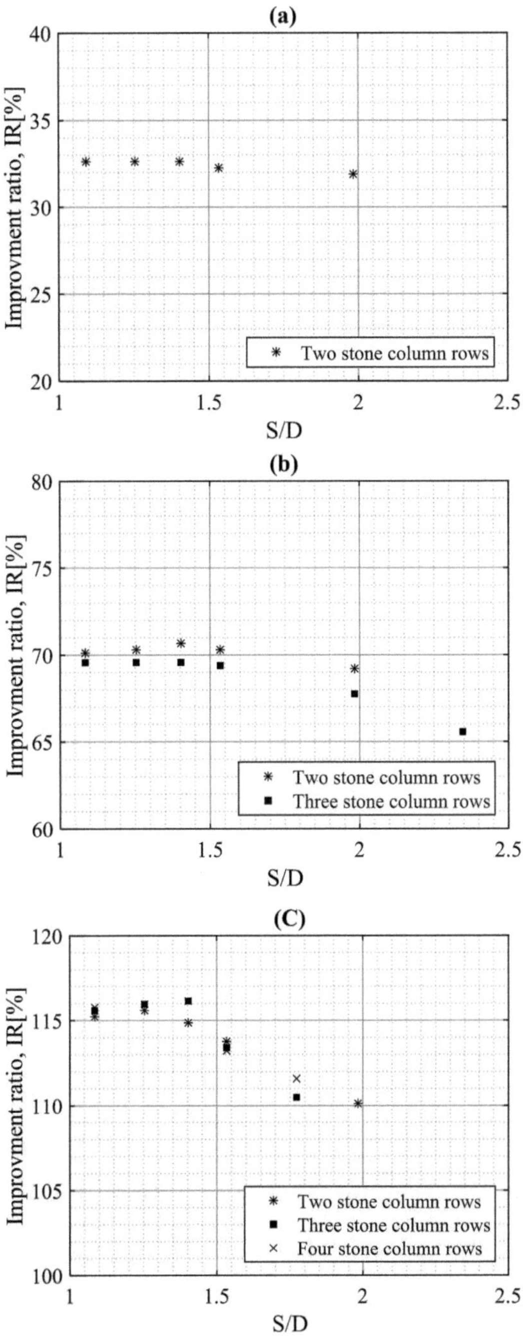

Fig. 4. Improvement ratio (IR) versus S/D for different replacement area, Cu = 5 kPa, $\Phi = 35^{o}$: (a) As = 10%; (b) As = 20%; (c) As = 30%

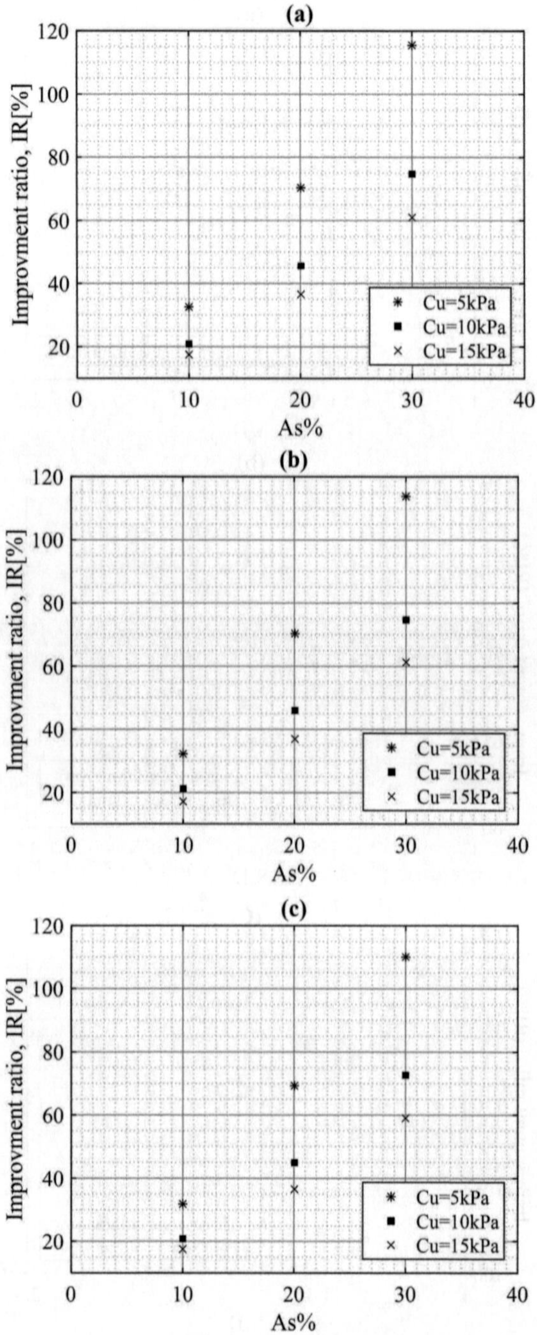

Fig. 5. Effect of shear resistance of clay on the improvement ratio for different replacement ratios, $\Phi = 35^{o}$: (a) S/D = 1.25; (b) S/D = 1.50; (c) S/D = 2.00

Table 3. Range of parameter used in the parametric study

Parameter	Range of value
Cohesion of clay soil, Cu [kPa]	5, 10, and 15
Unit weight of clay soil, γc [kN/m^3]	18
Angle of shear resistance of stone, ϕ_s [Degree]	35, 38, and 40
Unit weight of stone, γs [kN/m^3]	20
Stone column diameter, D [m]	0.3–1.60
Stone column diameter to spacing ratio	1.00–2.50
Replacement area, As [%]	10, 20, and 30
Footing width, B [m]	5

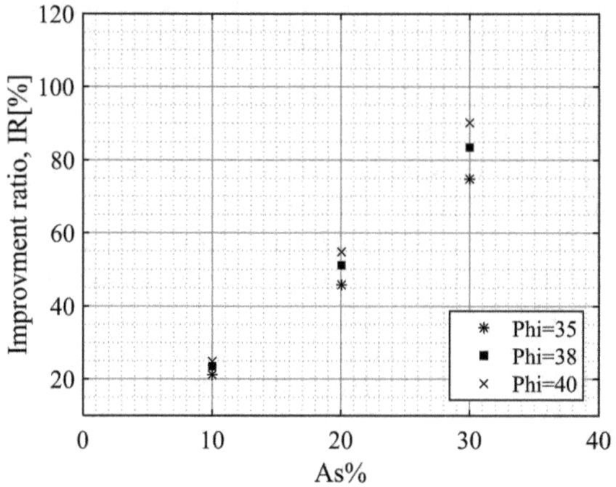

Fig. 6. Effect of shear resistance of stone on the improvement ratio for different replacement ratios, Cu = 10 kPa; S/D = 1.50

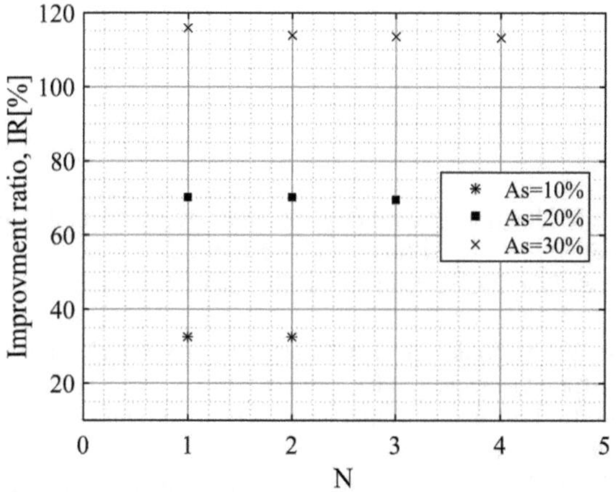

Fig. 7. The relation between number of stone column rows versus the improvement ratio, at S/D = 1.5, Cu = 5 kPa, Φ = 35°

in the improvement ratio particularly for low replacement ratio (<10%). However, there is a slight reduction in the improvement ratio for higher values of replacement ratio (30%). For a large number of stone columns, a small diameter was used to keep the replacement ratio constant. By reducing the column's diameter more load will be transferred to the clay soil (less strength material), which lead to the reduction in the improvement ratio (Black et al. (2007); Hanna et al. (2013))

3 Conclusions

Morgenstern-Price method of slices was used to calculate the bearing capacity of a soft clay reinforced with stone columns. The theory developed compared well with the available results in the literature. Parametric study was conducted on the parameters believed to govern the behavior of this system. The following was concluded:

1- Morgenstern-Price method of slices was successfully used to estimate the bearing capacity of reinforced clay soil.
2- (IR) significantly increases with the increrase of the replacement ratio.
3- (IR) reduces due to the increase of the spacing/ diameter ratio (S/D)
4- Based on the results of the present study, in order to optimize the benefit of the use of reinforced soft clay with stone columns is to use a ratio of columns spacing to columns diameter (S/D) equal to 1.5
5- For the same stone columns arrangement and shear resistance (diameter, spacing) the improvement ratio (IR) increases with a decrease of the shear strength of surrounding clay. However, the ultimate bearing capacity of the system is significantly increased due to the increase of the shear strength of the clay soil as well as the stone. Furthermore, the improvement ratio (IR) increases with the increase of the stone columns shearing resistance angle.

Acknowledgments. The financial support received from Concordia University is acknowledged.

References

Balaam, N., Booker, J.R.: Analysis of rigid rafts supported by granular piles. Int. J. Numer. Anal. Meth. Geomech. **5**(4), 379–403 (1981)
Bishop, A.W.: The use of the slip circle in the stability analysis of slopes. Geotechnique **5**(1), 7–17 (1955)
Black, J., Sivakumar, V., McKinley, J.: Performance of clay samples reinforced with vertical granular columns. Can. Geotech. J. **44**(1), 89–95 (2007)
Bouassida, M., Jellali, B., Porbaha, A.: Limit analysis of rigid foundations on floating columns. Int. J. Geomech. **9**(3), 89–101 (2009)
Castro, J.: Numerical modelling of stone columns beneath a rigid footing. Comput. Geotech. **60**, 77–87 (2014)

Etezad, M., Hanna, A., Ayadat, T.: Bearing capacity of a group of stone columns in soft soil. Int. J. Geomech. **15**(2) (2015)

Gibson, R., Anderson, W.: In situ measurement of soil properties with the pressuremeter. Civil Eng. Public Works Rev. **56**(658), 615–618 (1961)

Hanna, A., Etezad, M., Ayadat, T.: Mode of failure of a group of stone columns in soft soil. Int. J. Geomech. **13**(1), 87–96 (2013)

Hu, W.: Physical modelling of group behaviour of stone column foundation. (Ph.D. dissertation), University of Glasgow, UK (1995)

Hughes, J., Withers, N.: Reinforcing of soft cohesive soils with stone columns. Ground Eng. **7** (3), 42–49 (1974)

Lee, J.S., Pande, G.N.: Analysis of stone-column reinforced foundations. Int. J. Numer. Anal. Meth. Geomech. **22**(12), 1001–1020 (1998)

McKelvey, D., Sivakumar, V., Bell, A., Graham, J.: A Laboratory Model Study of the Performance of Vibro Stone Columns in Soft Clay. J. Geotech. Eng. **152**, 1–13 (2004)

Mitchell, J.K., Huber, T.R.: Performance of a stone column foundation. J. Geotech. Eng. **111**(2), 205–223 (1985)

Morgenstern, N., Price, V.E.: The analysis of the stability of general slip surfaces. Geotechnique **15**(1), 79–93 (1965)

Muir Wood, D., Hu, W., Nash, D.: Group effects in stone column foundations: model tests. Geotechnique **50**(6), 689–698 (2000)

Priebe, H.J.: The design of vibro replacement. Ground Eng. **28**(10), 31 (1995)

Rocscience: SLIDE 6.0—2D Slope Stability Analysis for Soil and Rock Slopes. Rocscience Inc. (2012)

Terashi, M., Kitazume, M., Minagawa, S.: Bearing capacity of improved ground by sand compaction piles. In: Deep Foundation Improvements: Design, Construction, and Testing. ASTM International (1991)

Turnbull, W. J., Hvorslev, M.J.: Special problems in slope stability. J. Soil Mech. Found. Div. (1967)

The Behavior of a Foundation Laterally Loaded at the Top over Highly Porous and Collapsible Soil

Roberto Kassouf[1(✉)], David de Carvalho[2], Paulo José Rocha de Albuquerque[2], and Nelson L. Fonte Jr.[3]

[1] Metropolitan College of Campinas - DeVry Metrocamp, Campinas, Brazil
Kassouf.engenharia@kassouf.com.br
[2] State University of Campinas, Campinas, Brazil
d33c@uol.com.br, pjra@fec.unicamp.br
[3] Head of Geoponto Engineering, Mogi das Cruzes, São Paulo, Brazil
lopesfontejr@bol.com.br

Abstract. In geotechnical engineering, the problem of laterally top-loaded piles occurs frequently. This type of foundation is often used in highly porous, collapsible soils, which are common in several regions of Brazil. Because of the limited information available in the literature, several load tests have been performed on piles subjected to this load. To analyze the behavior of piles in these collapsible soils, load tests were performed in steel piles (I), W 250 × 32.7 section (mm × kg/m), length 12 m, conventional bored piles (ϕ = 0.40 m; L = 12 m) and continuous helical auger piles (ϕ = 0.40 m; L = 12 m). All of them were tested at the same site. For each type of pile, a first load test was carried out with the soil in its natural condition of moisture content, followed by a second load test after the surface soil was flooded for 48 h. The results indicated a significant negative effect of flooding on the topsoil which consisted of sandy-silty clay with collapsible features down to the depth of 6 m. Load vs. horizontal displacement curves and soil coefficients of horizontal reaction were obtained. The results allowed the proposal of parameters for use in the soil under study. Before executing the load tests, laboratory and in-situ tests were performed to investigate the local subsoil.

Keywords: Laterally load tests · Piles · Porous soil · Collapsible soil

1 Introduction

The deep foundation designer, in addition to the executive-system-related aspects and the type of material employed, is concerned with the loading system, which can produce displacements due to axial load, transversal load and bending moments. Regarding lateral loading, the top of foundations is generally subject to these forces. This is the case of bridges, viaducts, transmission line towers, wind power generation towers, or along the piles shaft due earth's pressure. In countries with seismic activities, the construction code requires consideration of the horizontal load in foundation designs, thus minimizing the consequences of a possible earthquake.

W. Frikha et al. (eds.), *Soil Testing, Soil Stability and Ground Improvement*,
Sustainable Civil Infrastructures, DOI 10.1007/978-3-319-61902-6_9

When dimensioning foundations to resist lateral loads, the project criteria involve not only the ultimate horizontal loading capacity, but also the maximum pre-established displacement that can occur. Currently, there are many mathematical methods for prediction of the horizontal displacement of a pile. The common difficulty about these methods concerns the adoption of geotechnical parameters to be used in the calculations.

The main parameter used is the modulus of horizontal reaction (n_h), which is defined as the soil resistance along the foundation divided by its deflection at a point. Simplified mathematical models have been created for the analysis because the modeling of the transversal action problem is three-dimensional and extremely complex for routine solutions by project designers. The most commonly known and widespread theory for evaluation of these actions is the "Theory of Horizontal Reaction of the Soil", where the n_h factor represents the proportionality between the reaction and displacement acting on the soil mass. However, this factor is difficult to be theoretically estimated. Nevertheless, this factor can be "measured" using load tests to get a reliable value for horizontal resistance of the soil where the construction is to take place.

The reaction of the soil is a function of many factors, such as pile properties, soil stress *vs.* strain behavior, depth of the point analyzed, level of foundation displacement, etc. Because of the difficulty to establish a function covering all of these factors, the simplified Winkler's (1875) hypothesis is generally used, where the reaction of the soil is considered proportional to the displacement of the pile.

For horizontal loading, in the first few meters the surface soil has great influence on the load *vs.* lateral displacement behavior of the foundations. To predict the behavior of horizontally loaded foundations, theoretical approaches are available in the literature; however, parameters should be determined for the local soil before they can be used.

Surface soils with porosities above 50% cover vast areas in Midwest Brazil. Because of their large void volumes, these soils undergo great strain under load. In addition, many of these soils are collapsible, i.e., when the soils are under load and when a significant increase in the moisture content or soil saturation occurs, the structure collapses, which results in unacceptable displacement values for the buildings.

Given the lack of available information in the literature concerning horizontally loaded piles on highly porous collapsible diabase soil, this study was developed in order to review the performance of three types of piles in this type of soil. The tests were performed with the soil at their natural moisture content and after flooded on the surface.

Based on the horizontal loading tests, the effect of soil flooding on the load *vs.* horizontal displacement curve and the values for the horizontal reaction coefficient for both the natural and pre-flooded soil conditions were verified. The results obtained for the horizontal reaction coefficient were compared to the results for other types of foundations on similar soils.

The acting stresses and displacements on a pile under bending moments and horizontal loadings were determined using the theory of soil horizontal reaction, which is based on the model proposed by Winkler (1875). The soil behavior under horizontal forces is simulated by a set of independent, identical and equally spaced springs. Thus, the reaction of the soil is considered proportional to the displacement of the point being analyzed. This supposition simplifies the problem, considering that the ratio between the pressure of contact at the base of the foundation and its corresponding consolidation is the same for any supporting area. Using the model proposed by Winkler, the concept of the modulus of horizontal reaction, K, was introduced by Terzaghi (1955). It is defined as the ratio between the reaction of the soil (in units of force applied by the pile length) and the corresponding displacement (Eq. 1):

$$K = \frac{p}{y} \tag{1}$$

Where: K = the modulus of horizontal reaction (FL^{-2}), p = the applied pressure (FL^{-1}) and y = the horizontal displacement (L).

This notation presents the advantage of being independent of the diameter of the foundation. Therefore, the Eq. 2 can be rewritten as:

$$K = k_h.D \tag{2}$$

Where: k_h = the horizontal reaction coefficient (FL^{-3}) and D = the diameter of the foundation (L).

For pure sands, the elasticity modulus increases (approximately) linearly with depth. Therefore, the soil reaction to the load applied to the pile is assumed to increase linearly with depth (Eq. 3):

$$K = \frac{p}{y} = n_h.z \tag{3}$$

Where: n_h = the modulus of the horizontal reaction of the soil (FL^{-3}) and z = the depth (L).

Understanding the variation of K along the foundation is required for analysis of its behavior based on the theory of soil reaction. Refinements and sophistications in the reaction modulus function by depth are not justified since the errors in the results of calculations are minor compared with the ones involved in the estimation of numerical values of the modulus of soil reaction. Matlock and Reese (1960) agree with this assessment because the results are satisfactory and can be obtained for most practical cases as simple forms of variation of the reaction modulus with depth. Additionally, in practical problems, the uncertainty inherent to the estimation of soil behavior based on conventional tests is generally compatible with the minor errors that can be introduced by the depth using a simple form of the soil reaction modulus function.

Alizadeh and Davisson (1970) were the first researchers to present curves obtained for horizontal loads tests on sandy soils. They presented these curves in the form of n_h at the y-axis and the displacement y_0 at the x-axis. To create these curves, the authors used Eq. (4) by Matlock and Reese for the displacement for the application of only one horizontal load parallel to the ground surface, i.e.:

$$y_o = \left(\frac{HT^3}{E_pI_p}\right)\Delta_y \tag{4}$$

Where: T is the relative stiffness between the pile and the soil, and for soils with sandy behavior and normally consolidated clays. It is defined by Eq. 5:

$$T = \sqrt[5]{\frac{E_p I_p}{n_h}} \tag{5}$$

To calculate n_h, Eq. 6 is used:

$$n_h = \frac{4,42H^{5/3}}{y_o^{5/3}\left(E_p I_p\right)^{2/3}} \tag{6}$$

In the present study, the soil characteristics were analyzed using geotechnical laboratory tests and in-situ tests, with the aim of predicting soil behavior in terms of deformability, resistance and collapsibility. To determine the horizontal reaction coefficient (n_h), curves of horizontal reaction coefficients (n_h) *vs.* horizontal displacement at the surface (y_o) were calculated based on the load tests performed and adopting a range of values of horizontal displacement. Alizadeh and Davisson (1970) proposed the curves (n_h) *vs.* (y_o) using values in the 6.35–12.70 mm range. The curves of horizontal reaction coefficients *vs.* horizontal displacement were divided into two separate groups: load tests performed with soil in its natural moisture content and load tests performed with the soil pre-flooded on the surface.

2 Materials and Methods

The aims of this study were achieved based on geotechnical analysis of the soil under investigation, execution of piles, conduction of load tests on piles and analysis of the obtained data.

2.1 Geological and Geotechnical Characteristics

The soil is composed of diabase bodies which can also be found embedded in the Itararé Formation and in the Crystalline Complex in the form of sills and dykes. Such materials are pedologically classified as purple latosols, and are mineralogically constituted by quartz, ilmenite, magnetite, kaolinite, gibbsite, iron oxides and hydroxides. Thicknesses range from 5 to 30 m. (Zuquete 1987).

The load tests were carried out at the Experimental Site of the University of Campinas (Cavalcante et al. 2007), where the profile of the subsoil is diabase soil. It has a highly thick 6-m thick sandy-silty clay surface layer followed by a sandy-clayey silt layer down to the depth of 19 m. The water table is found at the depth of 18 m. The mean geotechnical profile of the subsoil was obtained from the results of tests on undeformed samples taken after opening two wells and performing Standard Penetration Tests (SPT) and Static Penetration Tests (CPT). The results of the laboratory tests and field tests performed can be found in Peixoto (2001), Fig. 1.

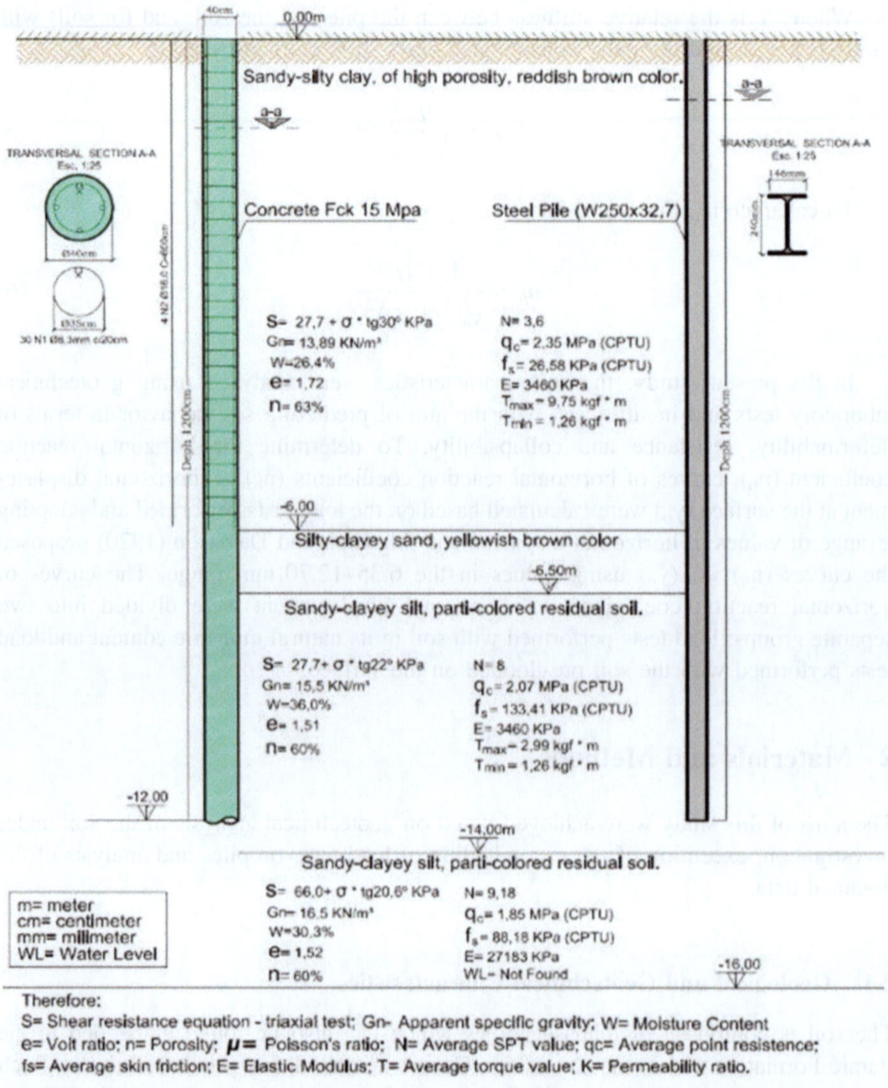

Fig. 1. Static Penetration Tests (CPT) and the results of the laboratory tests and field tests performed Peixoto (2001).

The characteristics of collapsibility of the subsoil of the Experimental Site were reviewed by Monacci (1995). The criterion used was the one proposed by Vargas (1978), which defines soils as collapsible when the coefficient of structural collapse, "i", is greater than 2%, considering that "i" is defined by the following equation:

$$\mathrm{i} = \frac{\Delta ec}{1 + ei} \tag{7}$$

In the equation above, $\Delta \mathbf{e_c}$ represents the variation in the void ratio due to collapse of the soil structure, and $\mathbf{e_i}$ is the index of voids before flooding.

The indexes of collapse at the depths of 0.75 m, 5.00 m and 8.00 m, were determined from simple oedometric tests, with soil flooded at determined pressures, which are presented in Table 1. Note that collapsibility in the layer of high porous soil is reduced as a function of depth.

Table 1. Collapse coefficients according to the applied voltage and the depth - Monacci (1995)

Depth 0,75 m		Depth 5,00 m		Depth 8,00 m	
σ (kPa)	i (%)	σ (kPa)	i (%)	σ (kPa)	i (%)
5	4,97	–	–	–	–
9,8	11,09	9,8	2,41	–	–
19,4	7,4	19,4	3,76	–	–

Gon (2011) performed oedometric tests with stresses of 100 kPa, 200 kPa and 400 kPa. The samples taken at 1 m of depth were observed to be collapsible, with a very high index for all flooding stresses. For the 2 m and 3 m deep layers, it was possible to observe the low collapsibility index for the 100 kPa and 200 kPa stress, which in turn had a high index for the 400 kPa stress. Based on the data obtained for the 100 kPa stress, only the 1 m, 4 m and 8 m depths were observed to be collapsible. However, for the 200 kPa stress, only the 2 m and 3 m depths were not collapsible. Finally, for the flooding stress of 400 kPa, all of the depths were observed to be collapsible, with the exception of the 8 m depth.

3 Load Tests

3.1 Piles Analyzed

3.1.1 Steel Piles

A 12 m long steel pile I was used; 250 × 32.7 section W (kg × mm/m); Moment of Inertia Ix = 4,937 cm^4; Young's Modulus E = 205,000 MPa and Area = 42.1 cm^2; in Gerdau (2015). The pile dimensions are shown in Fig. 2.

3.1.2 Continuous Flight Auger Piles and Bored Piles

These piles were constructed with 0.40 m nominal diameter and 12 m length. The concrete used in continuous flight auger piles were pumped, cement consumption of 400 kg/m^3 and aggregates (sand and gravel); slump ±240 mm. The characteristic compression resistance of concrete (fck) was 15 MPa, slump ±70 mm, using gravel and sand as aggregates. The longitudinal frame of both piles were 4 bars with

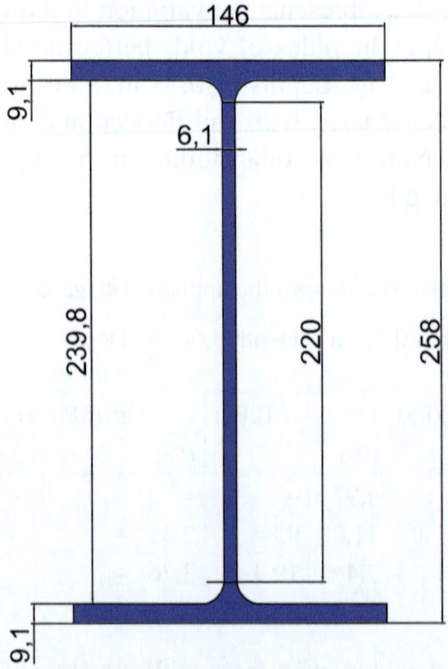

Fig. 2. Steel pile dimensions

16 mm diameter, CA-50 steel, 6 m long, plus a 12 m long Dywidag steel bar with 32.0 mm diameter positioned in the center of the pile. The stirrups were 6.3 mm every 20 cm. The Young's modulus of these piles was in average 20 GPa (Miranda Junior 2006).

3.2 Results and Analysis

Based on the results of the load tests, load *vs.* horizontal displacement curves were obtained.

The results for steel piles are shown in Fig. 3.

Figures 4 and 5 show the results for bored and continuous flight auger piles.

The load tests were conducted for the three types of piles with slow maintained loads. The pre-flooding of the soil was conducted for 48 h through an excavation around pile head with the head of 0.70 m × 0.70 m and 0.50 m depth

It is observed that:

Pre-flooding the topsoil caused drastic reduction of soil resistance to horizontal loading. Taking a 5 mm displacement as reference for the steel pile into the natural moisture content soil, the supported load is approximately 50 kN, whereas for the pre-flooded soil for this displacement, the supported load is approximately 3 kN. For bored piles in the natural moisture content soil for 5 mm, the supported load is about 48 kN, while for the pre-flooded soil for this displacement, the supported load is about 8 kN. For continuous flight auger piles, the supported load is about 45 kN in soil in its natural moisture content, whereas for pre-flooded soil for this displacement of 5 mm, the supported load is about 2 kN.

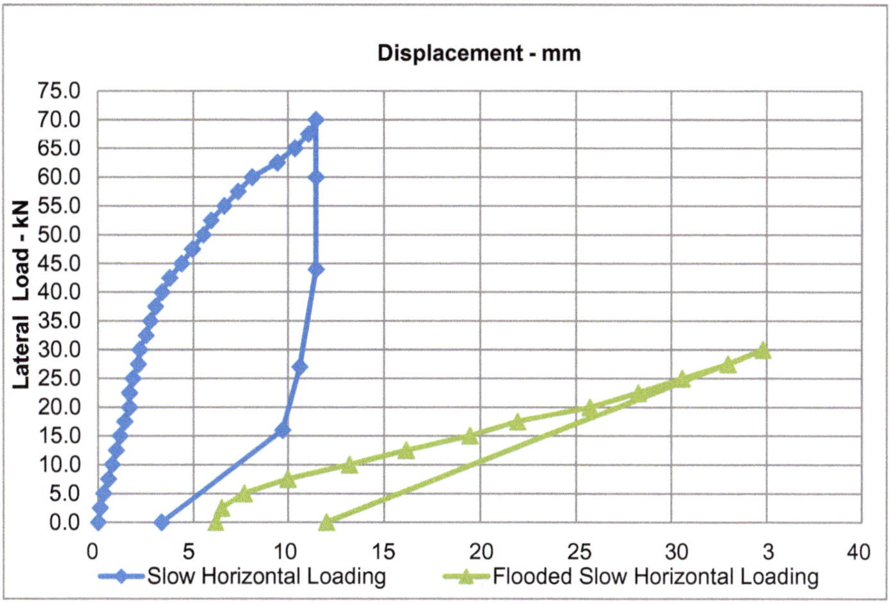

Fig. 3. Load vs. horizontal displacement curves on Steel Pile (W250 × 32,7) mm × kg/m; L = 12 m.

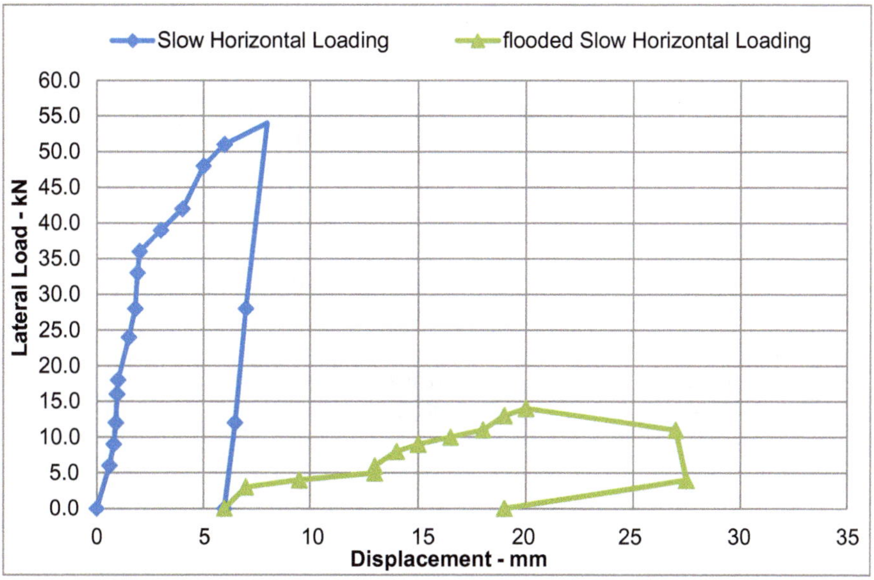

Fig. 4. Load vs. horizontal displacement curves - Bored pile (ϕ = 0,40 m; L = 12 m)

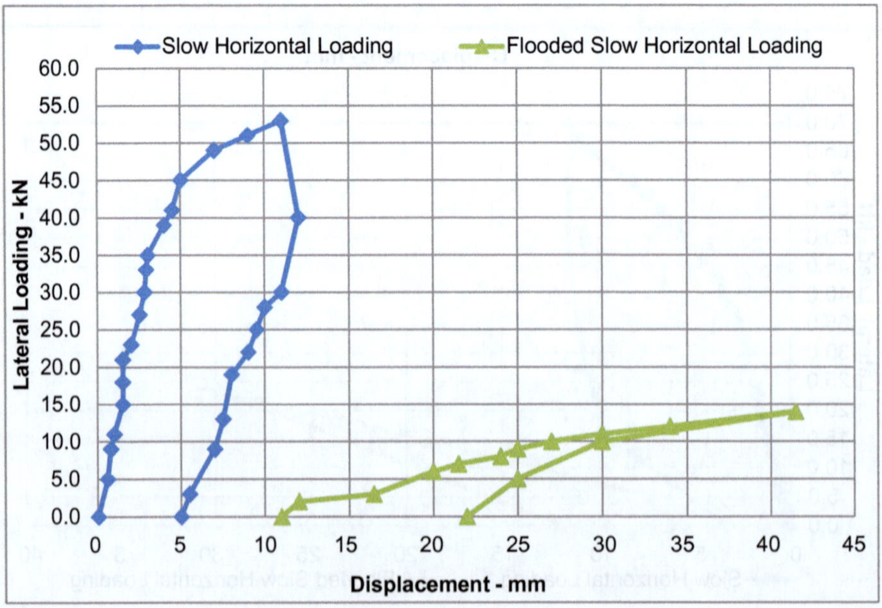

Fig. 5. Load vs. horizontal displacement curves - Continuous flight auger (ϕ = 0,40 m; L = 12 m)

Table 2. Values of horizontal loads, displacements and horizontal reaction coefficients. Source: The authors.

Type pile	Author Name	Local Brazil	Soil condition	n_h (MN/m^3) displacement 6 to 12 mm
Steel	Silva, Marcella B.M	Unicamp	Natural	7.71
			Flooded	0.40
Bored	Miranda Jr., Gentil		Natural	11.55
			Flooded	0.56
CFA	Miranda Jr., Gentil		Natural	9.86
			Flooded	0.24
Caisson	Kassouf, Roberto		Natural	12
			Flooded	6

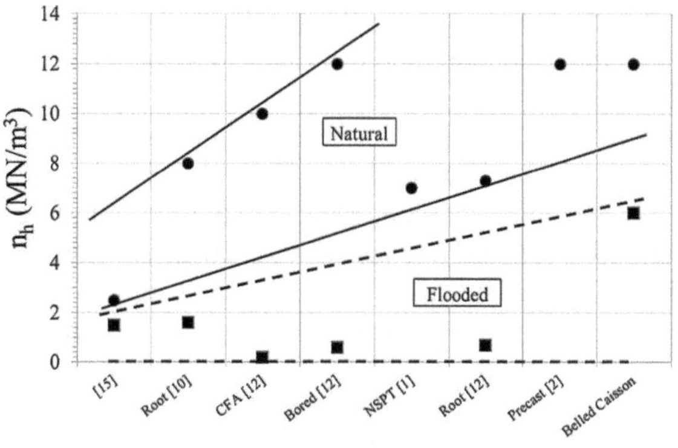

Fig. 6. Graph of nh vs. authors load tests.

4 Conclusions

The load tests have indicated that the soil has low bearing capacity to horizontal loading, even in its natural condition of moisture content. These results reinforce the need to perform load tests on these types of soil to get precise parameters for projects.

High variation in moisture content down to a few meters deep into the soil significantly reduces the piles ability to support horizontal loadings. Thus, in foundations subject to such loading in these types of soils, care should be taken to minimize this possibility, both in the case of leakage in piping and in the case of infiltration by rain water.

The permeability coefficients of this soil both in its natural condition and in the compacted condition are in the range between 5×10^{-4} cm/s and 5×10^{-7} cm/s, respectively.

The values of horizontal coefficients n_h obtained from the analysis of steel piles, bored piles and flight auger piles, that are presented in Table 2, differ from the literature for different types of deep foundations built on soil with similar behavior to the one from the present study (under natural and pre-flooded conditions) are presented in Fig. 6.

However, these values are within the range of values n_h found for sandy porous soils of the State São Paulo, Brazil. Miguel (1996) found 7.50 and 8.0 MN/m^3 for bored, Strauss and root piles, respectively. Ferreira et al. (2001) obtained values n_h of 7.4 and 11.0 MN/m^3 for bored piles in the city of Bauru. Souza et al. (2008) got the value of 5 MN/m^3 for bored piles in Ilha Solteira. For the Experimental Site under analysis, Miranda Junior (2006) obtained the average values n_h of 7.28 MN/m^3 for root piles, 9.86 MN/m^3 for continuous flight auger piles and 11.55 MN/m^3 for bored piles.

Carvalho et al. (1996) obtained the value of 11.90 MN/m^3 for pre-cast concrete piles.

Kassouf (2012) had to place close to the same type of soil the value obtained the value of 12 MN/m^3 for a caisson, for the same type of soil in a place close the Experimental Site.

The results of load tests conducted by Miranda Junior (2006) show that the reinforced concrete blocks in the pile head can increase load capacity of both natural moisture content soil and soil with significant variations of moisture content.

For collapsible soils, the analysis of the possibility of water saturation or large variations in moisture content of the surface soil up to few meters, during the construction service life, are of fundamental importance to determine the project parameters.

Acknowledgments. The authors thank the Unicamp-Universidade Estadual de Campinas, the FAPESP-Fundação de Amparo à Pesquisa do Estado de São Paulo and DeVry Metrocamp-Metropolitan College of Campinas.

References

Alizadeh, M., Davisson, M.T.: Lateral load tests on piles – arkansas river project. J. Soil Mech. Found. Div. ASCE **96**(SM5), 1583–1604 (1970)

Carvalho, D., Albuquerque, P.J.R., Claro, A.T., Ferreira, C.V.: Análise de estaca carregada transversalmente no topo, em solo residual de diabásio, Seminário de Engenharia de Fundações Especiais (SEFE 3), vol. 1, pp. 145–154 (1996)

Gerdau – Perfis Gerdau Açominas aplicados como Estacas Metálicas em Fundações Profundas. 8ª edição (2015)

Gon, F.S.: Caracterização geotécnica através de ensaios de laboratório de um solo de diabásio da região de Campinas/SP. M.S thesis. Department of Geotechcnics and Transportation, University of Campinas, Campinas, Brasil (2011)

Kassouf, R.: Análise de Prova de Carga em Tubulão a Céu Aberto Submetido a Esforço Horizontal em Solo não Saturado de Diabásio da Região de Campinas. Dissertação de Mestrado. Universidade Estadual de Campinas, SP (2012). 111p.

Matlock, H., Reese, L.C.: Generalized solutions for laterally loaded piles. J. Soil Mech. Found. Eng. Div. ASCE **86**(SM5), 63–91 (1960)

Miguel, M.G.: Execução e análise de provas de carga horizontal em estacas em solo colapsível. M.S. thesis. Departament of Geotechcnics, University of São Paulo, São Carlos, Brasil (1996)

Miranda, J.G.: Estacas Submetidas a Esforços Horizontais em Solos Colapsíveis do Interior de São Paulo nas Condições Natural, Melhorada e Inundada. Ph.D. Dissertation, Faculty of Agricultural Engineering, University of Campinas, Campinas (2006)

Monacci, M.G.: Estudo da Colapsibilidade de um Solo do Campo experimental da Faculdade de Engenharia Agrícola, Unicamp. Dissertação de Mestrado, Faculdade de Engenharia Agrícola. Unicamp, p. 130 (1995)

Peixoto, A.S.P.: Estudo do ensaio SPT – T e sua aplicação na pratica de engenharia de fundações – tese de doutorado, Faculdade de engenharia agricola, Unicamp (2001). 468p.
Terzaghi, K.: Evalution of Coefficients of Subgrade Reaction. Géotechnique **5**(4), 297–326 (1955)
Vargas, M.: Introdução à Mecânica dos Solos. McGraw -Hill do Brasil, São Paulo (1978)
Zuquete, L.V.: Análise e proposta metodológica sobre cartografia geotécnica para condições brasileiras, Tese de Doutorado, vol. 3. EESCUSP, São Carlos (1987)

Early Applications of DMT in Arabian Gulf Area – Three Case Studies

Sharif Emad(✉)

GTC LAB, Geotechnical Engineering, e.construct, Dubai, UAE
emad.sharif@econstruct.ae, emad.sharif@gtc-lab.com

Abstract. DMT was used for three main applications in famous projects in Dubai, UAE and in Duqm Port in Sultanate of Oman in the Arabian Gulf. The 1st two applications were conducted in man-made sandy earthfill embankments formed by dredging from the sea bed and stabilized by deep ground improvement. The recent correlations of S. Marchetti to incorporate the DMT stress history parameter K_D for CPT-based CRR was used to confirm post-compaction CPT-based liquefaction analysis in the Duqm Site, whereas DMT was used in Dubai site to confirm the over consolidated nature of upper sandy fill soil that was previously Vibro-Compacted (8–10 years). On the other hand, the design of large diameter flexible steel oil storage tanks is based on controlled & strict, edge settlement requirements. Without DMT, the estimates of soil modulus are made based on SPT or CPT tests resulting in conservative estimates, and suggesting the need for deep ground densification/improvement. Site specific correlation was developed for a large site between DMT based M and CPT tip resistance (qc) indicating at least 50% increase over the CPT based M.

1 Introduction

DMT was only recently introduced for practical use in Dubai and the Gulf Area in General. This paper describes three main early applications of DMT in main projects. The several man-made islands in Dubai and other places, present a convenient area of application of DMT and SDMT for both assessment of the deep man made earth fill, design and control of deep ground improvement procedures and assessment of liquefaction potential. Further, the construction of large diameter oil storage flexible steel tanks presents another important area where DMT can provide valuable information to accurately assess the settlement of tanks.

This paper presents some of the recent applications of DMT in this area.

1.1 Man-Made Islands/Embankments

Construction on artificial islands or embankments by dredging and filling is very common in the Arabian Gulf region for different purposes.

W. Frikha et al. (eds.), *Soil Testing, Soil Stability and Ground Improvement*,
Sustainable Civil Infrastructures, DOI 10.1007/978-3-319-61902-6_10

1.1.1 Background on Man-Made Islands & Embankments Construction and Stabilization

Several Man-made Islands were constructed along the shore line of Dubai. The first was Palm Jumeira Island having an approximate diameter of 8 Kms as shown in Fig. 1.

Fig. 1. Typical shapes of man-made islands

The photos in Fig. 2 show part of the Commercial berth site in Duqm Port in Oman, proposed to accommodate several industrial facilities for the port.

Fig. 2. Commercial Berth – Duqm New Port

The islands/embankments were constructed with silty sands dredged from adjacent sea bed and laid by hydraulic filling with heights ranging between 12–18 m above original sea level as shown in Fig. 3.

Fig. 3. Typical photos of sand dredging operations

1.1.2 Ground Densification and Verifications

Stabilization and densification of the upper sandy fill is essential to improve its Engineering performance and mitigate hazards such as liquefaction and high differential settlements and long term creep movements.

Common Methods used are Vibro Compaction/Replacement and Dynamic Compaction/RIC for shallow sand depths. Vibro Compaction is most widely and commonly used method although other methods were also used.

Verification of VC works is typically based on Pre and post-compaction CPT measurements and zone loading tests in some cases.

The use of Robertson (2009) CPT-based Profiling (SBT) is very effective to establish the localities of high silt content/high friction ratio which are not likely compactable.

1.1.3 Actual Performance

Several roads, buildings on shallow and deep foundations exist since more than 8–10 years by today and performing very well as indicated in Fig. 4.

Fig. 4. Typical light weight and heavy structures in PJ

No published data exists about actual performance of the foundation, however the observed evidence indicates very well and satisfactory performance of the compacted earth fill.

Testing showed remarkable stiffness and strength increase with aging. Fills tested after several years of laying are stiffer than young fills.

1.2 Case 1: Confirming CPT-Based Liquefaction Analysis in Port of Duqm in Oman

The Duqm area lies along the coasts of the Indian Ocean, about 600 km SW of Muscat, as shown in Fig. 5, the Capital City of the Sultanate of Oman.

Fig. 5. Map of Oman showing location of Duqm

The area is intended to accommodate one of the largest Oil Ports in the Middle East with all associated facilities as sea port, airport, dry dock, oil refinery, Storage Terminals, infra-structures like roads, rail way, bridges, buildings and residential villas etc. that are under construction or on the anvil.

1.2.1 Description of DMT/CPT Application

A main component of the port facilities is the Commercial Berth which is made up of thick reclaimed sandy fill by dredging from the sea bed and hydraulic filling. The berth is >250 m in width and more than 1500 m in length extending from the coastal line into the sea. The typical soil profile indicates a thick (typically 12–18 m), slightly silty to silty fine sand (reclaimed soil) on top of a thin sea bed layer of silt/Clay soil that is underlain by bedrock of very to extremely weak claystone.

The use of shallow foundations to support the many proposed light to moderately heavy structures within the area of the Commercial Berth requires satisfaction of both liquefaction mitigation and acceptable settlement performance criteria, including control of potential differential settlement.

The Berth was investigated with large number of widely spaced boreholes with SPT testing conducted at 1 to 1.5 m intervals in each borehole as part of technical feasibility and initial risk assessment and planning of the port. Later, a more detailed investigation was conducted within the site of a typical project that lies within the mid-zone of the Commercial Berth and which is proposed for cement silos and concrete mixing plant. The subsurface conditions are indicated in the generalized subsurface profile in Fig. 6, which shows the upper sandy fill of about 15 m depth, underlain by thin sea bed layer of clayey soil followed by bedrock of claystone, as indicated in Fig. 6.

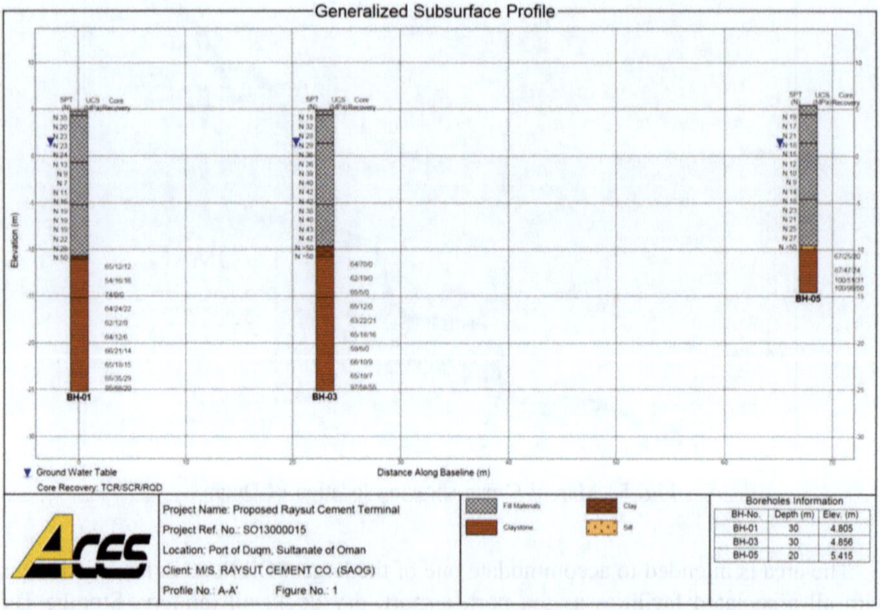

Fig. 6. Generalized subsurface profile – Duqm site

The SPT tests results obtained within 5 test borings, has indicated large stiffness variations and presence of weak zones. This is clearly indicated in Fig. 7 of SPT vs. elevation.

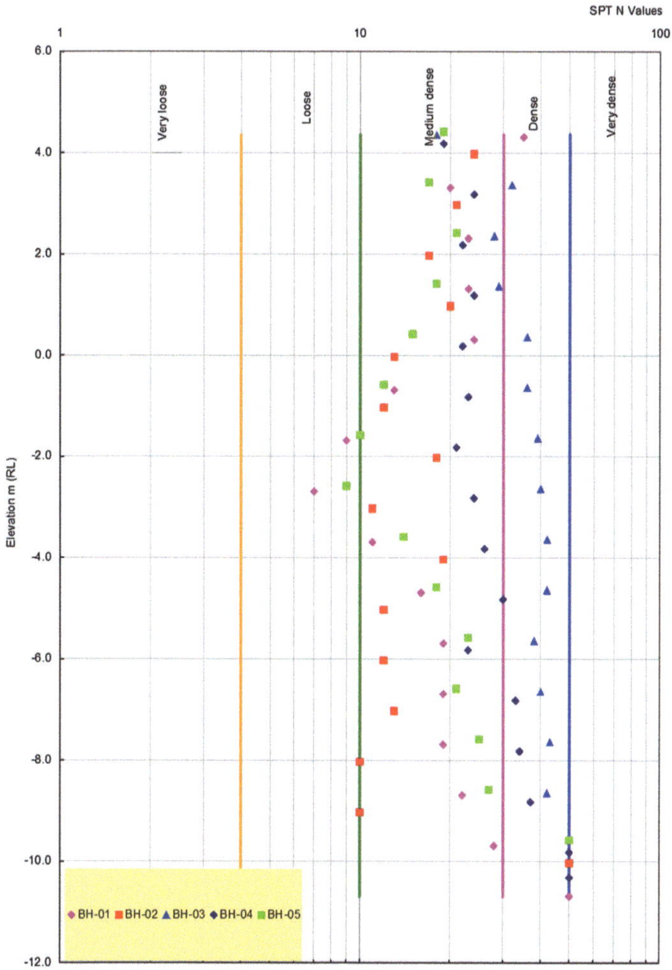

Fig. 7. SPT vs. Elevation – Duqm site

The obtained SPT results, suggested that deep ground improvement is necessary. Vibro compaction was found to be technically feasible, time and cost effective method to achieve the necessary targets. Pre compaction – CPT tests were conducted and used to assess the liquefaction potential and plan the VC works. The results showed the site is liquefiable and weak, and therefore VC works were conducted. 1st round VC: The typical layout of triangular grid of VC points is shown in Fig. 8.

Verification included post-compaction CPTU tests at three locations as indicated in the above layout. At each location, 2 pair of post-compaction CPT tests were conducted, at 1/3rd spacing from the poker point and at centre of spacing of a typical triangular grid. Typical post-compaction CPT results and CPT – based liquefaction analysis results are shown in Fig. 9.

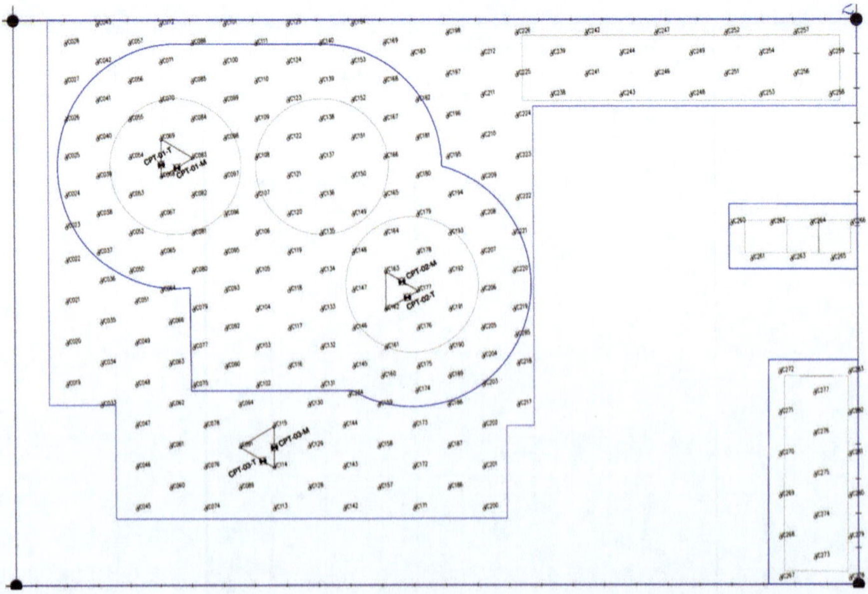

Fig. 8. Layout of VC points & post-compaction CPT tests locations – Duqm site

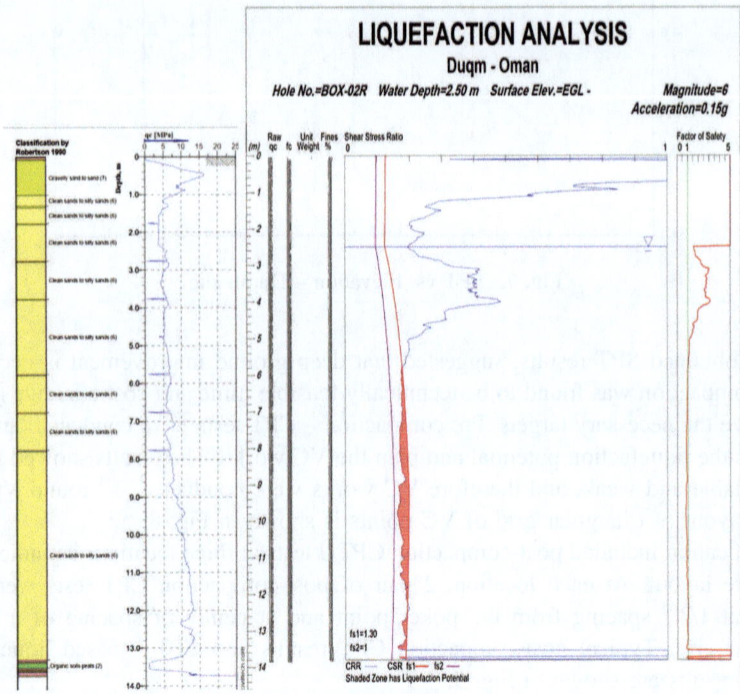

Fig. 9. Typical post-compaction– CPT result & liquefaction analysis Duqm site

Note: Liquefaction analysis is conducted according to NCEER (1997) recommendations with specialist software, which is commonly specified in most similar projects in GCC areas (e.g. Dubai). Design Earthquake effects as Magnitude (M) and Maximum ground acceleration (a_{max}) ranges of 5.5–6 and 0.10–0.15 g, respectively were considered as adopted by the Geotechnical Investigation Study, with amplification factor of 1.5 applied through the reclaimed soil above rock level.

Later, 3 post-compaction DMT tests were conducted, each at the centre of the test areas of post-compaction CPTs, to assess K_D effects on the CPT-Based liquefaction potential, and also for settlement performance of circular raft foundations proposed for the cement silos. Typical DMT results are shown in Fig. 10. Definitions of parameters measured and interpreted by DMT test are given in "Report of the ISSMGE, Technical Committee 16, 2001". The obtained results were repeatable and reliable.

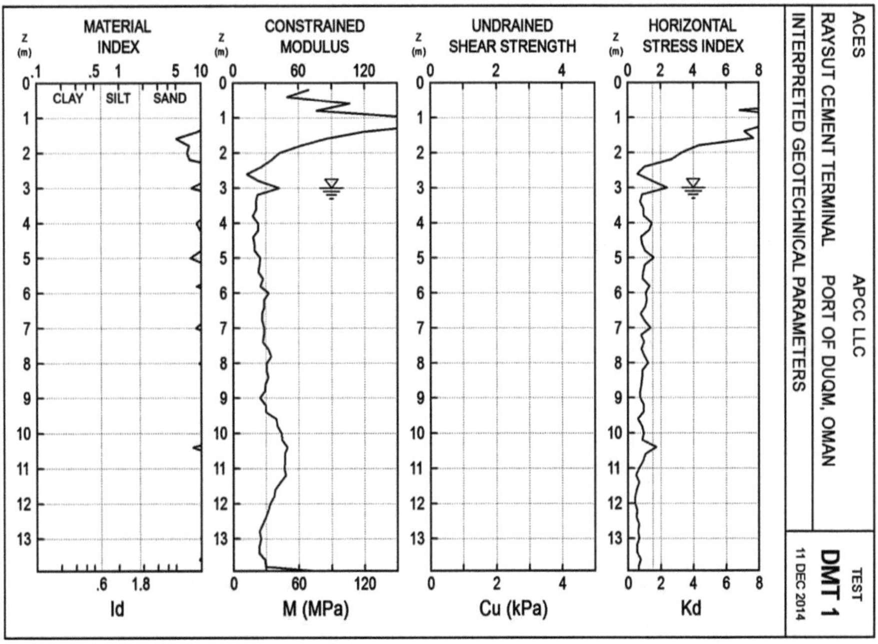

Fig. 10. Post-compaction Compaction DMT 1 (typical result) – Duqm site

Definitions of the different parameters incorporated with DMT test are shown in the below table extracted from The Flat Dilatometer Test (DMT) in Soil Investigation, Report of the ISSMGE, Technical Committee 16 on "Ground Property Characterization from In-Situ Testing", 2001.

SYMBOL	DESCRIPTION	BASIC DMT REDUCTION FORMULAE	
p_0	Corrected First Reading	$p_0 = 1.05 (A - Z_M + \Delta A) - 0.05 (B - Z_M - \Delta B)$	Z_M = Gage reading when vented to atm. If ΔA & ΔB are measured with the same gage used for current readings A & B, set $Z_M = 0$ (Z_M is compensated)
p_1	Corrected Second Reading	$p_1 = B - Z_M - \Delta B$	
I_D	Material Index	$I_D = (p_1 - p_0) / (p_0 - u_0)$	u_0 = pre-insertion pore pressure
K_D	Horizontal Stress Index	$K_D = (p_0 - u_0) / \sigma'_{v0}$	σ'_{v0} = pre-insertion overburden stress
E_D	Dilatometer Modulus	$E_D = 34.7 (p_1 - p_0)$	E_D is NOT a Young's modulus E. E_D should be used only AFTER combining it with K_D (Stress History). First obtain $M_{DMT} = R_M E_D$, then e.g. $E \approx 0.8 M_{DMT}$
K_0	Coeff. Earth Pressure in Situ	$K_{0,DMT} = (K_D / 1.5)^{0.47} - 0.6$	for $I_D < 1.2$
OCR	Overconsolidation Ratio	$OCR_{DMT} = (0.5 K_D)^{1.56}$	for $I_D < 1.2$
c_u	Undrained Shear Strength	$c_{u,DMT} = 0.22 \sigma'_{v0} (0.5 K_D)^{1.25}$	for $I_D < 1.2$
Φ	Friction Angle	$\Phi_{safe,DMT} = 28° + 14.6° \log K_D - 2.1° \log^2 K_D$	for $I_D > 1.8$
c_h	Coefficient of Consolidation	$c_{h,DMTA} \approx 7 \text{ cm}^2 / t_{flex}$	t_{flex} from A-log t DMT-A decay curve
k_h	Coefficient of Permeability	$k_h = c_h \gamma_w / M_h$ ($M_h \approx K_0 M_{DMT}$)	
γ	Unit Weight and Description	(see chart in Fig. 16)	
M	Vertical Drained Constrained Modulus	$M_{DMT} = R_M E_D$ if $I_D \le 0.6$ $R_M = 0.14 + 2.36 \log K_D$ if $I_D \ge 3$ $R_M = 0.5 + 2 \log K_D$ if $0.6 < I_D < 3$ $R_M = R_{M,0} + (2.5 - R_{M,0}) \log K_D$ with $R_{M,0} = 0.14 + 0.15 (I_D - 0.6)$ if $K_D > 10$ $R_M = 0.32 + 2.18 \log K_D$ if $R_M < 0.85$ set $R_M = 0.85$	
u_0	Equilibrium Pore Pressure	$u_0 = p_2 = C - Z_M + \Delta A$	In free-draining soils

Incorporating DMT Results to Support CPT-Based Liquefaction Analysis

In his recent paper, Marchetti (2016) described a method to incorporate the stress history parameter (K_D) of DMT into the Liquefaction Correlations. The final recommended relationship that shows the new correlation is shown in Fig. 11.

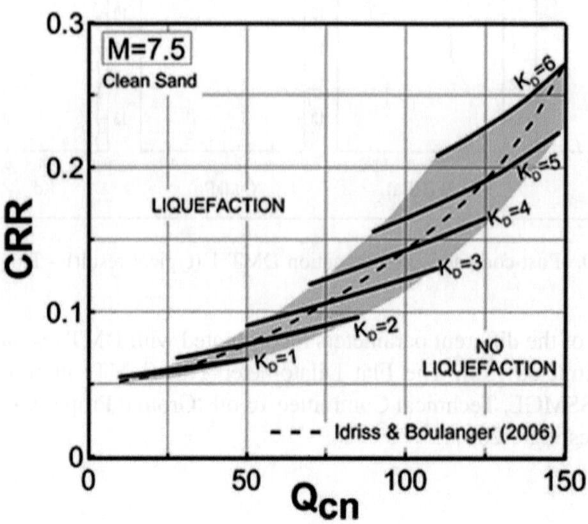

Fig. 11. Chart for estimating CRR in clean sand based on Q_{cn} & K_D (Marchetti 2016)

The relationship is based on Q_{cn}, K_D relationship as $Q_{cn} = 25 \times K_D$ given earlier by Robertson (2012), therefore, CRR is evaluated from measured CPT (Q_{cn}) and also from $Q_{cn} = 25 \times K_D$, and the average CRR is then evaluated as: Average CRR = $[(\text{CRR from } Q_{cn}) \times (\text{CRR from } K_D)]^{0.5}$. CRR is the cyclic soil resistance. The CRR values provided by Fig. 11 can be regarded as the sum of two parts. One part, the bulk of it, is the CRR estimated by the CPT consensus curve. The other part is the K_D correction. An advantage of Fig. 11 is that it shows in one graph the whole picture, facilitating the recognition of trends. For instance the graph indicates that for high values of K_D and Q_{cn}, corresponding to seasoned sands, the differences between the curves are higher. This later observation is of importance as the likely improvement effect of K_D on the value of soil cyclic resistance CRR can be quickly noticed. Therefore, for low K_D of <2, the effects may be ignored, whereas for higher K_D, then the effects are regarded as significant. This is also illustrated in Fig. 12 as given by Marchetti (2015).

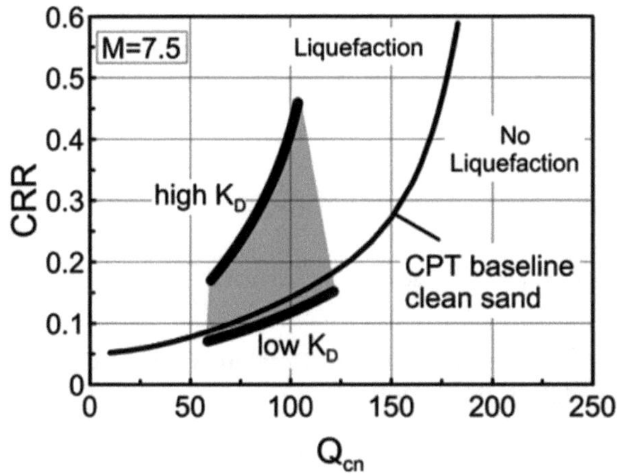

Fig. 12. CRR expressed as a function of Q_{cn} & K_D. Qualitative framework (Marchetti 2015)

An excel sheet was prepared to conduct the analysis for the above. The 3 DMT tests conducted in the middle of each test triangle has indicated low KD as indicated in the above tests results (from slightly <1 to 1 for tests 1 & 3 and around 2 to 3 for test 2). The conclusion was that K_D results did not impose any improvement on CRR, and the site was still classified as liquefiable and further round of VC was specified. The DMT results, however has also indicated the spatial variation as indicated by different ranges of K_D, which is in close agreement with the post-compaction CPT findings.

Following re-compaction works conducted by the contractor, other round of post-compaction CPT tests were only conducted that indicated satisfactory performance, as illustrated in Fig. 13. 2nd round of VC - Typical Post-compaction CPT result:

Liquefaction analysis was found satisfactory and the site was accepted, and the construction commenced.

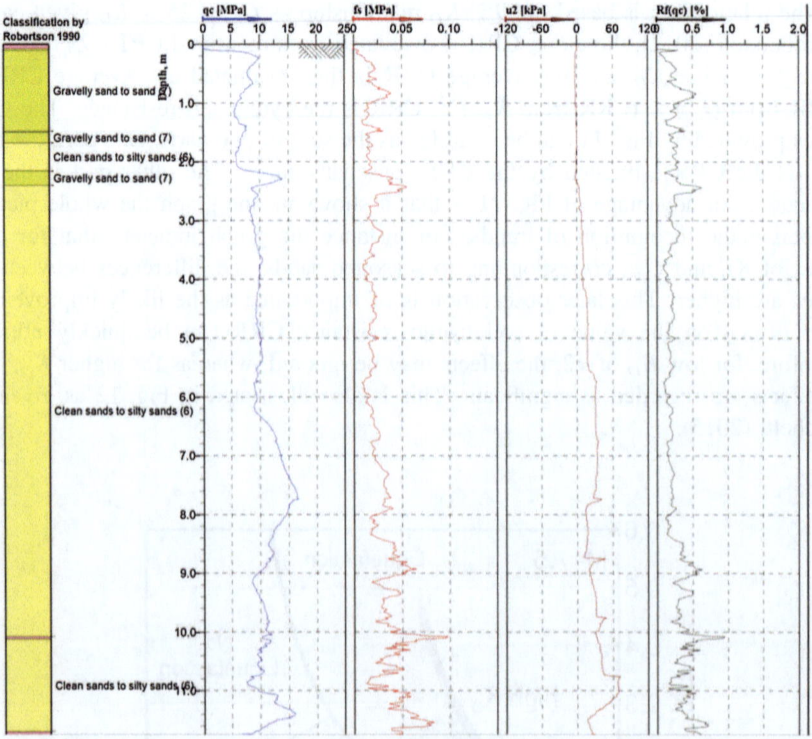

Fig. 13. Typical post-compaction compaction (2nd round) CPT – Duqm site.

1.3 Case 2: Confirming OC Nature of Previously Vibro-Compacted, Aged Sandy Fill in PJ in Dubai

Dubai is part of United Arab Emirates. It lies on the Arabian Gulf as shown in Fig. 14. Dubai area is distributed physiographically into main land and coastal areas in a ratio of 9:1. The main land is mostly occupied by Aeolian/desert sand dunes (80% of the area) with hard encrustations of local/in land sabkha's in areas with near surface water table and thin sheets of Aeolian sand overlying the local fans of gravels. The sand dune near coastal belt is enriched with carbonate source material from the sea-shells and carbonate rocks.

Dubai is Famous of its tall towers and several man made islands as indicated in Fig. 15.

Subsurface geological and geotechnical conditions in the island are described as follows. Results of Typical Test Borings in the Island would show the typical profile as illustrated in Fig. 16, composed of:

An upper layer of man hydraulically laid sandy fill of 12–18 m in depth sitting on a thin, weak sea bed layer of very silty SAND to sandy SILT (Carbonate content exceeds 70% and is >90% for many samples), followed by Cap rock of Calcarenite and Sandstones.

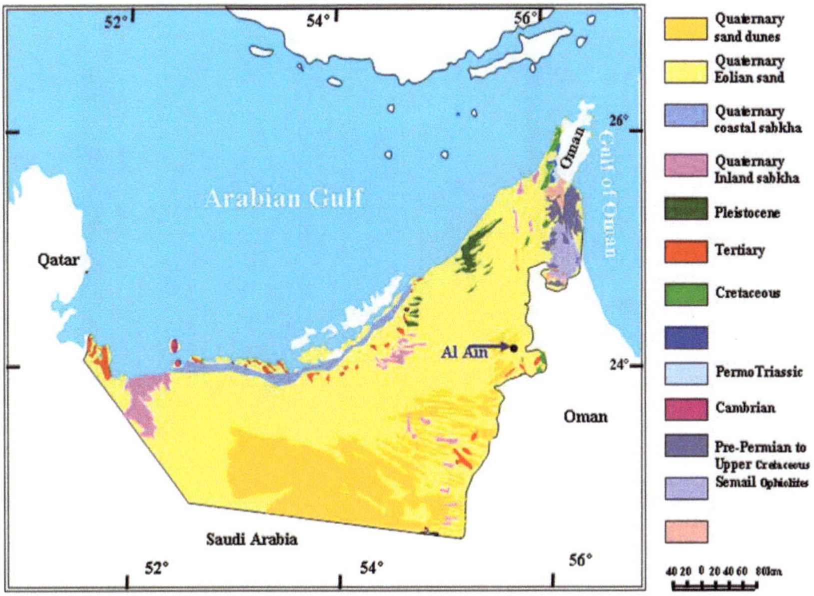

Fig. 14. Map of UAE indicating Dubai on the map

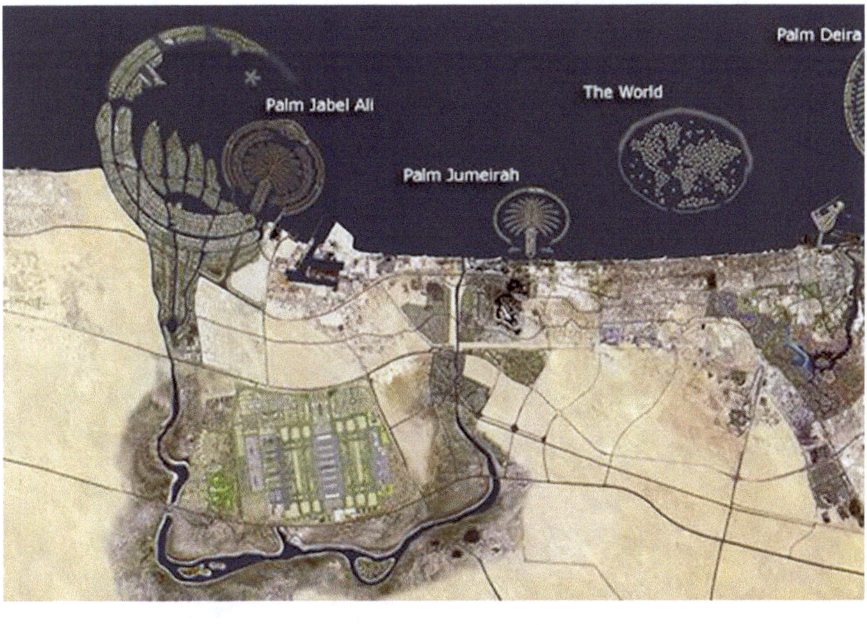

Fig. 15. Sattelite image of Dubai and coast line

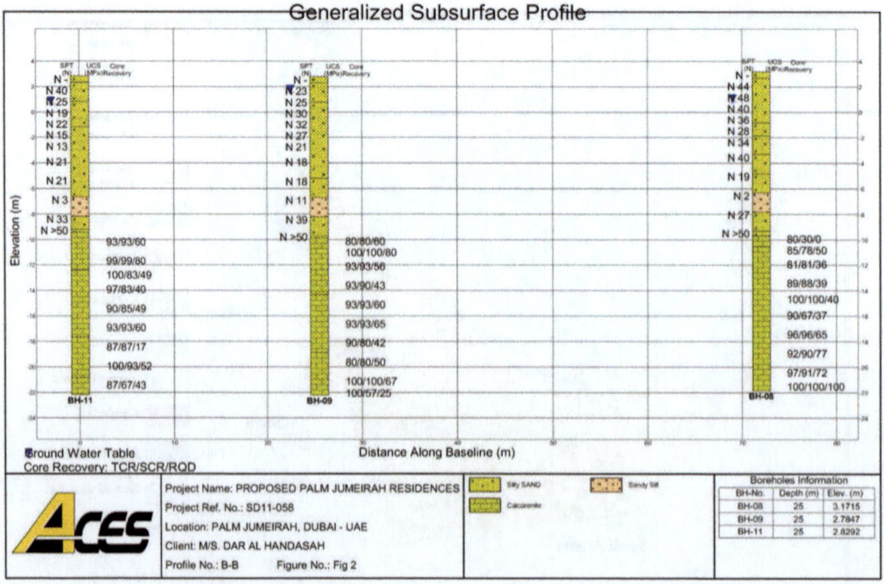

Fig. 16. Typical subsurface conditions within PJ Island

1.3.1 Application of DMT in Palm Jumeira

The site of a prestigious new project lies on the crescent Palm Jumeira. The site area is >150,000 sq. m with 4 towers of 160–190 m height range. The architectural design of the project as schematically indicated in Fig. 17, will make it a new landmark in Dubai.

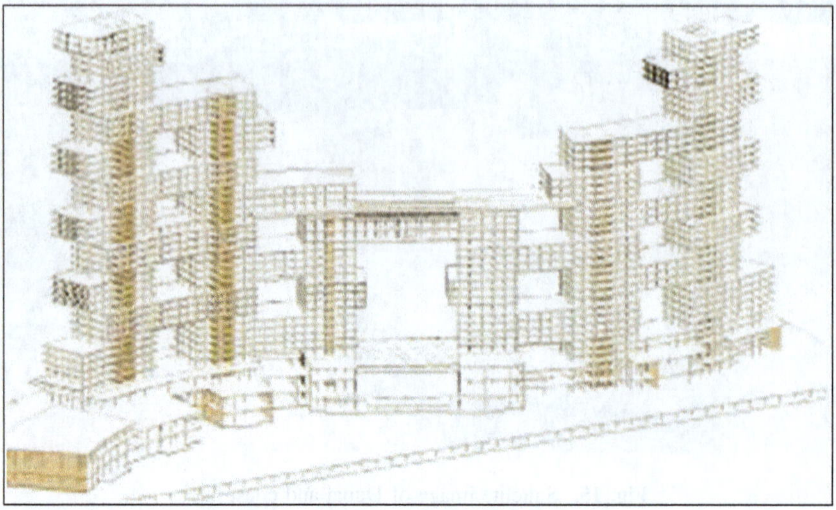

Fig. 17. View of the project architecture

The subsurface ground conditions are typical of the Palm Jumeira in general as indicated in the generalized subsurface profile presented in Fig. 18.

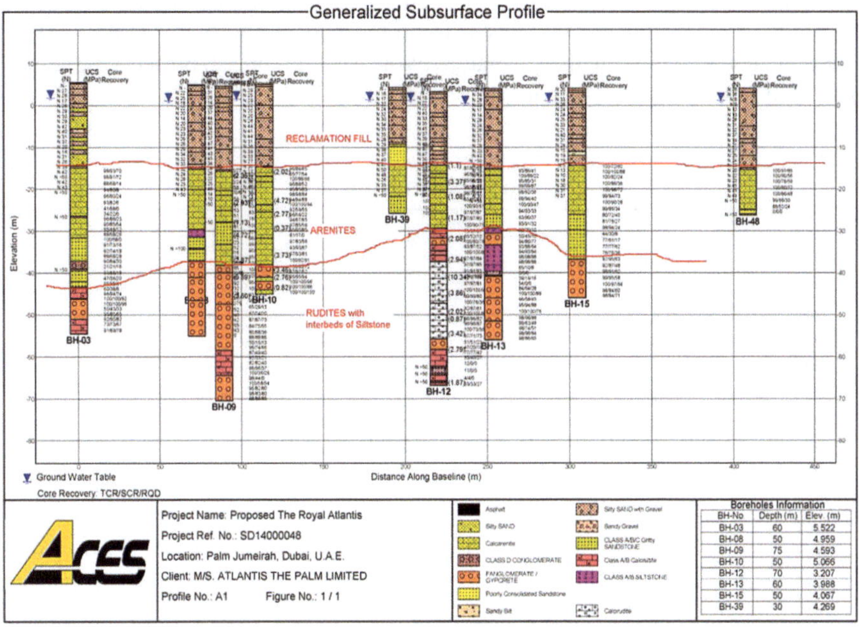

Fig. 18. Generalized subsurface profile along the site – Palm Jumeira Dubai

The upper silty SAND earthfill was previously deep compacted with Vibro Compaction since more than 8 years and hence the CPT tests results indicated dense conditions. Seismic studies included in-situ tests conducted to measure the shear wave and other dynamic properties of the site.

The typical MASW (multi-channel analysis of surface waves) results in the vicinity of the above CPT test result as shown in Fig. 19, indicate the shear wave velocity of the upper sand generally exceeds 400 to 500 m/s particularly within the upper 10 m.

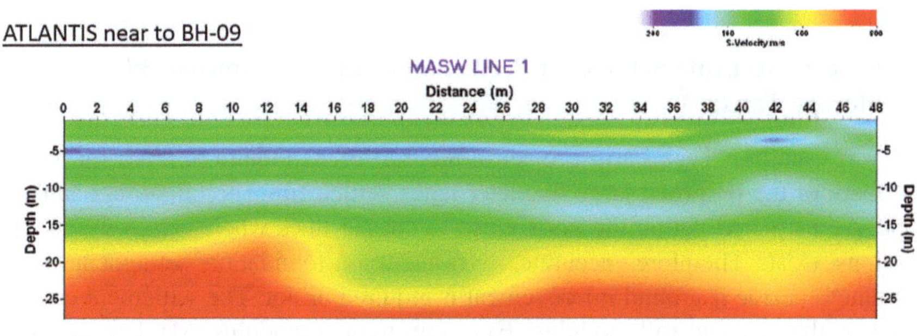

Fig. 19. Typical MASW profile – PJ Dubai

This was explained to be a result of the over-consolidated nature of the soil due to aging effects as reflected by the relatively high CPT results. However, a more specific justification was required to address the OCR which is difficult to assess by CPT tests results for sandy soil. Accordingly, few DMT tests were conducted at selected locations to establish the constrained modulus (M) profile and establish the DMT (M)/CPT (qc) ratio that would indicate the OC nature of the sand. Figure 20 presents the result of typical DMT (M) and CPT (qc) in Mpa as obtained.

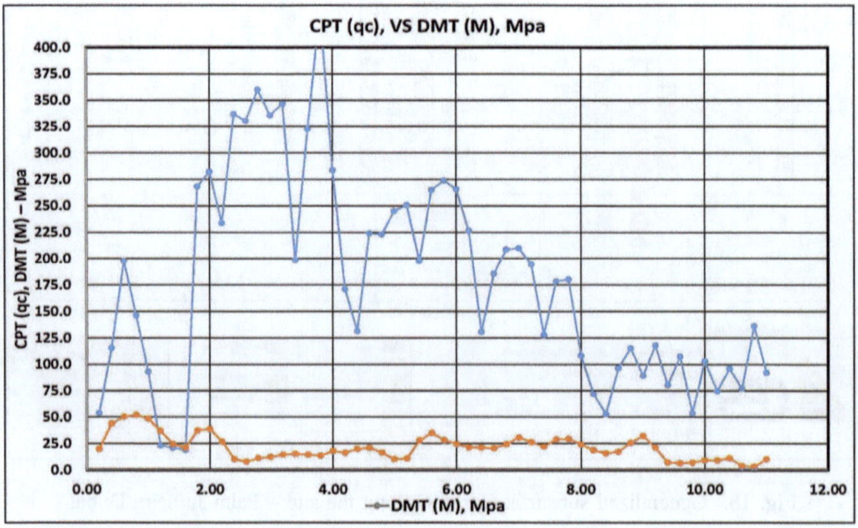

Fig. 20. DMT (M) – blue curve - and CPT (qc) – red curve - in Mpa vs. Depth

The M/qc ratio is presented in the below graph and the 1 m rolling average is also indicated.

M/qc ratio indicated in Fig. 21 shows that it is greater than 8 for the upper 8 m and exceeding 25 within the upper OC crust. This is a clear confirmation of the OC nature of the earthfill and provides the necessary justification of the MASW results. It further provided general good matching with CPT and MASW results.

1.4 Case 3: Accurate Settlement Prediction of Large Diameter Oil Storage Tanks

DMT was used in a Tank Farm Project in the Port of Al-Fujeira in UAE for accurate assessment of the proposed large diameter tanks settlement. The oil storage tanks are typically constructed within the ports close to coastal lines where relatively poor soil conditions exist. Therefore, accurate assessment of foundation settlement is very important to decide if ground improvement is required or not. The settlement estimates depend on the assessed soil modulus (E) or constrained modulus (M). For silty sandy soils that are generally encountered in that area, the estimates are generally based on

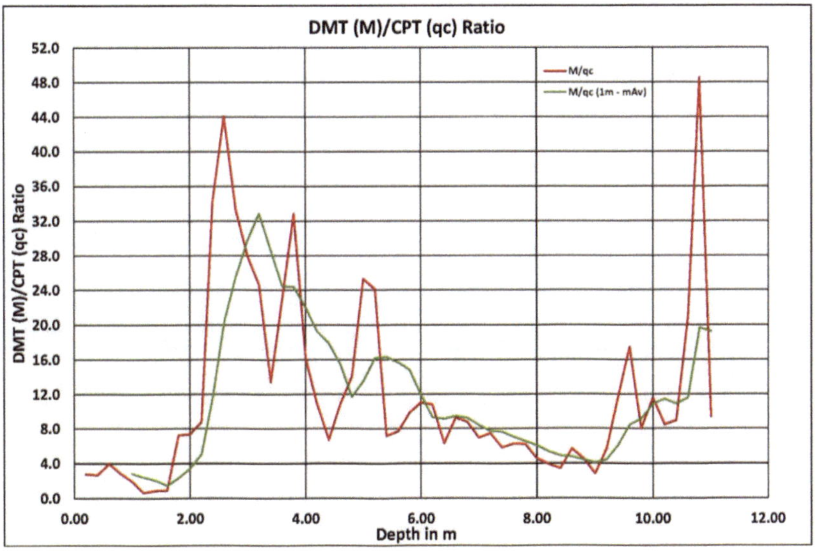

Fig. 21. M/qc ratio vs. depth

SPT and CPT (qc – tip resistance) tests results, which would generally result in low range values of E. The most commonly used relations are:

$E = N_{60}$ (Mpa) – (Ciria Report 43, 1995)
M = 4 qc for (qc < 10 MPa)
M = 2qc + 20 Mpa for (10 < qc < 50 MPa)
M = 120 Mpa for (qc > 50 MPa) – (Lunne et al. 1997)

The Eurocode 7, mentions (E = 2.5qc MPa), which is also consistent with the above estimate of M for NC conditions of sand. For OC sands, the following relation is typically recommended: M = 5 qc for (qc < 50 MPa), and M = 250 MPa for (qc > 50 MPa) (Lunne et al. 1997).

The relation between M, Shear Modulus G and Young's Modulus E according to Elasticity Theory, based on Poisson's ratio ν are given below:

$$M/G = 2(1 - \nu)/(1 - 2\nu), \text{ and } G/E = 2(1 + \nu)$$

The effect of over consolidation (stress history) is a main factor that is generally underestimated with CPT based M or E estimates.

1.4.1 Effect of Shear Strain/Loading Level

The foundation settlements shall be assessed for shear strain in the range of 0.1% (This is established based on observations made for well designed actual foundations). Therefore, with absence of reliable procedure to estimate OCR of sands from the results of CPT or SPT tests, then, the above is only illustration of the amount of conservatism or risk which the interpreter has to take to estimate M or E for his calculations!

For large diameter tanks (>50 m), the use of conservative estimates of Es would result in settlements exceeding the maximum permissible limit (API 650) and hence deep ground improvement and other expensive solutions would be needed. Figure 22 shows illustration of the sensitivity of settlement of the tank to the modulus E of the upper overburden soil layer.

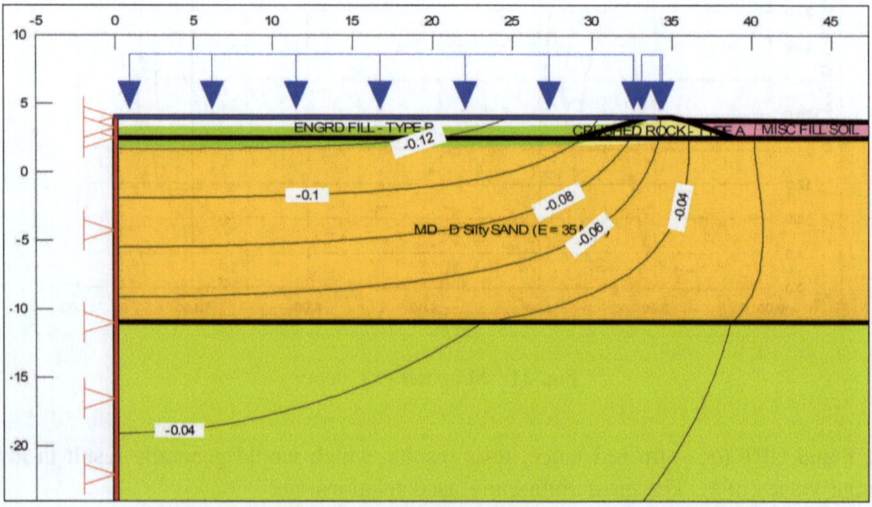

Fig. 22. E = 35 Mpa – settlement at tank edge = 60 mm

Typical settlement analysis of 68 m Diameter tank for range of E modulus of the upper overburden silty Sand for 300 kPa uniform Hydrotest pressure.

By adopting DMT based modulus of 60 Mpa, then the edge settlement was found to be 45 mm only.

Note: Axi-Symmetric FE analysis for settlement was conducted with Sigma/W software (www.geo-slope.com).

Local experience has indicated that actual tank settlements under hydrotest loading test are much less than the theoretical estimates of tank settlements based on the above SPT and CPT methods. The actual measured settlements are typically <50% of the theoretical estimates.

The cost of deep ground improvement with Vibro-Compaction or Stone Columns is several hundred thousands of dollars per each large diameter tank, in addition to several months of delay of the project.

It is also important to point out that the hydrotest conducted (API 650) at the completion of construction of each tank is a preload that causes most of the permanent settlement component to take place and subsequent unloading and reloading of the tank during its operational life time would follow the more stiff reloading/unloading modulus with in-significant movements as experienced with so many existing tanks of various diameters.

Therefore, the accurate and reliable assessment of the soil modulus is very significant for confidence of the estimated foundation settlement. DMT was therefore introduced to provide reliable and direct measurement of in-situ modulus of the soil and provide site – specific correlation to the other penetration tests as SPT and mostly CPT that are widely used.

Case Study: The proposed storage tank farm includes tanks having diameters of 68 m and 36 m for the small ones. All the tanks were designed for an average max. hydrotest load of up to 300 kPa. The main design criteria is that the shell plate settlement under max. load not to exceed 50 mm (preferable). However, settlements up to 100 mm may be tolerated by the owner, provided the settlements are uniform (differential settlements between edge points at 10 m arc length not to exceed 13 mm) and most of the settlements to be of short term nature (to take place during the hydrotest).

The investigation for this Tank farm project consisted of a number of geotechnical boreholes drilled up to 40 m depth, CPT and DMT tests in addition to geophysical survey with MASW (Multi Channel Analysis of Surface Waves), as given in Fig. 23.

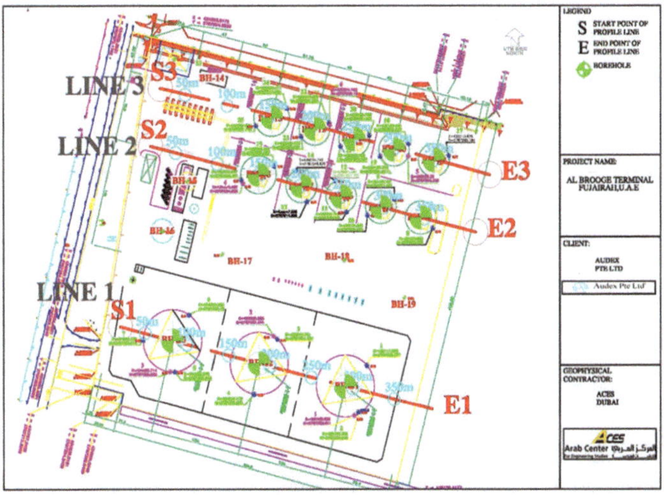

Fig. 23. Layout of 3 MASW lines covering all tanks locations

The typical geological cross section indicated the site is made up of thin upper rockfill layer of 1–4 m thickness that was constructed to provide a level platform with thickness increasing towards the sea side, underlain by medium dense to dense fine silty sand that is followed by bedrock of very weak to weak Sandstone and conglomerates. GWT exists at 2.5 m depth approximately. Generalized subsurface profile showing the main soil and rock layers is shown in Fig. 24. SPT results of all boreholes indicated Medium dense to Dense soil with few localized weak lenses/pockets. The DMT & CPT results, after 3 m of pre-drilling though rockfill cover, indicated OC sand followed by Medium dense silty Sand, as shown in Fig. 25.

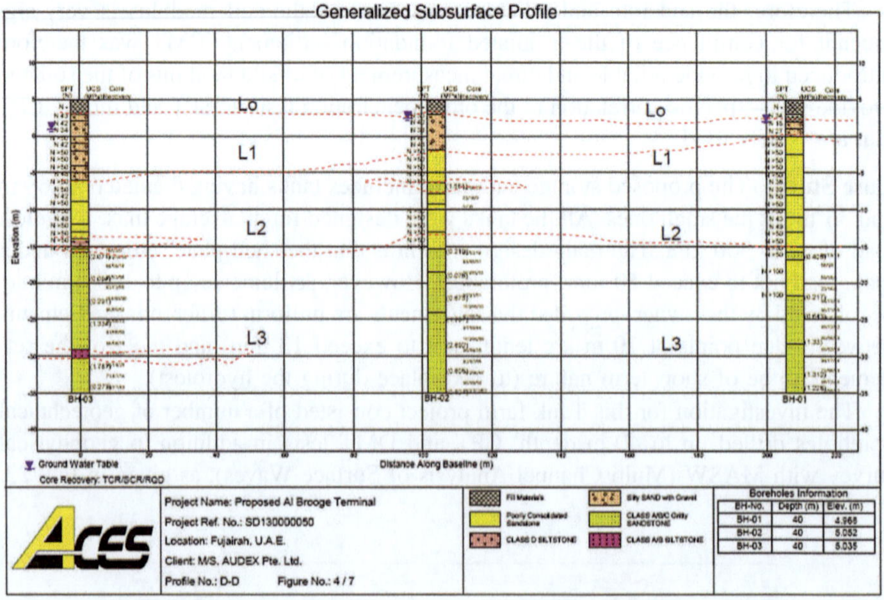

Fig. 24. Subsurface profile indicating general homogeneity – Fujeira site.

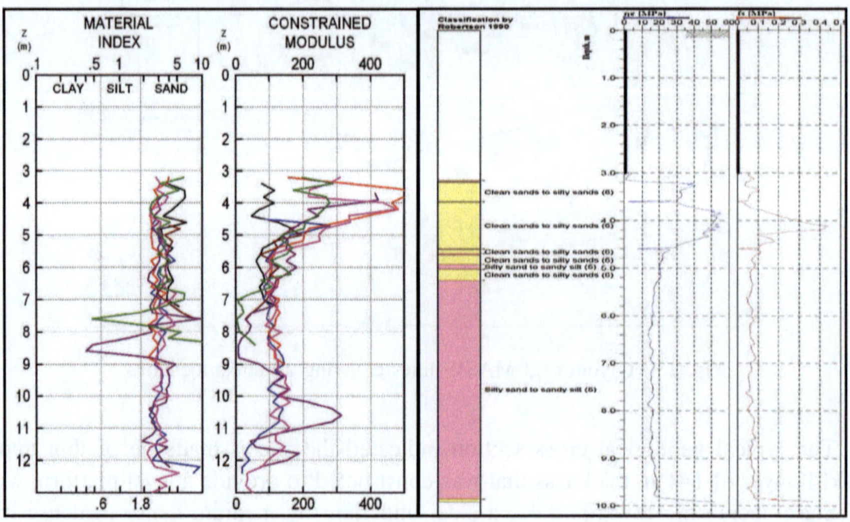

Fig. 25. Summary of all the 7 DMT tests & typical CPT test result with SBT profiling

The main parameter controlling the settlement estimates of the sand is obviously the modulus of deformation E. Based on SPT and CPT results, the average Es was estimated to be 35–40 Mpa, resulting in settlements exceeding 50 mm at the large diameter tank edges. Based on direct measurement of DMT, M average was slightly

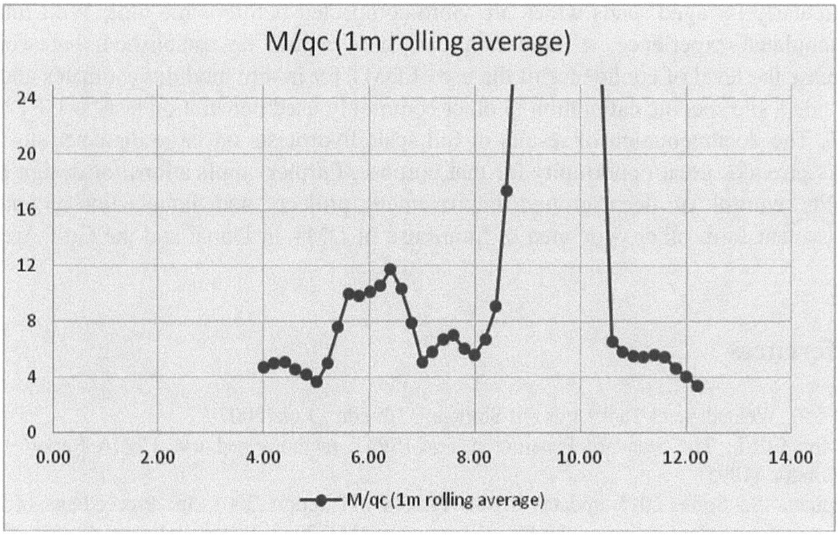

Fig. 26. M/qc ratio obtained (vertical axis) vs. depth (horizontal axis) in m

around 100 Mpa, and E is >60 Mpa. The typical DMT (M)/ CPT (qc) ratio is shown graphically in Fig. 26.

It shows that M/qc may safely be taken as 6 instead of 4 as adopted for NC sand, which means that for average qc = 15 Mpa, then M = 90 Mpa and E will be >60 Mpa. This is at least 50% greater than the typical CPT based estimates adopting NC sand conditions.

The final recommendation suggested that no deep ground improvement is necessary for settlement limitation. Vibro Compaction was only suggested for the main tanks to improve the stiffness within potential weak localities, as indicated by the CPT and some of the DMT results. The project construction phase has commenced and the next step will be the observation of the results of actual hydrotests that are planned for each of the tanks, which will provide the final confirmation on the foundation ground behavior. As stated previously, previous experience has indicated that the actual results of settlements are less than the theoretical predictions. This will increase the confidence in the methods adopted for the soil modulus estimates.

2 Conclusion

The use of DMT in Dubai and the Arabian Gulf Area is relatively recent. The early applications presented in this paper indicate that DMT would be of great value to identify the effects of previous stress history and over-consolidation of either man made and stabilized embankments or natural sandy soils, resulting in better understanding of actual foundation soil behavior and avoiding expensive foundation solutions that may not be necessary. Further, the use of DMT to support the confidence on assessment of the liquefaction potential by CPT or SPT-Based methods is of main significance, particularly for aged sands which are vibro-compacted before some time. With future accumulated experience, a more rich data base would be established that would increase the level of confidence of the use of DMT for in-situ modulus estimates and to provide a site specific calibration to other commonly used penetration tests as CPT and SPT. The documentation of results of full scale hydrotests on large diameter storage tanks provides great opportunity for that purpose. Further, applications for design and quality control of deep ground improvement projects and liquefaction potential assessment form other wide area of future use of DMT in Dubai and the Gulf Area.

References

API 650: Welded Steel Tanks For Oil Storage, 11th edn., June 2007

Clayton C.R.I.: The Standard Penetration Test (SPT): methods and use. CIRIA Report 143, London (1995)

Marchetti, S.: Some 2015 updates to the TC16 DMT report 2001. In: Proceedings of 3rd International Conference on the Flat Dilatometer DMT 2015, Rome, Italy, pp. 43–65 (2015)

Marchetti, S.: Incorporating the stress history parameter K_D of DMT into the liquefaction correlations in clean uncemented sands. J. Geotech. Geoenviron. Eng. **142**(2), 04015072 (2016)

NCEER: Proceedings of the NCEER workshop on evaluation of liquefaction resistance of soils. Technical report NCEER-97-0022, 31 December 1997

Robertson, P.K.: Interpretation of cone penetration tests – a unified approach. Can. Geotech. J. **46** (11), 1337–1355 (2009). doi:10.1139/T09-065

Robertson, P.K.: The James K. Mitchell lecture: interpretation of in-situ tests – some insights. In: Proceedings of 4th International Conference on Geotechnical and Geophysical Site Characterization, Porto de Galinhas, vol. 1, pp. 3–24 (2012)

Lunne, T., Robertson, P.K., Powell, J.J.M.: Cone penetration tests in geotechnical practice (1997)

The Flat Dilatometer Test (DMT) in Soil Investigation, Report of the ISSMGE, Technical Committee 16 on "Ground Property Characterization from In-Situ Testing" (2001)

Some Laboratory and Numerical Studies on the Behaviour of Stone Columns Installed in Mumbai Marine Clay

Vinay Bhushan Chauhan[1(✉)], Yashwant A. Kolekar[2], and Satyanarayana M. Dasaka[1]

[1] Civil Engineering Department, Indian Institute of Technology Bombay, Mumbai 400076, Maharashtra, India
chauhan.vinaybhushan@gmail.com, dasaka@civil.iitb.ac.in

[2] Civil Engineering Department, Government College of Engineering, Pune 411005, Maharashtra, India
yashwantkolekar@gmail.com

Abstract. Mumbai marine clay is problematic in nature for substructures and it needs to be strengthening before making it available for any construction activity. Out of many available geotechnical solutions, stone columns are quite handy in significantly improving the bearing capacity of marine clay, which also supplements by dissipating excess pore water pressure and reducing the settlements of clay beds under external loads. For the investigation on behavior of stone columns, marine clay was collected from the Uran site near Mumbai, India. Slurry consolidated clay beds were prepared and gravity loading mechanism was adopted to consolidate the clay beds in the laboratory. Laboratory tests were performed to ascertain the uniformity and repeatability of the clay beds by slurry consolidation. Augers of 50, 75 and 100 mm diameter were used to bore the holes in the clay bed and aggregates were used for stone column preparation by replacement method. The static displacement controlled tests were performed on the stone column reinforced clay bed. In the present study, the loading is applied on stone column alone, as it leads to the ultimate axial capacity of stone column improved ground. Clay bed reinforced with stone column exhibits improved load-settlement response, compared to unreinforced clay beds. The improvement was very significant at higher settlements. The failure load of reinforced clay bed is around 6 times more than the unreinforced clay bed. In order to further understand the behaviour of stone column reinforced ground, numerical modeling of stone column reinforced clay bed is conducted using FLAC^{3D}. Using the validated numerical model of the stone column, the effect of variation of the diameter of stone column on its performance was studied. From the numerical analysis and physical tests conducted in laboratory, it is observed that the stone columns with smaller diameter when subjected to vertical loading carried higher bearing pressures when compared to that of the larger diameter stone columns, which may be due to the greater confining and larger bulge formed at a depth of 2–3 times the diameter of stone columns.

W. Frikha et al. (eds.), *Soil Testing, Soil Stability and Ground Improvement*,
Sustainable Civil Infrastructures, DOI 10.1007/978-3-319-61902-6_11

1 Introduction

The forced intrusions in the form of aggregates/stones for the densification of the soft soil strata are commonly referred as granular columns or stone columns. Stone column construction involves partial replacement of weak subsurface soils by the vertical column of compacted granular material that usually penetrates the weak strata to reinforce the surrounding soft soil and accomplish the ground improvement, so that it can safely withstand the superimposed loads (Barksdale and Bachus 1983). These stone columns are quite handy in significantly improving the bearing capacity of marine clay, which also supplements by dissipating excess pore water pressure and reducing the settlements. The vertical axial load applied at the top of stone column also induces lateral movements in the top portion of the stone column, thereby displacing the surrounding soil leading to formation of a bulge up to a depth of 2 to 3 times the diameter of stone column below the ground. This bulge in turn, helps the stone column to develop additional load carrying capacity by generating passive pressure conditions in the surrounding soil, which provides additional confinement for the stone column through the development of hoop stresses. To ascertain the behavior of stone column, laboratory tests and numerical analysis using FLAC^{3D} is performed. Also, a parametric study using physical modeling tests on the effect of variation of diameter of stone column installed in the consolidated clay bed on bearing capacity is studied in the present study.

2 Geomaterials and Experimental Program

The marine clay used in for the present study is obtained from the sea coast of Dronagiri, near Uran, Navi Mumbai, India. Marine clay, thus procured, is placed in specially prepared large bins of size 1.5 m wide, 3.5 m long and 2.5 m deep, and is covered with tarpaulin sheets throughout the year to prevent from adverse weather conditions. The particle size distribution curves of marine clay and aggregate (used for stone column) are shown in Fig. 1(a). The pre-consolidation pressure of marine clay is determined according to the BIS: 2720 Part 15 (1986) and is found to be 18 kPa, as shown in Fig. 1(b). Physical properties of marine clay used for preparation of clay bed are presented in Table 1.

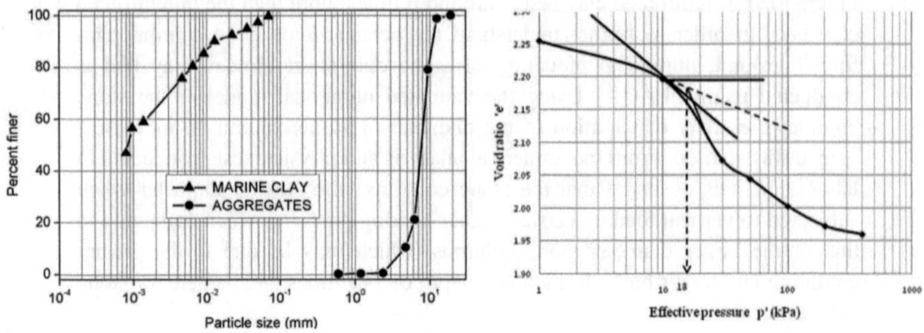

Fig. 1. (a) Grain size distribution of marine clay and aggregates (b) pre-consolidation pressure of marine clay

Table 1. Physical properties of marine clay

Properties	Values
Natural moisture content, w (%)	84
Field density, γ (kN/m^3)	14.96
In-situ vane shear strength, S_u (kPa)	8 to 9
Specific gravity of soil solids (G_S)	2.74
Liquid limit (in-situ conditions) (%)	101
Liquid limit (oven dried) (%)	51
Plastic limit (%)	40
Plasticity index (%)	61
Classification of soil	OH
Pre-consolidation pressure, p_c (kPa)	18
Free swell index (%)	31

For preparing the clay bed by consolidation process, amount of water equal to 1.5 times liquid limit of the soil is added, so as to have homogeneous sedimentation (Sridharan and Prakash 2003; Murugesan and Rajagopal 2007). The soaked clay is made into slurry with the assistance of mechanical mixer. Murugesan and Rajagopal (2007), Gneil and Bouazza (2009) and Shahu and Reddy (2011) consolidated the clay slurry to form clay beds for conducting studies on stone column reinforced clay beds. Murugesan and Rajagopal (2007), Shahu and Reddy (2011) adopted dead weight consolidation. To understand the behavior and performance of stone columns in the laboratory, unit cell concept is undertaken (Barksdale and Bachus 1983; Bae et al. 2002; Ambily and Gandhi 2007; Murugesan and Rajagopal 2007; El-Garhy et al. 2011; Shahu and Reddy 2011). This unit cell is an imaginary cylindrical volume of soil with concentric stone column and the boundary of the unit cell is the extent beyond which the effect of stone column loading cannot be felt when subjected to critical loading. Unit cell (shown in Fig. 2) is adopted in the present study, which consisted of a detachable collar at the top and cylindrical tank in the bottom part. The detachable collar placed above the unit cell has dimensions of internal diameter 350 mm, height of 250 mm and thickness 12 mm. To accomplish consolidation of large clay beds simultaneously, new simplified and economical gravity loading based lever arm set-up, as shown in Fig. 3, is devised in the present study.

Pressure equal to pre-consolidation pressure is applied to the clay slurry leading to removal of excess pore water to achieve vane shear strength in the range of 8 to 9 kPa and bring the consistency of the clay bed to that at the site and thereby simulating field conditions in the laboratory. However, Barksdale and Bachus (1983) reported that stone columns perform better in clays having undrained shear strength in the range of 15 to 50 kPa. So, the consolidation pressure is increased from 18 kPa to 36 kPa, which resulted in undrained shear strength in the range of 20 to 22 kPa, which is in the specified range to provide better performance of stone columns. On the completion of consolidation after 30 to 35 days, the detachable collar is removed and the clay protruding above the unit cell is trimmed and leveled.

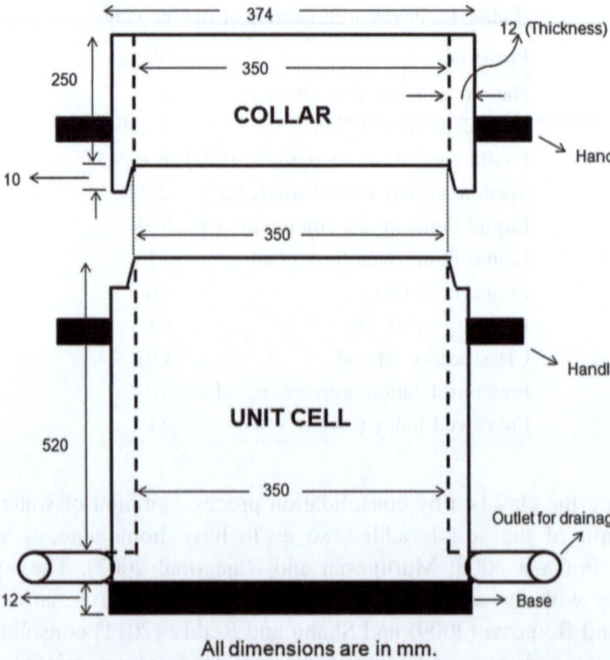

Fig. 2. Schematic layout of unit cell and collar with dimensions

Fig. 3. Specially designed loading frame for the consolidation of clay

3 Parametric Study

Bearing capacity of stone column is also a function of H/D (height/depth) ratio of stone column. Effect of diameter and H/D ratio of the stone column on the behaviour of stone column reinforced ground has been studied by various researchers. Murugesan and Rajagopal (2007, 2008, 2009, 2010) and Ambily and Gandhi (2004, 2007) adopted H/D ratio of unit cell greater than 1.0 and above, whereas, Malarvizhi and Ilamparuthi (2007) considered the H/D ratio of unit cell less than 1.0. Lovisa and Sivakugan (2015) observed that for doubly drained specimen with the H/D ratio up to 3, the wall friction does not play any role and same average degree of consolidation versus time charts can be considered. Bae et al. (2002) conducted study of the stone columns having H/D ratios as such as 5, 6, and 7. Malarvizhi and Ilamparuthi (2007) carried out similar studies and observed that with increase in H/D ratio of stone columns the settlement was reduced. So, H/D ratio for the present study is chosen 5, 6.67 and 10.

For preparation of stone column, replacement method is adopted to form fully penetrating stone columns in the consolidated clay bed using an auger (assisted by a centering plate) of diameters of 50 mm, 75 mm and 100 mm to prepare bore holes of corresponding size (Fig. 4). The aggregates, used for construction of stone columns in the marine clay bed, are procured locally. Aggregates passing through 12.5 mm and retained on 2 mm sieves are used for 100 mm and 75 mm diameter stone columns; whereas, for 50 mm diameter stone columns aggregates passing through 9.52 mm and retained on 2 mm sieves are used, following the recommendations of Nayak (1983). Aggregates are placed in each layer with the assistance of a hollow flexible pipe (aggregate placer) attached to the bottom of a funnel so that they reach the bottom and are not lost midway by getting stuck to the surrounding sticky marine clay. The aggregates are then compacted in 10 layers with a steel rod and unit weight of the stone column is maintained at 16 kN/m^3. A layer of sand of 2 mm is placed over the surface of completed stone column. Physical properties of aggregate and sand used in preparation of stone column is shown in Table 2.

Performance of stone column reinforced clay beds are undertaken in the geotechnical laboratory using 1-g small scale models. For static load tests on unreinforced clay bed, a loading plate of 100 mm diameter is centrally placed over prepared consolidated

Fig. 4. Hole formed by auger and hole filled up with aggregates

Table 2. Physical properties of aggregate and sand

Properties	Aggregate	Sand
D_{10} (mm)	4.5	0.65
D_{30} (mm)	7.0	0.80
D_{60} (mm)	8.5	1.2
Coefficient of uniformity, C_u	1.28	1.85
Coefficient of curvature, C_c	1.89	0.82
Classification as per USCS	Poorly graded gravel (GP)	Poorly graded sand (SP)
Internal friction angle (Degrees)	41°	–
Unit weight (kN/m^3)	16	–

clay bed and loading is applied at a constant displacement rate of 1.2 mm/min up to 50 mm displacement. Pressure-settlement response of clay beds consolidated at 18 kPa and 36 kPa is studied. For static load tests on reinforced clay bed, loading plates of 12 mm thick and diameter equals to that of the stone column of diameter 100 mm, 75 mm and 50 mm are studied in the present study. Loading procedure is similar to that of unreinforced clay bed. In the present studies, results from three tests carried out on the consolidated clay bed in unit cell subjected to 36 kPa consolidation pressure are presented.

4 Numerical Analysis

Clay beds with stone column having dimensions similar to laboratory test are analysed using finite difference analysis tool FLAC^{3D}. Elastic modulus of aggregate, clay and steel are taken as 45 MPa, 4 MPa and 180 GPa. Poisson's ratio of aggregate, clay and steel are taken as 0.3, 0.45 and 0.29. Figure 5 shows the numerical grid generated to simulate the unreinforced and reinforced clay beds placed inside steel tank. Clay bed and aggregate are modelled as an elasto-plastic material following Mohr-Coulomb failure criterion and tank is modelled as elastic material. The steel container consisting of clay bed and stone column is modelled numerically, and the model is fixed at bottom to represent the physical test conditions. Very fine grid is chosen for the simulation of unreinforced and reinforced clay bed model as shown in Fig. 5. The interface (shown in Fig. 6) between stone column and clay is modelled as linear spring-slider system with interface shear strength defined by the Mohr-Coulomb failure criterion. The normal stiffness (k_n) and shear stiffness (k_s) of the interface are taken as per guidelines of Itasca (2011). Interface parameters, k_n and k_s, have been assigned a value of 6×10^8 kN/m^2/m at the periphery of stone column. In the present analysis, diameter of stone column is varied as 50, 75 and 100 mm and displacement at a constant incremental displacement is applied up to 50 mm to observe the load settlement behaviour of the same and results are validated with physical test results.

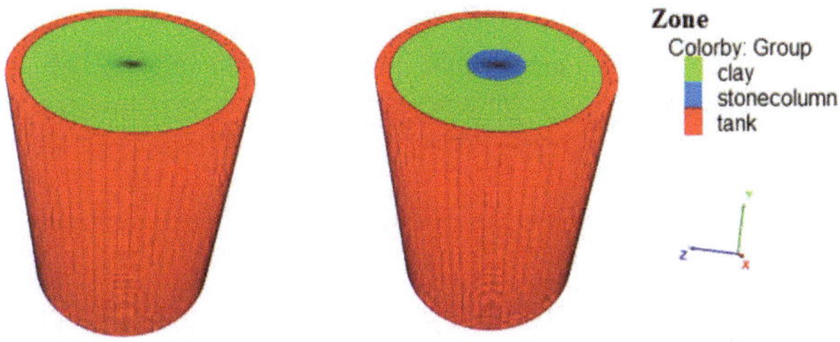

Fig. 5. Numerical grid for unreinforced and reinforced clay bed

Fig. 6. Interface between stone column and clay bed in FLAC^{3D}

5 Results and Discussion

5.1 Effect of Consolidation Pressure

With doubling of the consolidation pressure from 18 kPa to 36 kPa, there is 1.48 times increase in the bearing pressure at settlement of 50 mm for the unreinforced clay beds as shown in Fig. 7(a).

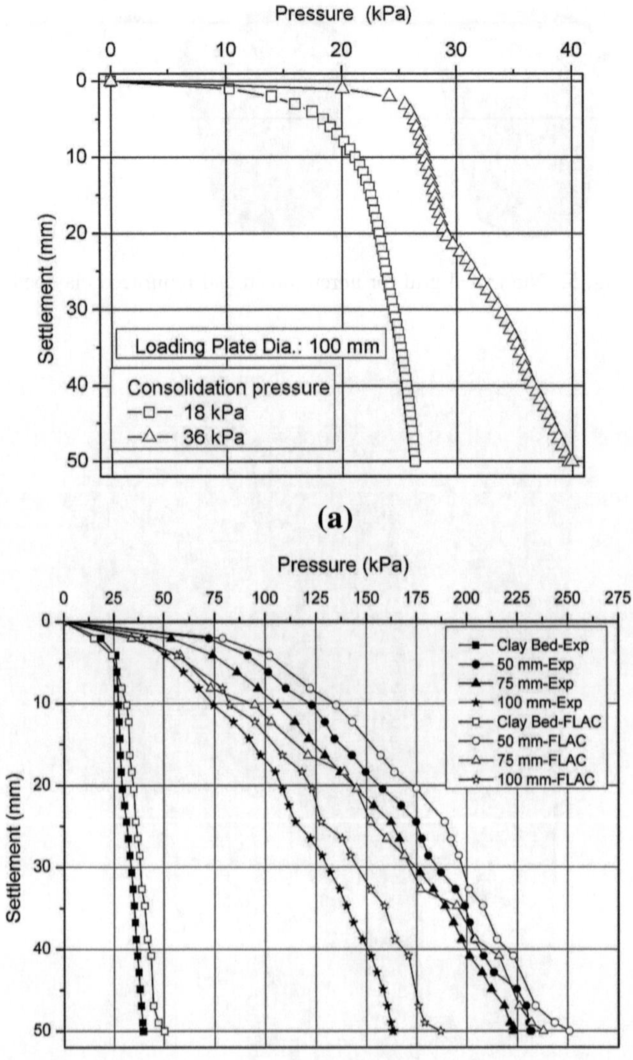

Fig. 7. Pressure-settlement response of (a) unreinforced clay beds consolidated at 18 kPa and 36 kPa (b) reinforced clay bed with different diameter of stone columns

5.2 Influence of Diameter of Stone Column

The diameter of stone column has significant effects the behaviour and performance of stone column reinforced clay bed. There is marked improvement in the bearing pressure of the stone column reinforced clay bed over the unreinforced clay bed. The increase in bearing pressure of clay bed with 100, 75, and 50 mm diameter stone columns is 4, 5.5 and 5.8 time that of unreinforced clay bed, respectively, as shown in

Fig. 7(b), at 50 mm settlement. It is observed that the experimental results for the clay bed and the clay bed reinforced with different diameter stone columns are in tandem with the numerical results. However, the numerical results are slightly on the higher side. From the numerical analysis, bulging contours of stone column in lateral directions (x and z) are presented for 100 mm diameter stone column in Fig. 8. It can be noted from these figures that the bulging of stone column extends up to a depth of 2D of stone column, where D is the diameter of the stone column. Similarly, lateral strain up to 15% is observed in the stone column, which resulted in creating passive pressure in clay bed, and led to the increase in load bearing capacity of stone column.

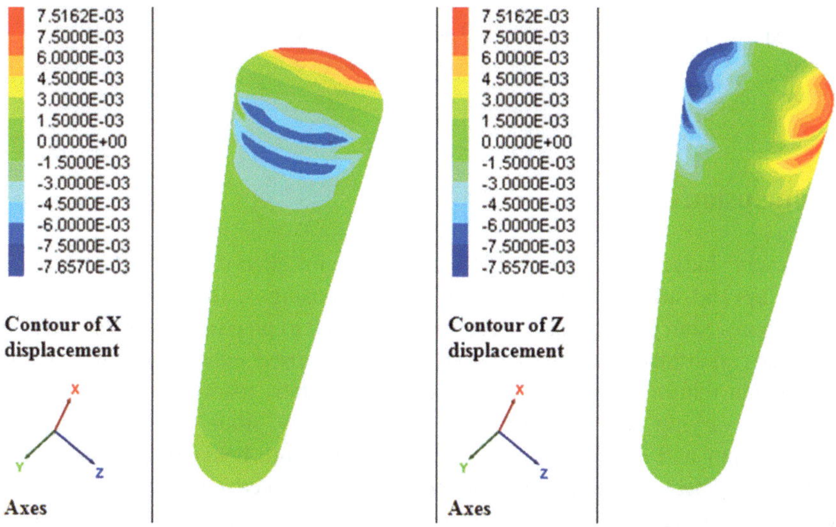

Fig. 8. Displacement contours of stone column of 100 mm diameter

5.3 Improvement Factor (I_f)

The performance improvement of the stone column reinforced clay bed is represented using a non-dimensional parameter – bearing pressure improvement factor (I_f), which is the ratio of footing pressure (q_c) for a reinforced clay bed at a given settlement to the footing pressure (q_o) for an unreinforced clay bed, at the same settlement, as defined below:

$$I_f = q_c/q_o \tag{1}$$

Improvement factor for the stone column reinforced clay bed is presented in Table 3. Briefly discuss the results shown in Table 3.

Table 3. Improvement factor for the stone column reinforced clay bed

Test description	Bearing pressure at 25 mm settlement (kPa)	Improvement factor $(I_f)_{25}$	Bearing pressure at 50 mm settlement (kPa)	Improvement factor $(I_f)_{50}$
Unreinforced clay bed	32	–	40	–
Reinforced clay bed 50 mm	201	6.3	233	5.8
Reinforced clay bed 75 mm	164	5.1	223	5.6
Reinforced clay bed 100 mm	115	3.6	164	4.1

6 Conclusions

In the present study, the behaviour of stone columns installed in very soft cohesive soil (marine clay) and subjected to static loads is studied using small scale models tested at 1-g conditions. Slurry consolidation method is adopted to prepare clay beds, simulating the natural conditions. The slurry consolidation is carried out in the unit cell itself. Stone column reinforced clay beds are subjected to static load to ascertain their performance by varying their diameter. In the present study, stone columns of diameters 50 mm, 75 mm and 100 mm are used to understand the behavior of stone column reinforced clay beds subjected to static loading. A validated numerical model is also developed in FLAC^{3D} to have better insight into the stone column behavior. The following conclusions are drawn from the present study.

1. With doubling of the consolidation pressure there is 1.48 times increase in the bearing pressure at settlement of 50 mm for the unreinforced clay beds.
2. With the increase in the diameter of stone column there is decrease in the bearing pressure of stone column which is attributed to the confining stresses, which is more in smaller diameter stone columns.
3. Clay bed reinforced with stone column exhibits improved load-settlement response, compared to unreinforced clay beds. The improvement was very significant at higher settlements. The failure load of reinforced clay bed is around 6 times more than the unreinforced clay bed, which is in tandem with the observations of other researchers. It is noted that only top portion of the stone column, i.e. depth ranging from 2 to 3 times diameter of stone column, undergoes bulging, and influencing the response of stone column to external loads.
4. The numerical simulation results show bulging at a depth of 2D which develops passive pressure conditions at the interface.

References

Ambily, A.P., Gandhi, S.R.: Experimental and theoretical evaluation of stone column in soft clay. In: Proceedings of ICGGE-2004, IIT Bombay, India, pp. 201–206 (2004)

Ambily, A.P., Gandhi, S.R.: Behaviour of stone columns based on experimental and FEM analysis. J. Geotech. Geoenviron. Eng. **133**(4), 405–415 (2007)

Bae, W.S., Shin, W.S., An, B.C.: Behaviours of stone column improved with stone columns. In: Chung, S., Kashiwagi, S., Hong, S.W. (eds.) Proceedings of the 12th International Offshore and Polar Engineering Conference, Kitakyushu, Japan, pp. 675–678, 26–31 May 2002

Barksdale, R.D., Bachus, R.C.: Design and construction of stone columns. Report No. FHWA/RD-83/026, Federal Highway Administration, Washington, D.C., USA (1983)

BIS: 2720 (Part-15): Methods of Test for Soils-Determination of Consolidation Properties. Bureau of Indian Standards, New Delhi (1985)

El-Garhy, B., Maraie, M., Youssef, A.: Behaviour of model footings resting on soft clay reinforced by floating granular piles: experimental study. Int. J. Geotech. Eng. **5**(4), 415–424 (2011)

FLAC 3D (5.0), Itasca (2011)

Gniel, J., Bouazza, A.: Improvement of soft soils using geogrid encased stone columns. Geotext. Geomembr. **27**(3), 167–175 (2009)

Lovisa, J., Sivakugan, N.: Tall oedometer: method to account for wall friction. Int. J. Geomech. **15**, 1–9 (2015). 04014045

Malarvizhi, S.N., Ilamparuthi, K.: Comparative study on the behaviour of encased stone column and conventional stone column. Soils Found. **47**(5), 873–886 (2007)

Murugesan, S., Rajagopal, K.: Model tests on geosynthetic encased stone columns. Geosynthetics Int. **14**(6), 346–354 (2007)

Murugesan, S., Rajagopal, K.: Performance of encased stone columns and design guidelines for construction on soft clay soils. In: Li, G., Chen, Y., Tang, X. (eds.) Proceedings of the 4th Asian Regional Congress on Geosynthetics, Shanghai, China, pp. 729–734 (2008)

Murugesan, S., Rajagopal, K.: Shear load tests on stone columns with and without geosynthetic encasement. Geotech. Test. J. ASTM **32**(1), 35–44 (2009)

Murugesan, S., Rajagopal, K.: Studies on the behaviour of single and group of geosynthetic encased stone columns. J. Geotech. Geoenviron. Eng. ASCE **136**(1), 129–139 (2010)

Nayak, N.V.: Recent advances in ground improvements by stone columns. In: Proceedings of Indian Geotechnical Conference, Madras, vol. 1, p. V-19 (1983)

Shahu, J.T., Reddy, Y.R.: Clayey soil reinforced with stone column group: model tests and analyses. J. Geotech. Geoenviron. Eng. **137**(12), 1265–1274 (2011)

Sridharan, A., Prakash, K.: Self-weight consolidation: compressibility behaviour of segregated and homogeneous fine grained sediments. Mar. Georesour. Geotechnol. **21**(2), 73–80 (2003)

Performance of Recycled Plastic Pin (RPP) for Slope Stabilization

M.S. Khan[1(✉)], M. Sahadat Hossain[2], M.A. Khan[2], and Mohammad Faysal[2]

[1] Department of Civil Engineering, Jackson State University, 1400 John R Lynch Street, Jackson, MS 39217, USA
J00797693@jsums.edu

[2] Department of Civil Engineering, The University of Texas at Arlington, 416 Yates Street NH 119, Arlington, TX 76019, USA
Hossain@uta.edu, {md.khan22,md.faysal}@mavs.uta.edu

Abstract. Surficial failures of highway slopes in clayey soils are quite common throughout the United States. These failures commonly occur for the slope constructed with expansive clay, especially after prolonged rainfall. These failures are also predominant in North Texas area and cause significant maintenance problems for the Texas Department of Transportation (TxDOT). As an alternative to the conventional slope stabilization technique, a green and cost effective slope stabilization method using the Recycled Plastic Pin (RPP) had been utilized and tested for its performance. RPPs are driven into the slope face to provide additional resistance along the slip surface, which increases the factor of safety against shallow slope failure. Current study summarizes the long term performance of a highway slope on expansive clay reinforced with RPP. The slope is located over US 287 near the St. Paul overpass in Midlothian, Texas. Surficial movement had taken place over the slope, resulting in cracks over the shoulder and near the bridge abutment. Three 50 ft. slope sections were stabilized using RPPs in March 2011. In addition, two 50 ft. unreinforced control sections were utilized between the reinforced sections to evaluate the performance of slope sections stabilized with RPP. After installation of the RPPs, the performance of the slope was monitored by using instrumented RPPs, inclinometers and topographic survey. The performance monitoring results indicate that, maximum deformation of the reinforced slope section is less than 3.75 cm (1.5 in.). However, more than 38 cm (15 in.) of vertical settlement was observed at the control sections during the last 5 years monitoring period. Also, few slope-sections just opposite side of reinforced slope at the same highway failed during the monitoring period. Based on the last 5 years monitoring data, it was summarized that RPP can be successfully utilized for slope stabilization.

1 Introduction

Shallow slope failure refers to surficial slope instabilities along highway cuts, fill slopes and embankments. Moderate-to-steep slopes and embankments underlain by expansive clay soils are susceptible to shallow landslides during intense and prolonged rainfall events. In many cases, the failure surface is parallel to the slope face. Shallow slope

W. Frikha et al. (eds.), *Soil Testing, Soil Stability and Ground Improvement*, Sustainable Civil Infrastructures, DOI 10.1007/978-3-319-61902-6_12

failures generally do not constitute a hazard to human life or cause major damage. However, they can be a hazard to infrastructure, by causing damage to guardrails, shoulders, road surfaces, drainage facilities, utility poles, and/or the slope landscaping (Titi and Helwany 2007). In some cases, shallow slope failures can affect regular traffic movement if debris flows onto highway pavements.

Typically, failure occurs because of an increase in pore water pressure and reduction in soil strength due to the progressive wetting of the near-surface soil. This condition is further exacerbated by moisture variations due to seasonal climatic changes, which cause cyclic shrinkage and swelling of the upper soils. Cyclic shrinkage and swelling leads to the change in void ratio of soil and lower the shear strength. Reduction of shear strength results in sloughing and shallow slope failures which are predominant in the North Texas. It poses a significant maintenance problem to the Texas Department of Transportation (TxDOT) (Wright 2005). The depth of the shallow slope failure varies with soil type and slope geometry, but generally ranges between 0.9 m (3 ft.) and 1.8 m (6 ft.). (Loehr et al. 2007). Shallow failures often cause significant hazards to guard rails, shoulders, and roadways, which, if not properly maintained, may require extensive and expensive repairs.

Reinforcement is an effective remediation technique for relatively shallow slope failure conditions. In situ reinforcement methods for stabilizing slopes and embankments include soil nails, drilled piers, micro piles, and RPPs (Thompson et al. 2006). Among the available methods, the RPPs have been recognized as a cost effective solution for slope stabilization (Loehr and Bowders 2007; Khan et al. 2015). RPPs were first being utilized in the state of Missouri and Iowa, as a sustainable option to stabilize highway slopes and few field studies were conducted. The study summarized that no further failure was noticed after the slopes were stabilized using RPPs and the performance of the stabilized zone was promising (Loehr and Bowders 2007).

Typically, RPPs are fabricated from recycled plastics and waste materials (polymers, sawdust, and fly ash) (Chen et al. 2007). It is a lightweight material and less susceptible to chemical and biological degradation than other reinforcement materials. The use of RPPs can reduce the waste volume entering landfills and provide additional demand for recycled plastic (Sommers et al. 2000). A typical RPP is composed of High Density Polyethylene, HDPE (55%–70%); Low Density Polyethylene, LDPE (5%–10%); Polystyrene, PS (2%–10%); Polypropylene, PP (2%–7%); Polyethylene-terephthalate, PET (1%–5%); and varying amounts of additives, i.e., sawdust, fly ash (0%–5%) (McLaren 1995). The use of glass and wood fiber additives significantly improves the modulus of elasticity for plastic lumber (Breslin et al. 1998). RPPs driven into the slope face may provide an additional resistance along the slip plane that restricts the progression of the slip surface and increases the factor of safety. However, limited field studies have been conducted on the potential use of RPPs in slope stabilization.

The current study presents the long term performance of RPP for a highway slope stabilization. Surficial movement and a cracked shoulder were observed on a highway slope located over highway US 287 in Texas. During March 2011, two 15.25 m (50 ft.) sections on the southbound side of US 287 were reinforced using RPPs. Another cracked zone formed over the shoulder of US 287 during the following year, and a new 15.25 m (50 ft.) section was reinforced considered on the southbound slope for the stabilization, using RPPs. In addition, two 15.25 m (50 ft.) control sections were kept between the reinforced sections to compare the performance. The reinforced and control sections of the slope were monitored after installation to evaluate the performance of the slope, using instrumented RPPs, inclinometers and surveying equipment.

2 Project Background

The slope is located near Highway US 287, near the St. Paul overpass in Midlothian, Texas. It is a 3(H): 1(V) fill slope with 30 ft. height. Surficial movement over the slope and shoulder cracks were observed on the shoulder near the crest of the slope in September, 2010. The cracks along the pavement shoulder might have occurred due to the surficial movement of the slope. The site location and the photo of the cracked shoulder are presented in Fig. 1.

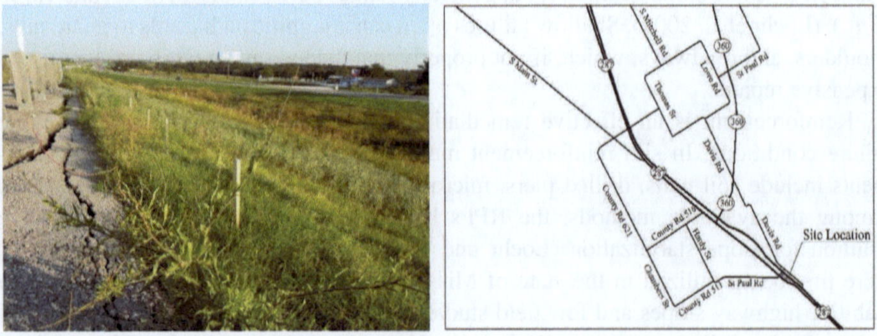

(a) (b)

Fig. 1. (a) Site location (b) Cracks along the shoulder of Highway US 287

The geology of the location is characterized by the Eagle Ford formation. The Eagle Ford, in the project vicinity, is composed of residual soils, consisting of clay and weathered shale (shaly clay), underlain by unweathered shale. The weathered shale is claylike and contains gypsum in-fills, jointed and fractured with iron pyrites. The unweathered shale is typically gray to dark gray and commonly includes shell debris, silty fine sand particles, bentonite and pyrite. The Eagle Ford formation consists of sedimentary rock in the process of degrading into a soil mass. This formation also contains smectite clay minerals and sulfates. The smectite clay minerals are highly expansive in nature and may be a contributing factor to the expansiveness of the soil. A previous study by Kibria and Hossain (2012) indicated that the dominant mineral of the soil is montmorillonite mineral, which has high shrink/swell.

A field exploration program was undertaken, which indicated that the slope was constructed with high plastic clayey (CH) soil. The liquid limits (LL) and plasticity indices (PI) of the samples ranged between 48 to 79 and 25 to 51, respectively. In addition, an increase in moisture below 5 ft. was observed, which ranged up to 20 ft. Based on the back analysis with PLAXIS 2D the factor of safety was 1.05 with fully soften shear strength (Khan et al. 2015).

3 Slope Stabilization Using RPP

The definition of the factor of safety of a slope is the ratio of resisting moment (M_R) to the driving moment (M_D), as presented in Eq. 1. RPPs installed at the slope provided an additional resisting moment (ΔM_R) along the slip surface, thereby increasing the resistance and factor of safety, as presented in Eq. 2. The schematic diagram of RPPs as slope reinforcement is presented in Fig. 2.

$$FS = M_R/M_D \tag{1}$$

$$FS = (M_R + \Delta M_R)/M_D \tag{2}$$

Where,
M_R = Resisting Moment along Slip Surface
M_D = Driving Moment along Slip Surface
ΔM_R = Additional Resisting Moment from Plastic Pin.

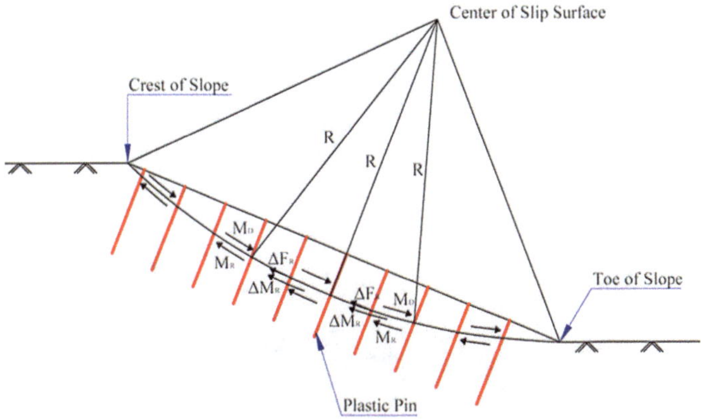

Fig. 2. Schematic diagram of slope reinforcing with RPP

Commercially available fiber reinforced RPP with square Section (4 in. × 4 in.) and 10 ft. and 8 ft. in length was utilized as reinforcement. RPP reinforcement with different spacing, (i.e., 3 ft., 5 ft. and 6 ft. c/c), was utilized in the Reinforced Section 1. Besides, a 4 ft. c/c spacing of RPP was utilized in Reinforced Sections 2 and 3. The layout of RPP is presented in Fig. 3(a). The calculated factor of safety from the FEM model (PLAXIS 2D) was 1.43, 1.48 and 1.54 for Reinforced Sections 1, 2 and 3, respectively (Khan et al. 2014).

RPPs were installed in Reinforced Section 1 and Reinforced Section 2 on March,

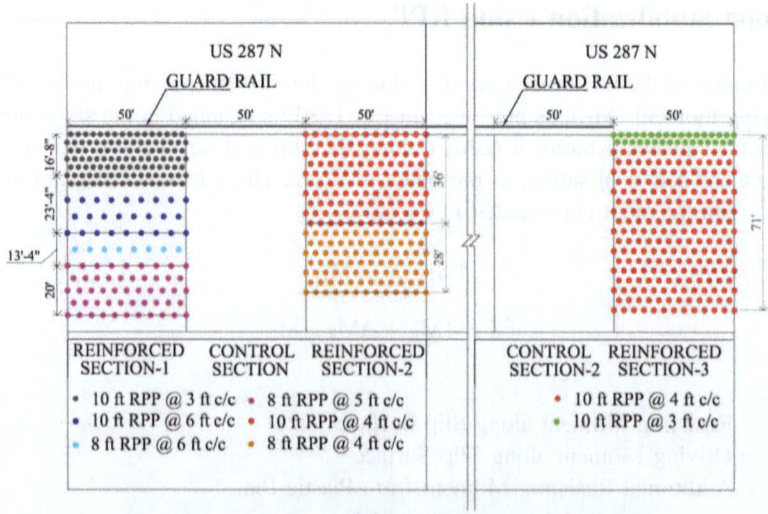

(a)

(b)

Fig. 3. (a) Layout of RPP at US 287 slope (b) Installation photo

2011, as presented in Fig. 3(b). Reinforced Section 3 was stabilized during March, 2012. A crawler-type drilling rig, having a mast-mounted vibrator hammer (model: Klemm 802 drill rig along with KD 1011 percussion head drifter), was utilized to install the RPPs. The crawler-type rig was suitable for the installation process over the slopes, as no additional anchorage was required to maintain the stability of the equipment, reducing labor, cost and time involved in the installation process. Based on the study, the average installation time for a 10 ft. long RPP was 4 min, and a total of 100 to 120 RPPs could be installed in one day.

4 Instrumentation and Performance Monitoring

Regular topographic survey and inclinometers were used to monitor the performance of RPP stabilized slope. To evaluate the performance of the reinforced slopes and to monitor the horizontal displacement, inclinometers were installed after the completion of field installation in Reinforced Section 1 and Reinforced Section 2.

4.1 Topographic Survey

To compare the performance between the reinforced section and control section, topographic surveys were performed on monthly basis. The purpose of this survey was to monitor the settlement of the crest of the slope.

4.2 Inclinometers

A total of 3 inclinometers, (designated as Inclinometer 1, 2 and 3) were installed in Reinforced Section 1, Control Section and Reinforced Section 2 to monitor the horizontal movement of the slope. The depth of each inclinometer casing was 30 ft. and it was installed perpendicular to the slope surface, 20 ft. below the crest. The layouts of the inclinometers are presented in Fig. 4. The inclinometers were monitored on a bi-weekly basis.

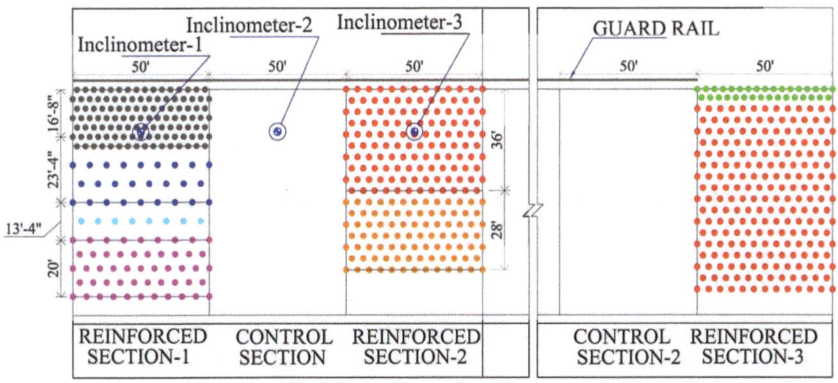

Fig. 4. Layout of inclinometer

5 Performance Evaluation

5.1 Topographic Survey

The total settlement over the crest of the southbound slope was measured during each survey, in between, the total settlement plot for September 2016 is presented in Fig. 5. The total settlement plot shows the control sections had significantly greater settlement at the crest when compared to the reinforced sections. The maximum settlements were 38 cm (15 in.) and 24 cm (9.5 in.) in Control Section 1 and Control Section 2, respectively, while Reinforced Section 1 had the lowest settlement of 7 cm (2.75 in.), followed by the Reinforced Section 3 and Reinforced Section 2, with settlements of 16 cm (6.25 in.) and 14 cm (5.5 in.), respectively. It should be noted that Reinforced Section 1 had the lowest spacing of RPPs (0.9 m, 3 ft. c/c) at the crest of the slope. Reinforced Section 2 and Reinforced Section 3 had 1.21 m (4 ft.) c/c spacing at the crest, which was higher and had higher deformation than Reinforced Section 1. Maximum incremental settlement on the reinforced section was less than 3.75 cm (1.5 in.) after five years of installation of RPPs.

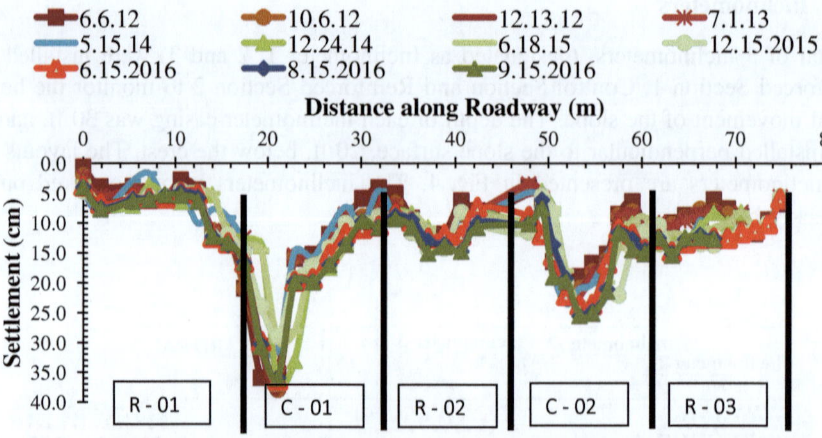

[R = Reinforced Section; C = Control Section]

Fig. 5. Settlement at the crest of slope (Southbound side)

5.2 Inclinometer

The inclinometers were monitored on a monthly basis. After each monitoring, the obtained data from the inclined plane of the inclinometer was normalized to the vertical plane and the horizontal movements of Inclinometer 1 and Inclinometer 3 are shown in Figs. 6 and 7 respectively. Inclinometer 2 malfunctioned; consequently, that data is not presented here.

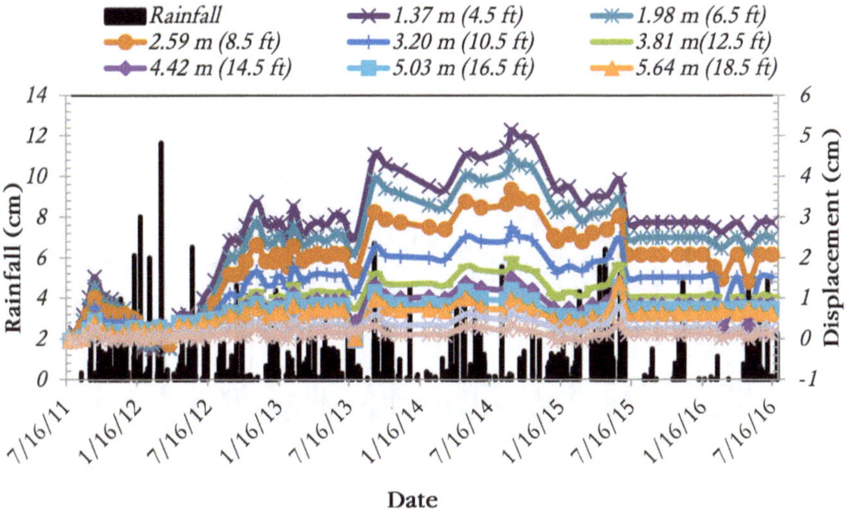

Fig. 6. Inclinometer - 1: cum displacement A-A with time

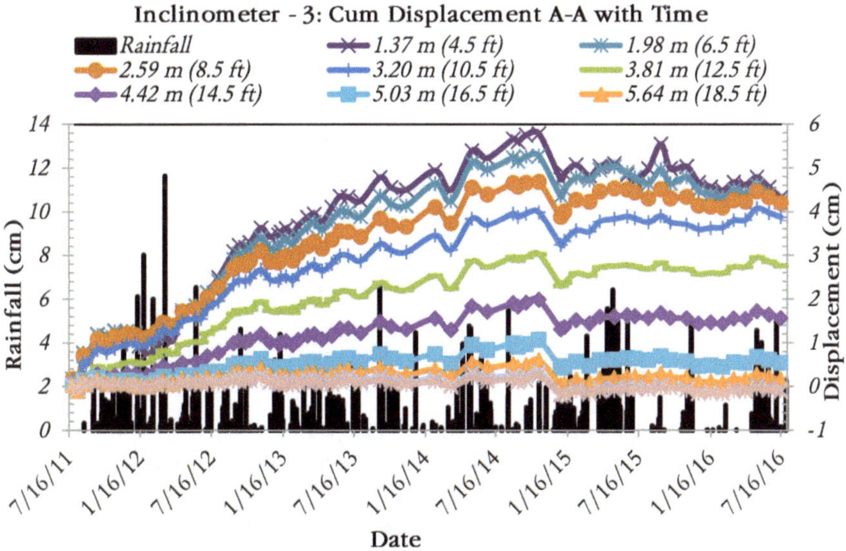

Fig. 7. Inclinometer - 3: cum displacement A-A with time

The comparisons of movement between Inclinometer 1 and Inclinometer 3 at different depths (1.37 m and 3.2 m) are presented in Fig. 8. It was observed that Inclinometer 3 had higher horizontal displacement (4.3 cm) than Inclinometer 1 (2.8 cm) at 1.37 m depth. A similar trend was observed at 3.2 m depth, where the observed displacement was 3.9 cm and 1.6 cm for Inclinometer 1 and Inclinometer 3, respectively. It should be noted that the maximum horizontal displacement took place at the surface of the slope and consequently reducing with deeper depth.

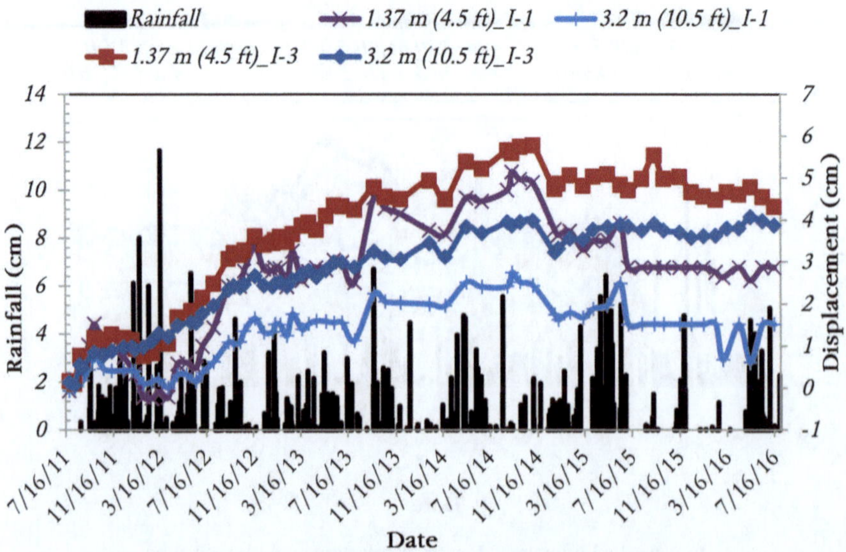

Fig. 8. Comparison of inclinometer - 1 & 3

Khan et al. 2014 conducted a numerical study on the slope stabilization using recycled plastic pin. During the study, the change in horizontal displacement of RPP at the crest of the slope is investigated using finite element method (FEM), with the change in RPP spacing. The FEM results indicated that, with the increment in RPP spacing from 0.2 m (2 ft.) to 2.4 m (8 ft.), the horizontal displacement of RPP increases from 2.5 cm (1 in.) to 12 cm (4.7 in.). In addition, the observed horizontal displacement from FEM analysis at the crest is in good agreement with the observed displacement at the field with 0.9 m (3 ft.) and 1.2 m (4 ft.) spacing.

5.3 Performance of the Northbound Slope

The northbound slope of highway US 287 was inspected visually on a monthly basis for last five years. A number of shallow slope failures were recorded on the control (northbound of US 287) slope and the failure locations are shown in Fig. 9. Photographs of the failures of Location 1 to Location 6 are also shown in Fig. 10. First time failures (Location 1 and Location 2) on the control slope were observed during September 2013, after a rainfall event. Soil was backfilled and compacted within the failure zone during the maintenance period. But the slope failed (Location 3) again during August, 2014 after a period of a heavy rainfall. Soil was backfilled again within

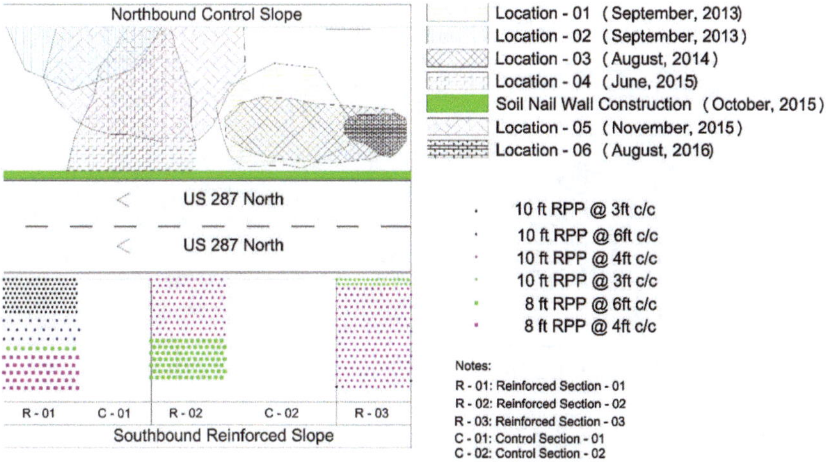

Fig. 9. Schematic diagram of slope failure within northbound control slope.

the failure location which was a temporary solution to prevent the movement of the slope. A global failure (Location 4) of the slope was observed on June, 2015 after a massive rainfall of 125 mm during that month. A soil nail wall was constructed to control the movement of the pavement and the whole slope was reconstructed with some geogrid reinforcement on October, 2015. Even after the reconstruction, the slope failed (Location 5) after a month of heavy rainfall. Most recent failure was recorded on August, 2016 and levelled as Location 6.

During the failures of the northbound slope, some sharp increment of horizontal displacement was observed at Inclinometer 1. The sharp increments of the horizontal displacement might indicate that the southbound slope had the tendency to slip due to the rainfall which provided additional load over the recycled plastic pin and may cause the additional movement. However, the RPP provided resistance and no sign of failure was observed at the reinforced section during the time of failure of the northbound slope. Moreover, no failure was also reported at the control section of the southbound slope. It should be noted that the control sections are located between the reinforced sections. During the failure of the northbound slope, the control section should tend to slip which might get resistance from the reinforced section. As a result, no failure was observed at the control section of the southbound slope.

Fig. 10. Failure locations (Location 01–06) at the northbound of the US 287 slope

6 Summary

A highway slope located on the southbound side of highway US 287 near the St. Paul overpass in Midlothian, Texas was stabilized using RPPs. Three 15.25 m (50 ft.) sections were selected and reinforced, using RPPs, after a crack, caused by slope movement, was observed on the shoulder. Additionally, two 15.25 m (50 ft.) control sections were utilized between the reinforced sections to compare the performances between the unreinforced and reinforced slopes. The field performance of the slope was monitored using topographic survey and inclinometers. Performance monitoring results for last 5 (five) years are summarized below.

- The unreinforced control sections of the southbound slope had significant settlement at the crest of the slope, as much as 38 cm whereas the incremental settlement on the reinforced section was less than 3.75 cm.
- Maximum cumulative displacement for inclinometer 1, at a depth below 1.37 and 3.20 m were 2.80 cm and 1.60 cm respectively.
- Maximum cumulative displacement for inclinometer 3, at a depth below 1.37 and 3.20 m were 4.30 cm and 3.90 cm respectively.
- Both inclinometers showed the same trend that, cumulative displacement decreases with the depth from the slope surface.
- Having the same geographic and climatic condition, the southbound control slope failed several times at different locations.

Based on the current study, it can be concluded that the overall performance of the reinforced sections was better than that of the control sections. The northbound control slope failed several times during the monitoring period whereas no visual depression or failure was observed within the southbound slope. Since the shallow failure is a major issue for slopes constructed with high plastic clayey soil; RPPs could provide a cost effective and sustainable solution to mitigate this problem.

References

Breslin, V.T., Senturk, U., Berndt, C.C.: Long-term engineering properties of recycled plastic lumber in pier construction. Resour. Conserv. Recycl. **23**(1998), 243–258 (1998)

Chen, C.W., Salim, H., Bowders, J., Loehr, E., Owen, J.: Creep behavior of recycled plastic lumber in slope stabilization applications. J. Mater. Civ. Eng. **19**(2), 130–138 (2007)

Khan, M.S., Hossain, S., Kibria, G.: Stabilisation using recycled plastic pins. J. Perform. Constructed Facil. 229–234 (2017). doi:10.1201/9781315206202-11. Online publication date: 14 Jun 2017

Khan, M.S., Hossain, M., Lozano, N.: Numerical study of slope stabilization using recycled plastic pin. In: Geo-Congress 2014 Technical Papers: Geo-characterization and Modeling for Sustainability, pp. 3092–3101. ASCE, February 2014

Kibria, G., Hossain, M.S.: Investigation of geotechnical parameters affecting electrical resistivity of compacted clays. J. Geotech. Geoenviron. Eng. (2012). doi:10.1061/(ASCE)GT.1943-5606.0000722

Loehr, J.E., Bowders, J.J.: Slope Stabilization using Recycled Plastic Pins – Phase III. Final report: RI98-007D, Missouri Department of Transportation, Jefferson City, Missouri (2007)

Loehr, J.E., Fennessey, T.W., Bowders, J.J.: Stabilization of surficial slides using recycled plastic reinforcement. Transp. Res. Rec. J. Transp. Res. Board **2**, 79–87 (2007). No. 1989

McLaren, M.G.: Recycled plastic lumber and shapes design and specification. In: Proceedings 13th Structures Congress, ASCE, vol. 1, pp. 819–833 (1995)

Sommers, L., Loehr, J.E., Bowders, J.J.: Construction methods for slope stabilization with recycled plastic pins. In: Proceedings of the Mid-continent Transportation Symposium, Iowa State University, Ames, Iowa, 15–16 May 2000

Thompson, M.J., White, D.J.: Design of reinforcement with small-diameter piles. In: Proceedings of Advances in Earth Structures: Research to Practice, GSP no. 151, pp. 1–12 (2006). Reston/VA

Titi, H., Helwany, S.: Investigation of Vertical Members to Resist Surficial Slope Instabilities (No. WHRP 07-03). Wisconsin Department of Transportation, Madison, WI (2007)
Wright, S.G.: Evaluation of Soil Shear Strengths for Slope and Retaining Wall Stability Analyses with Emphasis on High Plasticity Clays. Federal Highway Administration, Washington, D.C. (2005). FHWA/TX-06/5-1874-01-1

Evaluation of Liquefaction Potential of New Caledonian Nickel Ores

Samar Daoud[1(✉)], Imen Said[1], Samir Ennour[2], and Mounir Bouassida[1]

[1] ENIT, Université de Tunis El Manar – Ecole Nationale d'Ingénieurs de Tunis, LR14ES03-Ingénierie Géotechnique, Tunis, Tunisia
samar.daoud1989@gmail.com, imensaid2@gmail.com, mounir.bouassida@fulbrightmail.org

[2] Mining Engineering, MECATER, Rue Ibn Zohr, Immeuble Alizé, 2015 Le Kram, Tunisia
samir.ennour@mecater.com

Abstract. Although Nickel ores are generally characterized as plastic to very plastic silts they do liquefy under specific conditions. In fact, it was reported that New Caledonian Nickel ores (NCNo) can undergo liquefaction during their maritime transportation under wave motions. The purpose of this research is to develop a new empirical method to evaluate the liquefaction potential of these soils. First, the study consists on interpreting physical and mechanical characterization tests carried on Nickel ore samples originating from different sites of New Caledonia. This analysis permitted to conclude on the heterogeneity of NCNo and enabled the identification of three major types namely, laterites, earthy and grainy saprolites. The classification of these groups was based on the grain size distribution and the Atterberg limits tests. Afterwards, a review of the existing liquefaction potential evaluation methods is elaborated. The application of these methods to the case of NCNo has not allowed concluding on the liquefaction susceptibility of this material. Finally, a new method specific to each NCNo type was introduced based on the physical and mechanical test results.

1 Introduction

New Caledonian Nickel ores (NCNo) are obtained from open-cast mines and shipped to smelters located in China, Japan, Australia, Korea, etc. Several case histories of ore carriers capsize have been reported due to liquefaction of the transported material. According to The International Maritime Solid Bulk Cargoes Code (IMSBC 2008), these kind of cargoes are classified as Group A materials where several laboratory tests namely the Flow Table, the Penetration test and the Proctor/Fagerberg test are recommended to predict the risk of liquefaction before transport.

With reference to Koromila et al. (2013), cargoes that are at risk of liquefaction are those containing at least some fine particles and moisture and further mined or stored in exposed areas which allow the soaking up of large amounts of water. Such cargoes at the time of loading are typically in granular state and look like dry. However, whilst at

W. Frikha et al. (eds.), *Soil Testing, Soil Stability and Ground Improvement*,
Sustainable Civil Infrastructures, DOI 10.1007/978-3-319-61902-6_13

sea, they are subject to agitation due to the engine vibration; ship's rolling as well as swell impact. The oscillatory ship movement leads to resettling of the cargo particles and compaction of the inter-granular spaces. This compaction raises the water pressure sharply, forcing the particles apart, potentially leading them to lose direct contact. The cargo loses its shear strength and thus conditions are created for the material to behave like a liquid, i.e. to liquefy (IMSBC 2008). As a result, this latter can lead to a "free surface effect" that may upset the carrier's balance and causes it to capsize in heavy swells.

The existing guidelines for identifying potentially liquefiable soils are generally divided into two major types. The first one covers the laboratory tests enabling to obtain accurate results especially when carried out on undisturbed samples. In fact, many experimental tests carried out in laboratories are enough efficient since they are able to reproduce similar solicitations and field conditions to evaluate the susceptibility of a type of soil to liquefaction. The most famous tests are the direct shear test, the cyclic triaxial shear test, shaking table test… The second method is the empirical approaches developed based on various combinations of characterization test results, namely the Atterberg limits, the grain size characteristics, and the natural water content. These methods developed by researchers like Tsuchida (1970), Seed and Idriss (1982), Andrews and Martin (2000), are deduced from field observations of liquefied and non-liquefied soils after famous earthquakes. By means of these approaches, it is possible to determine the likelihood of liquefaction triggering in a particular soil under a given earthquake. Thus, the choice of this alternative is justified by its simplicity and cost-effectiveness.

Unfortunately, the efficiency and accuracy of these methods are still questioned since they do not provide reliable results covering all types of soil and solicitations. Indeed, although researchers had made efforts on identifying the susceptible liquefiable soil types based on mechanical and physical characterization tests, there is still no qualitative and quantitative parameter in the Geotechnical field that could be used as an evaluation tool of liquefaction susceptibility.

The final aim of this study is the elaboration of a method to evaluate the liquefaction potential of NCNo prior to their shipping in order to avoid further incidents. First, the study consists of interpreting physical and mechanical characterization tests carried on Nickel ore samples originating from different sites of New Caledonia. This analysis enabled the classification of the tested specimens into different groups and the identification of three major types of NCNo. Afterwards, an evaluation of the liquefaction potential of NCNo by means of the existing approaches is elaborated. Finally, a new method to assess the liquefaction susceptibility adapted for each type of NCNo is introduced.

2 Data Base Discussion and Interpretation

The first step in assessing the risk of liquefaction is to identify potentially liquefiable soils according to their physical characteristics (grain size distribution, water content, Atterberg limits…) and mechanical properties.

As part of Rheolat project, a series of physical and mechanical tests were conducted on numerous Nickel ore samples originating from different sites and mines of New Caledonia. Grain size distribution, Atterberg Limits and Cyclic triaxial shear tests were carried out by CERMES laboratories. Flow Table Tests were elaborated by SLN*.

In this section, the different test results are presented, discussed and interpreted in order to classify the tested NCNo samples into groups and in final aim to evaluate their liquefaction potential.

2.1 Physical Characterization of NCNo

Grain Size Distribution Test Results

Particle size analysis were conducted on 24 samples of NCNo to allow determination of the soil nature corresponding to standard XP CEN TS/ISO 17892-4.

The obtained curves by combining dry sieving and sedimentometry results, point out the heterogeneity of NCNo. The representation of grain size distributions on the same graph in Fig. 1 enables distinguishing three major grading ranges. It is concluded that each of these latters corresponds to a type of NCNo.

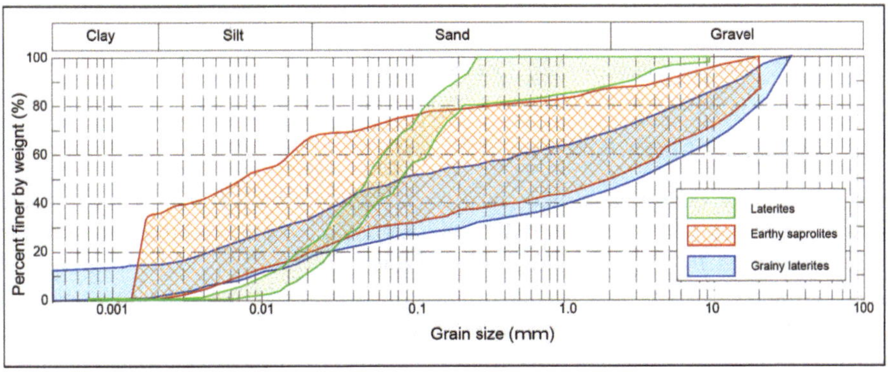

Fig. 1. Grading ranges of the three identified NCNo groups

Group 1: Laterites which are silty sands having most particles smaller than 80 μm.
Group 2: Earthy saprolites that are sandy silts containing a small fraction of coarse particles (>2 mm) not exceeding 30%.
Group 3: Grainy saprolites which are grainy sands having a high fraction of coarse particles (>30%).

Atterberg Limits Test Results

The Atterberg limits test is also carried out by CERMES laboratories on the same samples cited above corresponding to standard NF P 94-051. It is used in order to characterize the plasticity of a soil since it can influence the soil liquefaction properties.

The interpretation of the Atterberg limit test by representing the results on the Casagrande graph confirms the repartition of the tested samples into three distingable groups as shown in Fig. 2. It is also concluded that all NCNo are highly plastic silts.

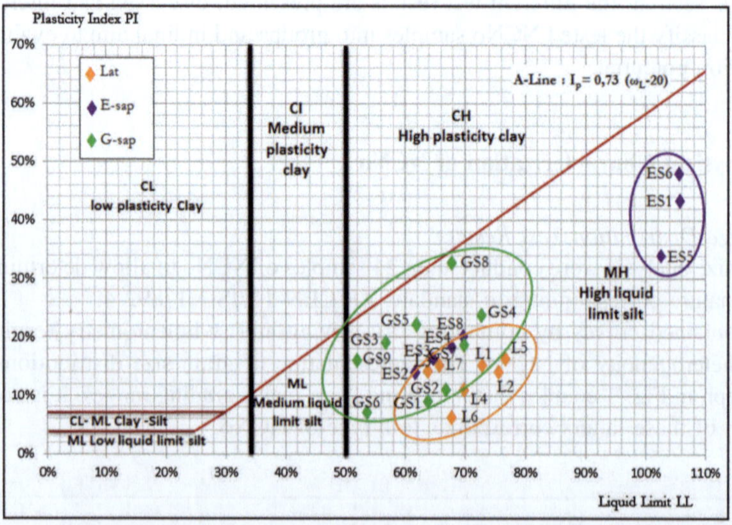

Fig. 2. Representation of Nickel ore tested samples on the Casagrande Diagram

2.2 Mechanical Characterization of NCNo

First, a series of consolidated undrained tests with measure of pore water pressure (CU + u) were carried out by CERMES laboratories on samples representative of the three different NCNo groups. Second, Flow Table tests have been elaborated by SLN laboratories on NCNo. The results of these mechanical testes are presented and interpreted in this paragraph. The critical water content is defined as the value above which the sample gets liquefied. At each test, the ratio of this value on the Liquid limit of the tested specimen is determined in order to be used in the development of the new approach (Table 1).

Cyclic Triaxial Shear Test Results CTT
The Cyclic Triaxial Tests (CTT) results allows to differentiate between a liquefiable and a non-liquefiable material when reaching one of the two following conditions; either the excess pore water pressure reaches the value of the consolidation stress $\Delta u = \sigma'c$ or the peak-to-peak axial strain within a cycle is equal to 5%. In this paragraph, the results of the saturation degree variation as well a comparison of the three samples responses to cyclic shearing are discussed in order to evaluate the NCNo susceptibility to liquefaction.

Figure 3 illustrates the effect of the degree of saturation (S_R) variation on the liquefaction properties in the case of laterites samples. The obtained shapes of the curves are in accordance with those found by Pecker (1984) when analyzing the effect of the

Table 1. CTT results on NCNo samples

Samples	<80 µm (%)	>2 mm (%)	IP (%)	W_c/LL
Laterites				
L1	100	–	15	0.86
L2	92	–	16	0.7
L6	87.5	–	6	0.63
L7	86.6	–	15	0.65
Mean value	**88.47**	–	**12.56**	**0.72**
Earthy saprolites				
ES1	78	–	43	0.71
ES2	55.6	–	48	0.56
ES4	93.8	–	51	0.57
ES5	33.8	–	61	0.58
Mean value	**66.85**	–	**45**	**0.67**
Grainy saprolites				
GS1	40	40	9	0.7
GS2	37	48	11	0.67
GS3	47	34	19	0.55
GS4	36	42	24	0.46
GS9	42	33	16	0.64
Mean value	**39**	**40.62**	**17.7**	**0.64**

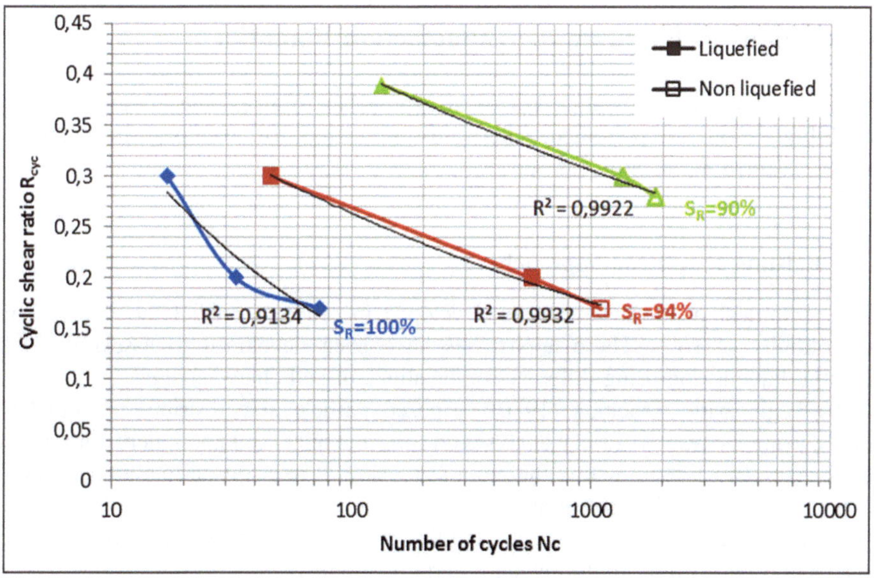

Fig. 3. Impact of the degree of saturation variation on the liquefaction properties in the case of Laterite samples

degree of saturation on the cyclic shear resistance CSR of a soil. It is mentioned that the required number of cycles to trigger liquefaction when the sample is partially saturated is higher than that when the sample is fully saturated. Therefore, the decrease of the degree of saturation improves the cyclic shear resistance of soils. This result is in accordance with the statement of Sherif et al. (1977) who proved that liquefaction resistance for soils increases with decreasing degree of saturation.

CTT were conducted on the three samples under the same $R_{cyc} = 0.2$. It is noted that sample representing group 2 needs a higher number of cycles to liquefy in comparison with the two other groups. This difference may be attributed to the high plasticity of earthy saprolites (PI = 43) which is the origin of the liquefaction delay. Actually, as reported by Alan F. Rauch, plastic fines usually create sufficient adhesion between the grains to limit the ability of larger particles to move into a denser arrangement and therefore to prevent liquefaction.

In contrast, sample from group 1 was the first to get liquefied compared to the two other groups. Hence, laterites are assumed to be less resistant to liquefaction than earthy and grained saprolites. This difference could be explained by the high percentage of fines in this sample (fraction < 80 μm = 100%). In fact, Finn et al. (1988), Lade and Yamamuro (1997), Zlatovic and Ishihara (1995) reported that liquefaction resistance decrease as the fine content increase. Moreover, according to Liang et al. (2000), it is proven that the coarse particles take an important part in soil to resist liquefaction and this is the case of grainy saprolites. Indeed, soils with a higher percentage of gravels tend to mobilize higher strength during shearing, and dissipate excess pore pressures more rapidly than sands.

3 Methods to Evaluate the Liquefaction Potential

In this section, the widely used empirical approaches to assess the liquefaction susceptibility of soils are presented. Afterwards, an attempt to develop a new empirical method adapted to NCNo is introduced.

3.1 Existing Empirical Approaches

Many empirical methods have been developed in order to predict the liquefaction susceptibility of soils. These approaches have been developed to assess the soil potential liquefaction using different combinations of soil characteristics.

To begin with, Tsuchida (1970) proposed particle size distribution boundary curves, using the results of sieve analyses on soils that did or did not liquefy during past earthquakes. Thus, representing the grain size curve of any soil on this graph, allows deducing its liquefaction potential.

Besides, Seed and Idriss (1982) stated that cohesive soils which are susceptible to liquefaction must fulfill the following criteria (combined):

- Percent finer than 0.005 mm < 15%
- Liquid Limit (LL) < 35%
- Water Content > 0.9 × Liquid Limit

If soils with these characteristics, plot below the A-line on the Plasticity chart, the best means of determining their cyclic loading characteristics is by additional test. Otherwise, clayey soils may be considered non-vulnerable to liquefaction.

In addition, the widely used Chinese criteria as presented by Koester (1992) on Fig. 4 showed that based on the soil representation on the Casagrande graph potentially liquefiable and non-liquefiable soils can be distinguished.

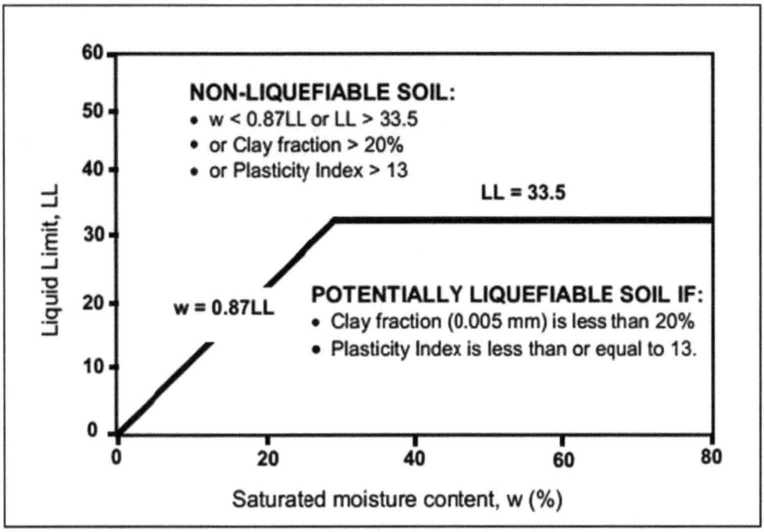

Fig. 4. Chinese Criteria Adapted to ASTM Definitions of Soil Properties (Perlea et al. 1999)

Furthermore, Andrews and Martin (2000) developed an empirical approach based on LL and percentage of fraction passing 2 μm presented in the following table.

Last but not least, Seed et al. (2003) recommended an assessment chart shown in Fig. 8 which is divided into three zones; Zone A where soils are considered potentially susceptible to liquefaction if Wc > 80% LL; Zone B where soils are considered potentially liquefiable with detailed laboratory testing recommended if Wc > 85% LL and Zone C where soils are considered generally not susceptible to classic cyclic liquefaction, although they should be checked for potential sensitivity.

3.2 Development of a New Empirical Approach for NCNo Case

Based on the existing widely used empirical approaches to evaluate the liquefaction potential of soils, a new method is proposed in this research which is more adapted to the case of NCNo. Most of the existing criteria are based on field observations where some soils got liquefied and some not under certain earthquakes. In this study, since it is not the case of a seismic solicitation, the results of CTT reproducing the transport conditions, are exploited to distinguish liquefiable soils from non-liquefiable ones and therefore a criterion for liquefaction is deduced based on the soil's physical and mechanical characterization.

4 Results

4.1 Evaluation of NCNo Liquefaction Potential Based on Existing Empirical Methods

The liquefaction susceptibility of NCNo samples was evaluated by means of the existing empirical approaches widely used in para-seismic analysis.

First, the grading ranges of each NCNo type was added to Tsuchida (1970) chart. It is seen from Figs. 5, 6 and 7 that all NCNo samples curves pass through both the 'Potentially Liquefiable' and 'Most Liquefiable Soil' boundaries. Therefore, according to this criterion and based on the representative grading ranges of each NCNo type, it can be concluded that the NCNo are susceptible to liquefaction under certain circumstances. Further, it should be mentioned when comparing the three obtained graphs that the most liquefiable type of NCNo is the Group 1 since its grading ranges falls exactly between the boundaries of potentially liquefiable soils. However, for Group 2 and 3 of NCNo only a part of their grain size distribution curves are plotted inside the boundaries defined by Tsuchida (1970).

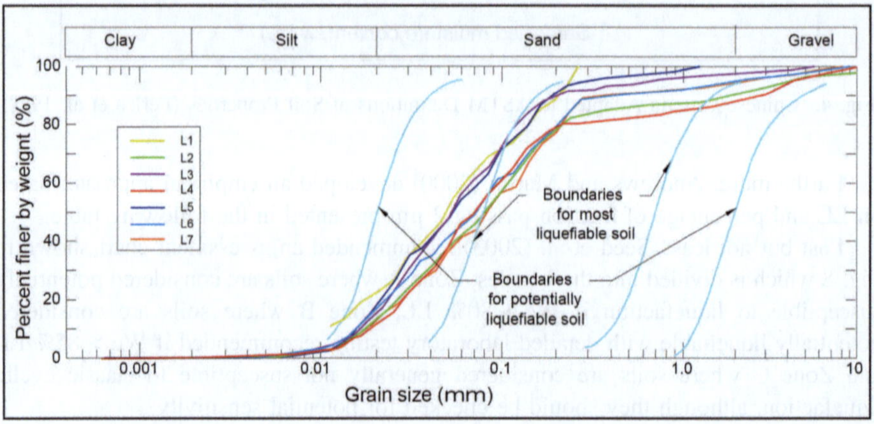

Fig. 5. Representation of the grain size distributions of Laterites NCNo on the Tsuchida graph

However, conforming to Seed and Idriss (1982) approach, NCNo are not prone to liquefaction since all their Liquid limits are above 35% and for all tested samples the water content does not exceed 0.9 × Liquid Limit (Cf. Table 1).

Moreover, according to the widely used Chinese criteria as presented by Koester (1992) NCNo are classified as non-liquefiable soils since all Liquid limits are above 33.5% and all the water content does not exceed 0.87 × Liquid Limit (Cf. Table 1).

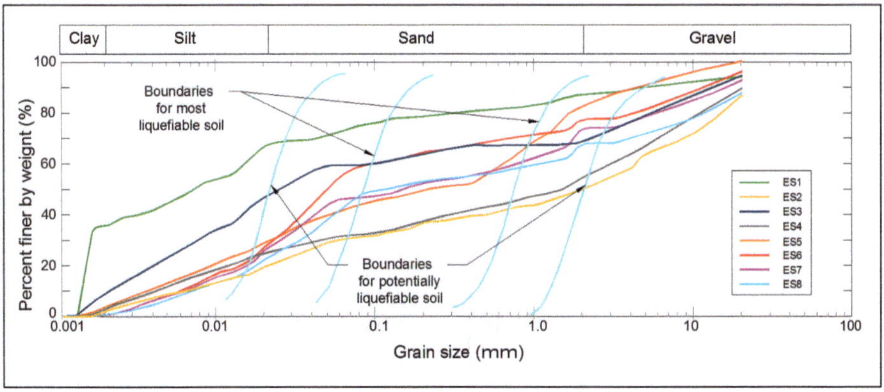

Fig. 6. Representation of the grain size distributions of Earthy Saprolites NCNo on the Tsuchida graph

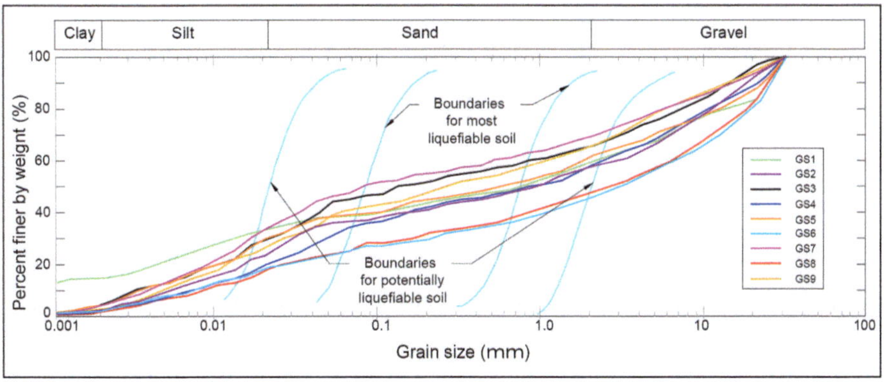

Fig. 7. Representation of the grain size distributions of Grainy Saprolites NCNo on the Tsuchida graph

In addition, following Andrews and Martin (2000) method, the three types of lateritic soils require further studies since their LL values are higher than 32% and their Minus 2 μm fraction is generally less than 10% (Cf. Table 2).

Table 2. Liquefaction susceptibility criteria by Andrews and Martin (2000)

	LL < 32	LL ≥ 32
Minus 2 μm fraction < 10%	Susceptible to liquefaction	Further studies required [Consider plastic non-clay sized grains]
Minus 2 μm fraction ≥ 10%	Further studies required [Consider plastic non-clay sized grains]	Not susceptible to liquefaction

Besides, the representation of NCNo on the graph recommended by Seed et al. (2003) showed that all the tested samples are plotted in zone C and thus are considered non susceptible to classic cyclic liquefaction, although they should be checked for potential sensitivity (Fig. 8).

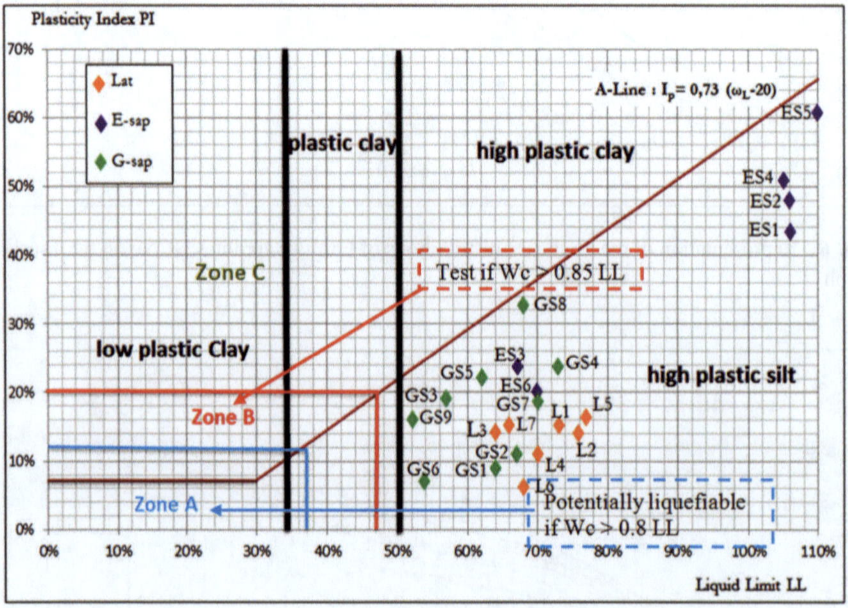

Fig. 8. Representation of the tested lateritic soils on Seed et al. (2003) recommended chart

Although these methods are well known and widely used, their usability is in doubt since they do not allow asserting on the liquefaction potential of NCNo. In fact while some of these methods consider that the tested Nickel ore samples are non-liquefiable, others claim they are rather susceptible to liquefaction or even liquefiable soils. This disagreement may be due to first, the fact that most of these empirical approaches were developed primarily for sands and not particularly for this type of soils (highly plastic silts). Second, these methods are based on observations of soil behavior under seismic loading which is doubtless a different solicitation comparing to the one due to swell effect.

For these reasons, in the following paragraph an adapted method to NCNo based on their physical and mechanical characterization is proposed in order to assess their liquefaction potential during maritime transport.

4.2 New Proposed Method for Evaluating the Liquefaction Potential of NCNo

As mentioned before, the existing empirical approaches used in para-seismic analysis are not adopted to the New Caledonian Nickel ores. Therefore a new method to evaluate the liquefaction potential of this type of soil when exposed to the swell solicitations is developed. This approach is mainly based on the Seed and Idriss (1982) method and Seed et al. (2003) criteria adding appropriate modifications to fit with the tested samples according to Cyclic Triaxial Test (CTT) results (Cf. Table 1). Moreover, taking into account the heterogeneity of NCNo, a criterion for each ore type is elaborated considering its specificity compared to the others (Cf. Table 3). These approaches are applicable for fully saturated samples.

Table 3. New approach for liquefaction assessment of New Caledonian Nickel ore under the swell effect

New approach for liquefaction assessment of New Caledonian Nickel ore under the swell effect		
Laterites	Earthy saprolites	Grainy saprolites
• Infra 80 μm > 80% • IP ≥ 10% • Wc > 0,6 LL	• Infra 80 μm ≥ 50% • MBV ≥ 5 • IP ≥ 30% • Wc > 0,55 LL	• Supra 2 mm > 30% • IP ≥ 12% • Wc > 0,5 LL

5 Conclusions

First, the interpretation of physical characterization tests carried on NCNo allowed identifying three major types of this material:

- Group 1: Laterites which are very fine soils having most particles smaller than 80 μm. Based on their characterization tests, they are classified as highly plastic silty sands.
- Group 2: Earthy saprolites which are sandy silts containning a small fraction of coarse particles (>2 mm) not exceeding 30%. According to the United Soil Classification System (LCPC-USCS), this soil is a high plastic silt having a high clay content (MBV > 5).
- Group 3: Grainy saprolites are grainy sands having a high fraction of coarse particles (>30%).

Second, the discussion of CTT results conducted on the three types of NCNo proved that these latters behave differently under the same solicitations. In fact, it is mentioned that earthy saprolites may be the less liquefiable type of NCNo. This high resistance to cyclic shearing may be attributed to the high plasticity of this ore (PI = 43) which is believed to contribute to the liquefaction delay. On the other hand, laterites appears to be the most susceptible type of NCNo to liquefaction since they acquires the less number of cycles to reach the criteria. This difference was explained by the high percentage of fines in this sample (fraction < 80 μm = 100%). For the grainy saprolites, it is proven that the coarse particles take an important part in soil to resist liquefaction.

Afterwards, the evaluation of the liquefaction potential of NCNo by means of the existing methods hadn't provided reliable results to conclude. In fact, according to Tsuchida (1970) chart, all three types of NCNo appear to be susceptible to liquefaction. However, conforming to Seed and Idriss (1982) approach as well as the Chinese criteria as presented by Koester (1992), NCNo are not prone to liquefaction since they are characterized by high vales of Liquid limits and for all tested samples the water content does not exceed 0.87 × Liquid Limit.

Finally, a new method adapted to NCNo was developed to assess the susceptibility of each type of ore to liquefaction. This approach has been inspired from the existing criteria and based on the physical and mechanical test results.

Acknowledgments. This research has been developed with the financial support of CNRT (Centre National de Recherche Technique "Nickel et son Environnement") and SEM (Syndicat des Producteurs-Exportateurs et Exportateurs de Minerai de Nickel de NC) in the scope of Rheolat I and II projects 2011–2016. We thank them both for providing insight and expertise that greatly assisted this research. We would also like to show our gratitude to the CERMES-Navier laboratories who has conducted the physical characterization and the Cyclic Triaxial tests.

References

Andrews, D.C.A., Martin, G.R.: Criteria for liquefaction of silty soils. In: Proceedings of 12th WCEE, Auckland, New Zealand (2000)

Finn, W.D.L.: Dynamic analysis in geotechnical engineering. In: Proceedings, Earthquake Engineering and Soil Dynamics II – Recent Advances in Ground Motion Evaluations. Geotechnical Special Publication, ASCE, vol. 20, pp. 523–591 (1988)

IMSBC: Adoption of The International Maritime Solid Bulk Cargoes Code (2008), Annex 3, Resolution MSC.268 (85) (2008)

Koester, J.P.: The influence of test procedure on correlation of Atterberg limits with liquefaction in fine-grained soils. Geotech. Test. J. **15**(4), 352–360 (1992). ASTM

Koromila, I.A., et al.: Experimental investigation of cargo liquefaction and impact on the stability of a bulk – carrier. In: Proceedings of the 13th International Ship Stability Workshop, pp. 1–7 (2013)

Lade, P.V., Yamamuro, J.A.: Effects of non-plastic fines on static liquefaction of sands. Can. Geotech. J. **34**(6), 918–928 (1997)

Liang, R.W., et al.: Effect of clay particle content on liquefaction of soil. In: 12th World Conference on Earthquake Engineering, Auckland, New Zealand (2000)

Pecker, A.: Dynamique des sols «Collection géotechnique», 259 p. Presses de l'École nationale des ponts et chaussées, Paris (1984)

Perlea, V.G., Koester, J.P., Prakash, S.: "How Liquefiable are Cohesive Soils?" In: Proceedings of the Second International Conference on Earthquake Geotechnical Engineering, Lisbon, Portugal, vol. 2, pp. 611–618 (1999)

Seed, H.B., Idriss, I.M.: Ground motions and soil liquefaction during earthquakes, p. 134. Earthquake Engineering Research Institute, Berkeley (1982)

Seed, R.B., et al.: Recent advances in soil liquefaction engineering: a unified and consistent framework. EERC-2003–06. Earthquake Engineering Research Institute, Berkeley (2003)
Sherif, M.A., et al.: Saturation effect on initial soil liquefaction. J. Geotech. Eng. Div. **103**, 914–917 (1977). American Society of Civil Engineers
Tsuchida, H.: Prediction and countermeasure against liquefaction in sand deposits. Abstract of the Seminar of the Port and Harbour Research Institute, Ministry of Transport, Yokosuka, Japan, pp. 3.1–3.33 (1970)
Zlatovic, S., Ishihara, K.: On the influence of non-plastic fines on residual strength. In: Proceedings of IS-Tokyo 1995, First International Conference on Earthquake Geotechnical Engineering, vol. 1, pp. 239–244 (1995)

Effect of Variability of Soil Parameters in the Behavior of Shallow Foundations

Tahar Messafer(✉)

Civil Engineering Department, University M'Hamed Bougara Boumerdes, Boumerdès, Algeria
tmessafer@hotmail.co.uk

Abstract. This work is divided into two folds. The first stage was to carry out a statistical study of geotechnical soil parameters obtained for a housing project. This comprised analysis of the mean, standard deviation, coefficient of variance, histograms, cumulative densities and normal distribution laws. In addition correlation using linear regression analysis was carried out between pairs of soil parameters obtained from laboratory tests. Then autocorrelation functions were developed for pressuremeter modulus.

The second stage was to develop a probabilistic approach to design shallow foundations. This assumed that strengths parameters such as cohesion and angle of shearing strength are variable. The results were compared to traditional techniques based on Terzaghi methods such as DIN and DTU and also to semi probabilistic approach such as Eurocode.

Keywords: Statistical analysis · Correlation · Laboratory tests · In situ test · Foundation analysis · Probabilistic approach

1 Introduction

The design of civil engineering structures requires a good knowledge of the subgrade. The first stage of a project is a geotechnical investigation as it allows the engineer to select representative values of soil characteristics necessary for the design. However, it is impossible to define in any point of a site the soil properties because the determination of representative parameter values is generally carried out on the basis of a few samples taken almost at random and in situ tests executed following a more or less wide mesh.

This is why the development of methods of statistical analysis and probability for the characterization of physical and mechanical properties of soils should solve the problem of variability of soil parameters. Then the use of these statistical methods in the probabilistic foundation design might be more advantageous compared to traditional methods used at present.

2 Site Presentation

The project is located in a mountainous region (Medea) in Algeria and includes the completion of 500 and 200 housing units as part of a 2000 housing program. The investigation in situ comprised the completion of 11 boreholes of 15 ml, 6 boreholes of 20 ml,

W. Frikha et al. (eds.), *Soil Testing, Soil Stability and Ground Improvement*,
Sustainable Civil Infrastructures, DOI 10.1007/978-3-319-61902-6_14

5 pressuremeter tests of 10 ml, 60 dynamic penetration tests and 2 piezometers 20 ml each deep.

Analysis of the results indicated the presence of two main layers which are:

Layer 1: silty clay with a thickness of about 4 m
Layer 2: gray marl

3 Statistical Analysis of Geotechnical Parameters

Data analysis will focus on the results of laboratory and in situ tests for model one layer (Site) and model two layers (clay and marl). The results compare values of the min, max, average, standard deviation and coefficient of variance (CV) for the 2 models.

Several geotechnical parameters were analyzed such as the density, water content, degree of saturation, grading analysis, Atterberg limits, compressibility parameters, compression index, swelling coefficient, cohesion and angle of friction. As well as the results of pressuremeter tests.

Only limits results obtained for models one layer and two layers are provided on Table 1 and these show that the two layers are both plastic.

Table 1. Summary of models Atterberg limits single layer, two layers

Paramètres	Model	Nbre sampl retained	Min values	Max values	Average values	Stand déviation	CV (%)
WL (%)	Site	36	45	59	54,24	4,13	7,62
	Clay	26	46	59	54,08	4,21	7,79
	Marl	10	45	59	54,10	4,18	7,72
IP (%)	Site	36	22	29	27,00	2,09	7,76
	Clay	26	23	29	26,96	2,16	8,02
	Marl	10	22	29	25.50	2,11	8.27

4 Analysis of Variability

Histograms and cumulative distribution of geotechnical parameters have been developed. Results for plasticity index (Ip) are shown on Fig. 1. A graphical representation of classes is given in histograms shown in Fig. 2.

From the results it can be seen that the Ip histogram for clay is identical to the site histogram. Hence to calculate the frequency or density of the site, clay data can be taken. Adopting Fig. 1, the cumulative distribution of Ip is shown in Fig. 3.

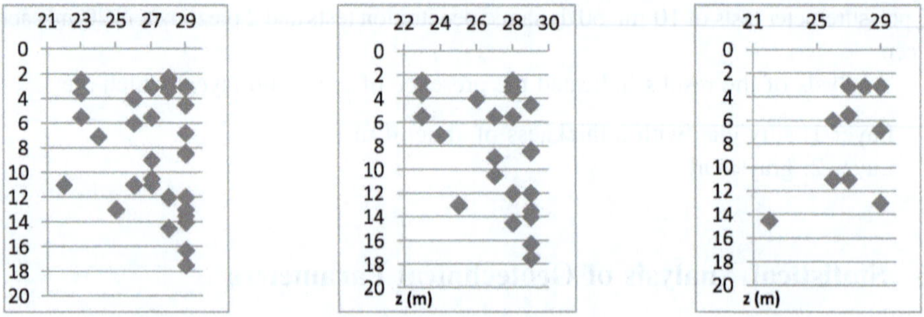

a) Site b) Clay c) Marl

Fig. 1. Distribution de Ip with depth for model 1(a) and model 2(b and c)

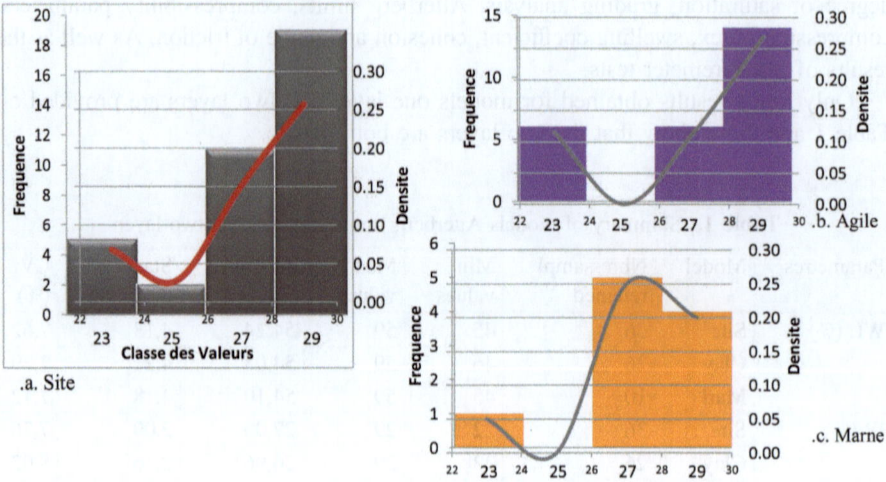

Fig. 2. Histograms of Ip

Figures 4, 5 and 6 show Gauss normal laws f(z) and the corresponding probabilities for Ip for the whole site and layers of Clay and Marl respectively.

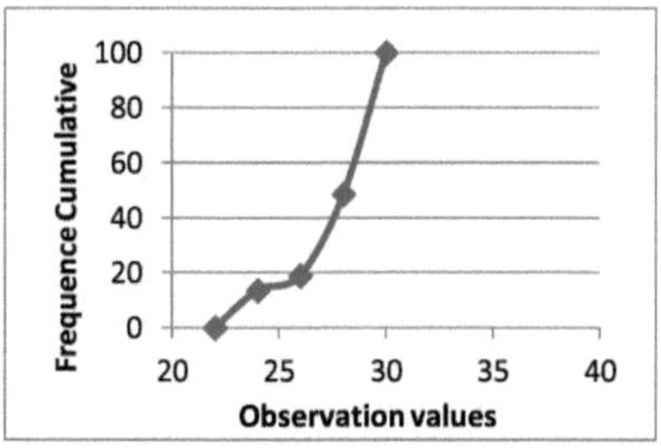

a) Site

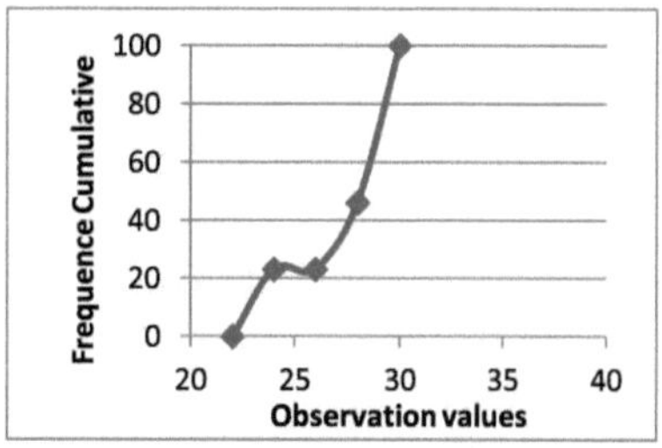

b) Clay

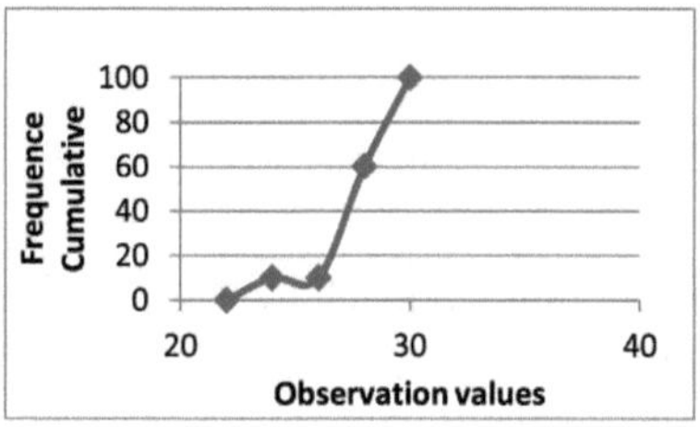

c) Marl

Fig. 3. Cumulative distribution of Ip

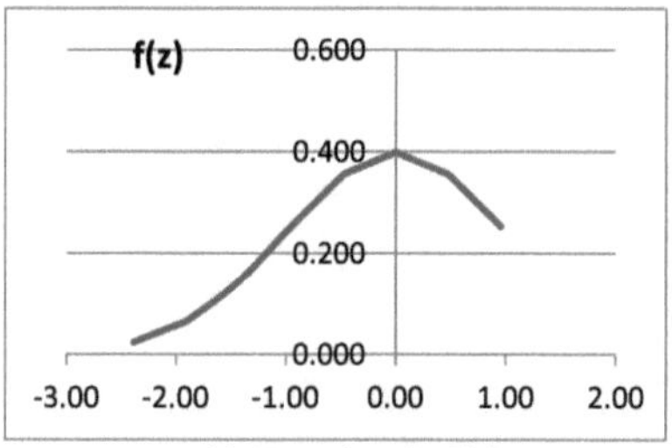

(a) Fonction de f(z)

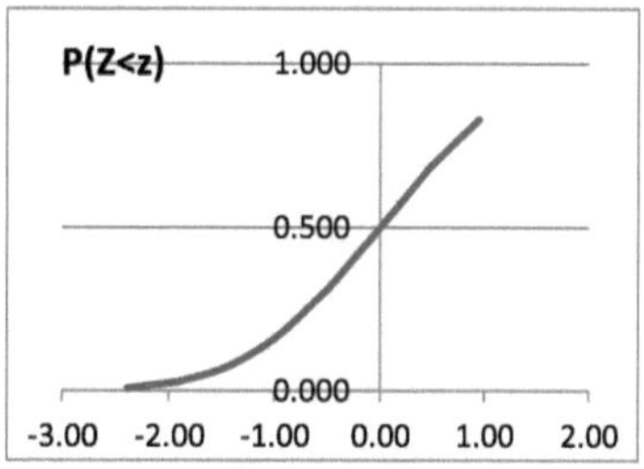

(b) Probabilité

Fig. 4. Normal distribution de Ip pour le Site

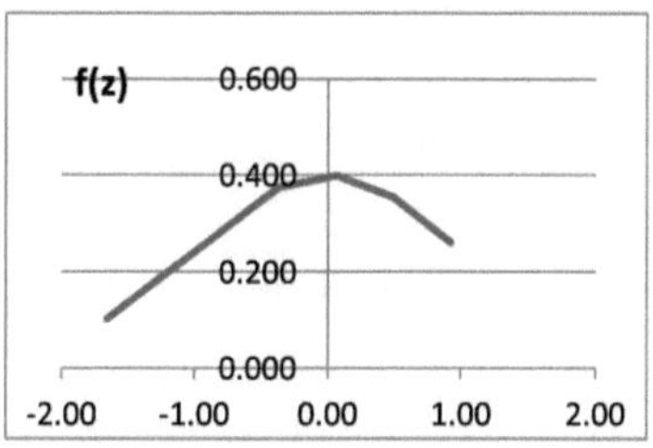

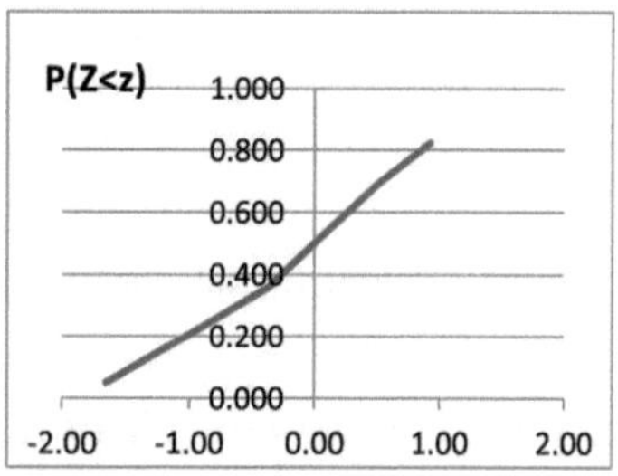

Fig. 5. Normal distribution of Ip for clay

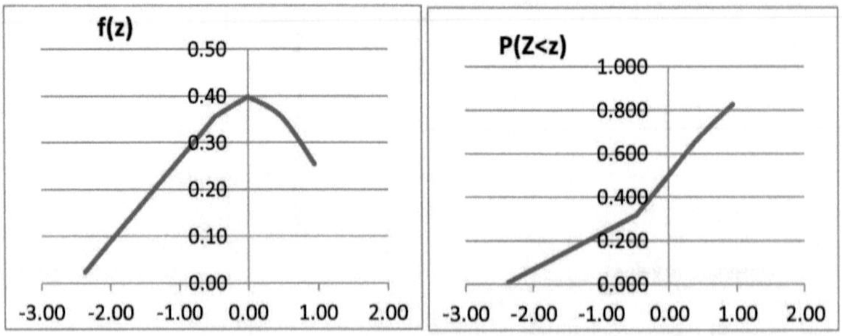

Fig. 6. Normal distribution of Marl

5 Correlations and Lineair Regressions

Figure 7 shows results of water content and Atterberg Limits correlations. The result gave good correlations between WL and IP for the whole site, clay and marl layers reflecting similarity of site materials origins. These results are similar to correlations obtained by several authors, for example by Magnan [1]: WL = 1.04 Ip + 27.02 R = 0.90

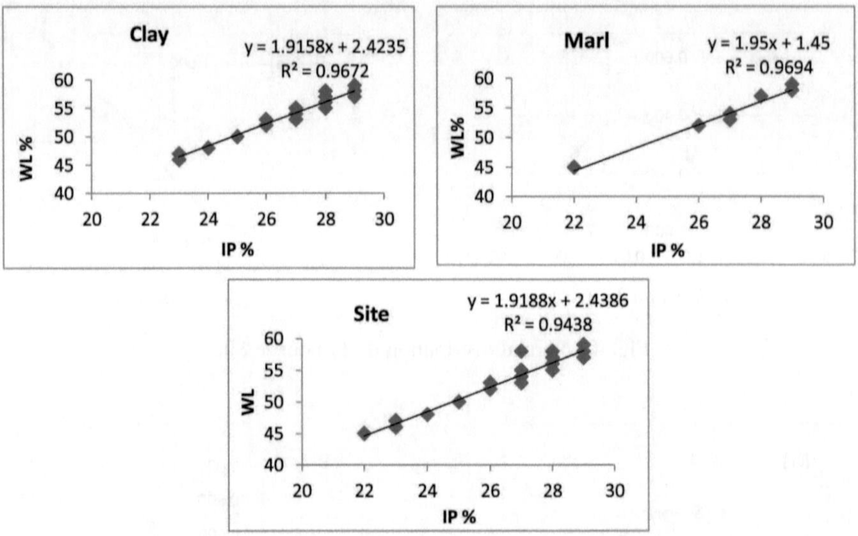

Fig. 7. Relations between water content and Atterberg limits

6 Spatial Auto-correlation

Analysis of spatial variability of physical and mechanical properties of the site was performed for the pressuremeter survey. The analysis was carried out in the vertical direction as it contained enough regularly spaced data. The purpose of this analysis is to determine the auto-correlation function which describes the representation given in the vertical direction and also to determine the remote auto correlation (or fluctuation scale) which determines the degree of dispersion of data.

The auto-correlation is a measure of the dependence of two or more points in a data set. From measurements y_i given x_i equidistant from a property, the estimated value of the auto-correlation between two values distant by x can be defined as follows, Salloum [2]:

$$\rho_\tau = \frac{\sum_{i=1}^{n-\tau}(y_i - \bar{y}_i)(y_{i+\tau} - \bar{y}_{i+\tau})}{\sum_{i=1}^{n}(y_i - \bar{y}_i)^2} \quad \tau = 0, 1, 2, \ldots, n-1 \qquad (1)$$

Where n is the number of measurements of the soil property and τ is the shift of the data set. $\bar{y}_i$ and $\bar{y}_{i+\tau}$ are values of the measured trends at x_i and $x_{i+\tau}$ respectively.

Among the auto-correlation function models, exponential functions of the following form: $C.e^{-\frac{|\tau|}{a}}$ can be used (Imanzadeh [3]):
τ represents the distance between two points of the ground where it is desired to determine the correlation and a is the auto correlation distance.

A simple but approximate method for calculating the vertical scale of fluctuation is given by Vanmarcke [4] as follows:

$$\delta_v \approx 0.8\bar{d} \qquad \text{with} \qquad \bar{d} = \frac{1}{n-1}\sum_{i=1}^{n-1} d_i \qquad (2)$$

Where δ_v is the vertical fluctuation scale and $\bar{d}$ is the average value of distances limited by the intersections of the trend function with the function $\xi(z)$ of soil property.

6.1 Drift Average Values (Linear Regression)

Observation of measured values distributions of soil parameters as a function of depth leads to visually distinguish two layers: a 4 m thick followed by a layer of 6 m as shown in Fig. 8. This represents measured values of pressuremeter module (EM) in a pressuremeter test.

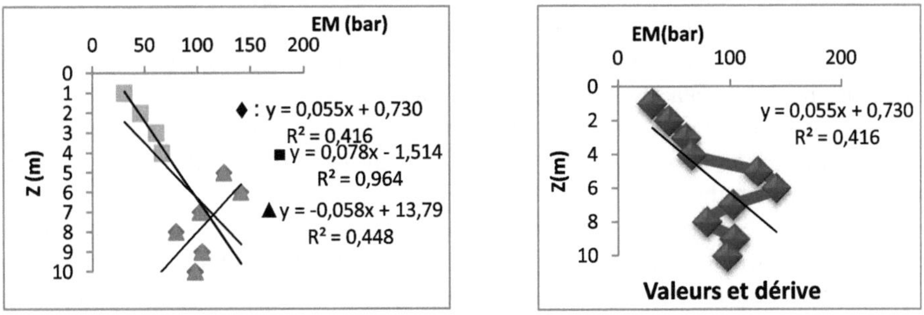

Fig. 8. Spatial variability of pressuremeter module (EM)

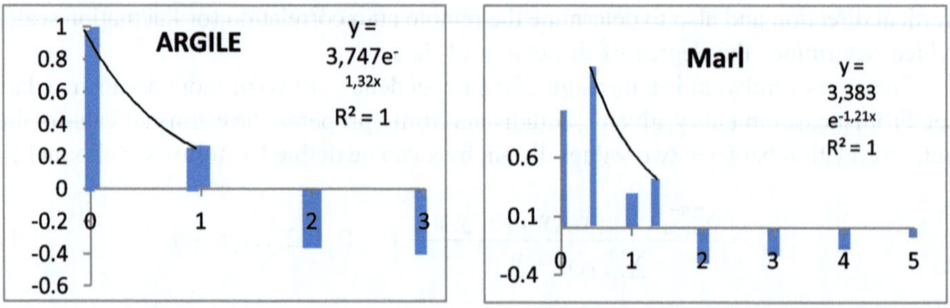

Fig. 9. Autocorrelation diagrams and EM autocorrelation functions in clay and marl

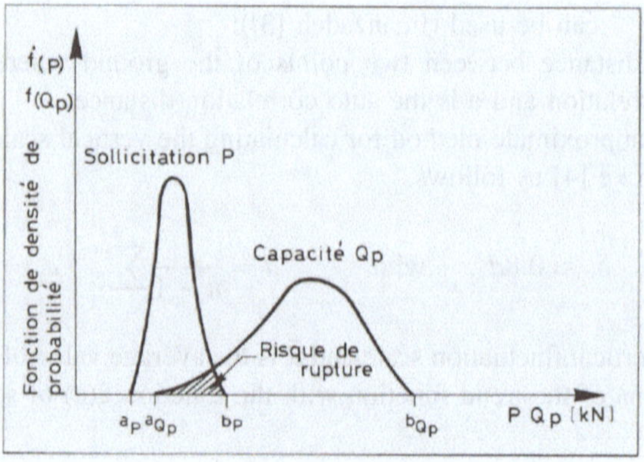

Fig. 10. Capacity-Solicitation model

6.2 Functions and Auto-correlation Distances

Figures 9 and 10 show the auto-correlation charts and autocorrelation functions of EM in the vertical direction for each layer separately (Marl and clay) and for the whole site.

For each auto-correlation diagram, an exponential function of the form, $y = C.e^{-\frac{|x|}{a}}$, has been adjusted on the first three or four values of the coefficient of auto-correlation, distance of auto correlation and fluctuation scale $\theta = 2a$.

Table 2 shows the results of the function of autocorrelation adjusted for pressuremeter modulus data in the clay and the marl.

The auto-correlation function in the vertical direction of EM was determined for clay and marl. The best results were obtained with the exponential function (Figs. 9 and 10) for which the value of R^2 is the highest.

The R^2 value obtained was found greater than 0.903, and the vertical autocorrelation distance was found around 0.52 m in clay and 0.59 m in the marl. This means that the pressuremeter modulus vary in the clay layer in an identical manner to the marl.

Table 2. Autocorrelation function, fluctuation scale θ wide field and the coefficient of determination R^2 in clay and marl and the whole site

Paraméter	Nature of soil	Auto-corrélation function $\rho(\tau)$	Auto-corrélation distance a	Fluctuation scale θ	R^2
EM	Clay	$y = 3{,}747e^{-1{,}32x}$	0.266	0.52	1
	Marl	$y = 3{,}383e^{-1{,}21x}$	0.295	0.59	1
	Site	$y = 5{,}188e^{-1{,}38x}$	0.72 m	1.44 m	0,903

The vertical distance of autocorrelation obtained for EM varies between 0.5 and 1.44. This indicates that these values are not dispersed.

7 Fondation Calculation

A probabilistic approach enables the study of the risk of failure by taking into account the variability of geotechnical parameters and also the variability of the pressure acting on the foundations. In what follows, traditional calculation results (Terzaghi, DTU, DIN) and Eurocode will be compared to probabilistic results, considering a normal distribution law.

7.1 Calculation of Bearing Capacity from Laboratory Tests

In the case of a strip footing, the bearing capacity under a vertical centered load is given by the following general relation (superposition method of Terzaghi):

$$q_l = \frac{1}{2}\gamma_1 B N_\gamma(\varphi) + cN_c(\varphi) + (q + \gamma_2 D)N_q(\varphi) \tag{3}$$

With:

q_l and q: bearing capacity and vertical lateral pressure respectively

γ_1, γ_2: Volumetric weight of the soil under the base of the foundation and laterally respectively

c: Cohesion of soil under the base of the foundation

B and D: Width of the foundation and ancrage depth of the foundation, respectively

$N_\gamma(\varphi)$,$N_c(\varphi)$ and $N_q(\varphi)$: Bearing Factors of the foundation

Table 3 contains the formulas of these three factors according to the French, German and Eurocode conventional concepts.

Table 3. Bearing capacity factors

Concept	Bearing factors		
	N_γ	N_c	N_q
Conventionnel French (DTU13.12)	$1{,}85(Nq - 1)tan\varphi$	$(N_q - 1)cot\varphi$	$e^{\pi \tan\varphi} \cdot \tan^2\left(\frac{\pi}{4} + \frac{\varphi}{2}\right)$
Conventionnel German	$2(Nq - 1)\tan\varphi$		
Eurocode 7-1	$2(Nq - 1)tan\varphi$		

7.2 Allowable Bearing Capacity of a Foundation (According to Terzaghi)

$$q_a = \gamma_1.D + \frac{q_l - \gamma_1.D}{F} \tag{4}$$

7.3 French Conventional Concept (DTU13.12)

$$q_{réel} \leq q_{ad} \tag{5}$$

$q_{réel}$, q_{ad}: real stress applied to the foundation and allowable pressure respectively

$$q_{ad} = \gamma D + \frac{\gamma \frac{B}{2} N_\gamma + \gamma D(N_q - 1) + cN_c}{F_s} \tag{6}$$

7.4 Conventional German Concept DIN 1054

$$\frac{V_b}{V} \geq Fs \quad \text{With} \quad V_b = q_L.A \tag{7}$$

A: area of the foundation and Fs: global safety factor
V_b and V: Limit load and external load applied on the foundation

7.5 Eurocode 7.1 (Semi-probabilistic Calculation for Safety)

Partial safety factors are applied to the soil parameters (γ, c, φ). The bearing capacity calculation is calculated as follows:

$$R_d = Q_L(\varphi'_d, c'_d, \gamma_d) \tag{8}$$

With, $c'_d = \frac{c'_k}{\gamma_{c'}}$, $\tan \varphi'_d = \frac{\tan \varphi'_k}{\gamma_{\varphi'}}$ $\gamma_d = \gamma_k$
The variation of γ specific weight is very low, for this reason we take $\gamma = 1$.
$\gamma_{\varphi'}, \gamma_{c'}$: Partial safety factors applied to φ' and c′ respectively
Indices k and d mean characteristic value and design value respectively

7.6 Probabilistic Method

The few published work on this subject (Mounji et al. [5], Chew et al. [6]) has shown the importance of taking account of soil variability. In this work, the capacity (Q) and solicitation (S) are considered random variables each having a mean and standard deviation.

The difference between the two random variables (Z = Q–S) is called margin of safety and is also a random variable. A limit state is reached when Z = 0 and failure occurs when Z < 0, and the probability of failure can be defined as (Boutahir et al. [7]):

$$P_f = P_f[(Q - S) \leq 0] = P_f(Q \leq S) \tag{9}$$

Where R and S follow Gauss normal distributions, the safety margin (Z) also follows a normal distribution.

$$f(Z) = \frac{1}{s_Z\sqrt{2\pi}} \cdot e^{-(Z-\mu_Z)^2/2s_z^2} \tag{10}$$

Whose mean and standard deviation are respectively:

$$\mu_Z = \mu_R - \mu_S \quad \text{and} \qquad s_z = \sqrt{s_R^2 - s_S^2}$$

The hatched area of Fig. 10 being the failure probability P_f calculated with the integral:

$$P_f = \int_{-\infty}^{0} f_Z(Z).dZ \tag{11}$$

Bearing capacity q_l, based on random variables φ, c and γ
The function $q_l(\varphi, \gamma, c)$ can be developed into a Taylor series. If we take the terms of the lowest order non null, we can calculate the mathematical expectation $\bar{y}$ and the variance S_y^2 of a function $y = f(x_1, x_2 \ldots x_n)$, (Hahn and Shapiro [8]), with x_i random variables defined by their average μ_i and their variance S_i^2:

$$\bar{y} = f(\mu_1, \mu_2, \ldots, \mu_n) + \frac{1}{2}\sum_{i=1}^{n} \frac{\delta^2 f}{\delta_{x_i}^2}(S_i^2) \tag{12}$$

$$S_y^2 = \sum_{i=1}^{n} \left(\frac{\delta f}{\delta x_i}\right)^2 (S_i^2)$$

The coefficient of variation of y will be:

$$V_y = \frac{S_y}{\bar{y}}$$

In the case of q_P (φ, γ and c), expressions (12) and (13) become:

$$\mu_q = \bar{q}_P = cN_c\lambda_c + \gamma tN_q\lambda_q + \gamma\frac{B}{2}N_\gamma\lambda_\gamma\frac{1}{2}\left(c\lambda_c\frac{\delta^{2N_c}}{\delta\varphi^2} + \gamma t\lambda_q\frac{\delta^{2N_q}}{\delta\varphi^2} + \gamma\frac{B}{2}\frac{\delta^{2N_\gamma}}{\delta\varphi^2}\lambda_\gamma\right)S_\varphi^2 \tag{13}$$

$$S_q^2 = \left(c\lambda_c\frac{\delta N_c}{\delta\varphi} + \gamma t\lambda_q\frac{\delta N_q}{\delta\varphi} + \gamma\lambda_\gamma\frac{B}{2}\frac{\delta N_\gamma}{\delta\varphi}\right)^2 S_\varphi^2 + N_c^2\lambda_c^2 S_c^2 + \left(tN_q\lambda_q + \frac{B}{2}N_\gamma\lambda_\gamma\right)^2 S_\gamma^2 \tag{14}$$

7.7 Calculation Results

7.7.1 Traditional Method

The geotechnical investigation conducted on the site helped fix a founding level of 2 m depth below ground level. Widths calculations at this depth for a strip footing and for $\gamma 1 = \gamma 2 = 20$ kN/m3 and solicitation S = 300 kN/ml are shown in Table 4.

Table 4. Widths B with Terzaghi method for Fs = 2 and 3

φ'[°]/c'[kPa]	13/40	15/61	15/49	14/51	15/44	15/53	16/53	14/58	15/51
Fs = 2	1,09	0,7	0,83	0,86	0,90	0,79	0,73	0,78	0,81
Fs = 3	1,60	1,04	1,23	1,27	1,32	1,17	1,09	1,16	1,20

7.7.2 DTU, DIN and Eurocode Methods

The results are given on Table 5.

Table 5. Widths B for DTU, DIN and Eurocode 7.1 methods

φ'[°]/c'[kPa]	DIN 1054		DTU 13.12		Eurocode 7
	Fs = 2	Fs = 3	Fs = 2	Fs = 3	
13/40	1,09	1,60	1,01	1,40	1,58
15/49	0,83	1,23	0,79	1,11	1,21
15/44	0,90	1,32	0,85	1,18	1,30
16/53	0,73	1,09	0,70	0,99	1,07
15/51	0,81	1,20	0,77	1,09	1,18

The results show that the widths obtained from DIN approach are slightly higher than those of the DTU. However the results of Eurocode are much closer to the results of DIN for FS = 3.

7.7.3 Probabilistic Method

The foundation is a strip footing width of B and parameters shown on Table 6.

Table 6. Mechanical properties of the soil

											μ	s	V	s^2
φ'	(°)	13	15	15	14	15	15	16	14	15	15	0,87	0,06	0,75
γ_1	kN.m-3	20,9	20,9	20,6	20,8	20,6	20,1	20,6	20,5	20,0	20,6	0,32	0,02	0,10
c'	kN.m-2	40	61	49	51	44	53	53	58	51	51,1	6,43	0,13	41,36

It is considered that the solicitation (S) and the capacity (Q) are random and follow a normal probability law and thus the safety margin (Z = Q–S) also follows a normal distribution.

The results of long-term load bearing capacities are summarized on Table 7. Values μ_q and s_q used for the calculation of risk are determined from Table 8.

Table 7. Probability of a value being smaller than z for a standard normal distribution

	Bearing capacity															
	B = 0,5 m				B = 0,75 m				B = 1 m				B = 1, 5 m			
	ql	Ql	Z	F(z)	ql	Ql	Z	F(z)	ql	Ql	Z	F(z)	ql	Ql	Z	F(z)
	200	100	−2,08	0,05	200	150	−2,07	0,05	200	200	−2,06	0,05	200	300	−2,05	0,05
	300	150	−1,57	0,12	300	225	−1,57	0,12	300	300	−1,57	0,12	300	450	−1,56	0,12
	400	200	−1,06	0,23	400	300	−1,06	0,23	400	400	−1,07	0,23	400	600	−1,08	0,22
	500	250	−0,55	0,34	500	375	−0,56	0,34	500	500	−0,57	0,34	500	750	−0,60	0,33
	538,96	269,48	−0,35	0,38	544,11	408,08	−0,34	0,38	549,25	549,25	−0,33	0,38	559,55	839,32	−0,31	0,38
	600	300	−0,04	0,40	600	450	−0,06	0,40	600	600	−0,08	0,40	600	900	−0,12	0,40
	659,10	329,55	0,26	0,39	665,92	499,44	0,27	0,38	672,74	672,74	0,28	0,38	686,39	1029,59	0,30	0,38
	690,12	345,06	0,42	0,37	696,08	522,06	0,42	0,36	702,03	702,03	0,43	0,36	713,94	1070,91	0,43	0,36
	714,00	357	0,54	0,34	720,82	540,61	0,55	0,34	727,64	727,64	0,55	0,34	741,29	1111,94	0,56	0,34
	730,83	365,42	0,63	0,33	737,46	553,09	0,63	0,33	744,08	744,08	0,63	0,33	757,33	1136,00	0,64	0,33
	753,90	376,95	0,75	0,30	760,57	570,43	0,75	0,30	767,24	767,24	0,75	0,30	780,57	1170,86	0,75	0,30
	760,39	380,19	0,78	0,29	766,25	574,69	0,78	0,30	772,12	772,12	0,77	0,30	783,86	1175,79	0,77	0,30
	810,96	405,48	1,04	0,23	818,84	614,13	1,04	0,23	826,72	826,72	1,04	0,23	842,48	1263,71	1,05	0,23
	848,32	424,16	1,23	0,19	855,24	641,43	1,22	0,19	862,16	862,16	1,22	0,19	876,01	1314,02	1,21	0,19
μ	607,61	303,81	0,00		611,81	458,85	0,00		616,00	616,00	0,00		624,39	936,58	0,00	
s	196,16	98,08	1,00		198,98	149,23	1,00		201,81	201,81	1,00		207,52	311,29	1,00	

Table 8. Calculation *of* μ_q et de s_q for différent B

Sol	Clay			
	B = 0,5 m	B = 0,75 m	B = 1 m	B = 1,5 m
$c\,Nc$	658,6	658,6	658,6	658,6
$\gamma\, t\, Nq$	91,4	91,4	91,4	91,4
$\gamma\, B/2\, N\gamma$	15,0	22,5	30,0	45,0
q_l	765,1	772,6	780,1	795,1
$1/2\; c\; d^2\, Nc/d\varphi^2\;\; s_\varphi^2$	3 867,8	3 867,8	3 867,8	3 867,8
$1/2\,\gamma\, t\; d^2\, Nq/d\varphi^2\; s_\varphi^2$	1 216,5	1 216,5	1 216,5	1 216,5
$1/2\gamma\; B/2d^2\; N\gamma/d\varphi^2\; s_\varphi^2$	561,9	842,9	1 123,9	1 685,8
Somme	5 646,3	5 927,2	6 208,2	6 770,1
moyenne μ_q (kPa)	6 411,3	6 699,8	6 988,3	7 565,2
$c\;\; dN\, c/d\varphi$	2 294,4	2 294,4	2 294,4	2 294,4
$\gamma\, t\, dNq/d\varphi$	530,9	530,9	530,9	530,9
$\gamma\, B/2\, dN\gamma/d\varphi$	131,2	196,8	262,3	393,5
Somme	2 956,5	3 022,1	3 087,7	3 218,8
$(..\text{somme}..)^2\; s_\varphi^2$	6555589,07	6849668,34	7150199,79	7770619,2
$Nc^2\;\; s_c^2$	6867,97	6867,97	6867,97	6867,97
$(t\; Nq + B/2\, N\gamma)^2 s_\gamma^2$	2,67	3,056	3,47	4,38
s_q^2	6562459,70	6856539,37	7157071,23	7777491,55
Ecart type s_q	2561,73	2618,5	2675,27	2788,81
variance V_q	0,4	0,39	0,38	0,37

First Case: Assume that the load is constant of intensity S $= 300\ \text{kN/m}$, the probability of failure for widths B of 0.5 m; 0.75 m; 1.0 m and 1.5 m are summarized on Table 9. Figure 11 shows the method of calculating the probability of failure for a width equal to B of 1 m.

Table 9. Risk of failure for constant loads

Widths B (m)	0,5	0,75	1,0	1,5
Risk of failure (%)	59,3	29,8	18,2	10,1

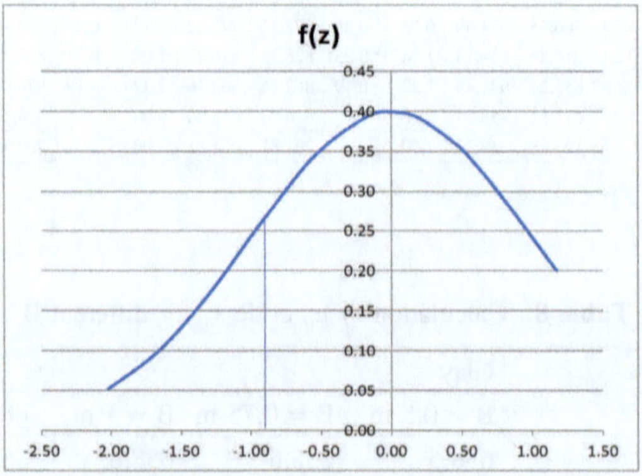

Fig. 11. Risk of failure for B = 1 m

Second case: We assume that the stress (S) and the capacity (Q) are random with a normal distribution, the load to be transmitted to the ground is characterized by the following values.

Table 10. Risk of failure for variable loads and capacities

Widths B (m)	0,5	0,75	1,0	1,5
Probability de failure (%)	1,4	0,9	0,7	0,5

Results for probability of risk calculation are shown in Table 10 for different widths B. Figure 12 shows the distributions of Q and S in the case where the width B is 1 m.

$$P_f = 1 - F\left(\frac{\mu_S - \mu_Q}{\sqrt{s_S^2 + s_Q^2}}\right) = 1 - F(+2,46) = 0,69\% \tag{15}$$

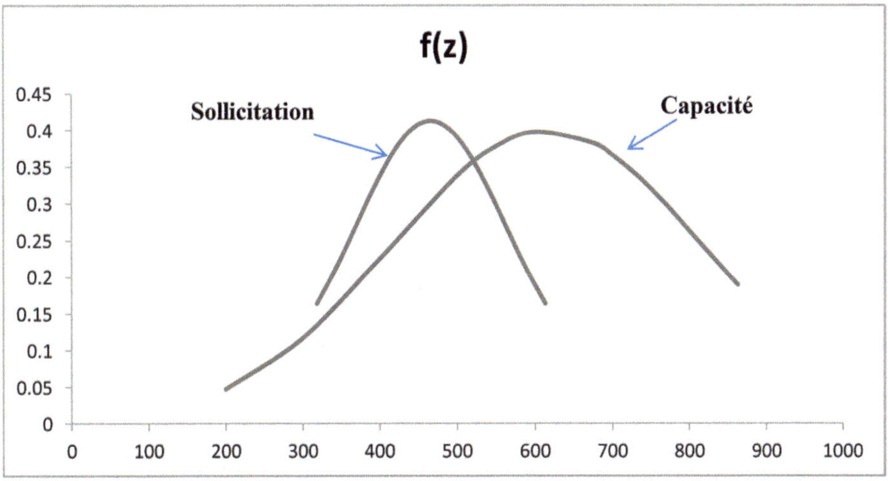

Fig. 12. Diagram of normal distribution of capacity and solicitation for B = 1 m

The results show that the risk of failure in the case of a variable load is negligible for widths ranging between 0.5 to 1.5 m. However in the case of a constant load, probability of failure varies from 60 (%) in the case of a 0.5 m width of foundation to 10 (%) in the case of a width of 1.5 m. In general for widths larger than 1 m the probability of failure is less than 20 (%). With traditional methods (Terzaghi, DIN and DTU) to obtain B values between 1 and 1.5 m, an Fs equal to 3 is required.

8 General Conclusions

The geotechnical investigation showed that the soil is composed of two essential layers namely a clay layer with thickness of 4 m followed by a layer of marl. In order to compare the results we decompose the soil in 2 models. The first model is the whole site as one layer (site) and the 2nd is a two-layer soil (clay and marl).

The results of the geotechnical parameters were presented in the form of histograms, cumulative distributions and normal laws. Regression equations were established between pairs of parameters and gave results comparable to those in the literature.

We presented an analysis of the spatial variability of the pressuremeter modulus EM. We found that if we assume that EM is an exponential function, the autocorrelation distance is approximately 0.5 m indicating that EM values are not dispersed.

Foundation calculations were carried out by traditional probabilistic methods. We showed that in the probabilistic approach bearing capacity is a random parameter because it is based on parameters c and φ which are themselves random. Compared to traditional methods the probabilistic approach is a powerful calculation tool but requires to be utilized among practitioners.

References

1. Magnan, J.P, Baki. A., Pouget, P.: Analyse statistique de la variabilité des propriétés physiques et mécaniques d'un sol, Bulletin de liaison Laboratoire Centrale des Ponts et Chausse, No. 186 (1993)
2. Salloum, N.: Evaluation de la variabilité spatiale des paramètres géotechniques du sol à partir de mesures géophysiques: application à la plaine alluviale de Nahr-Beyrouth (Liban), Thèse de doctorat, Université de Grenoble, Liban (2015)
3. Imanzadeh, S.: Effects of uncertainties and spatial variation of soil and structure properties on geotechnical design. Cases of continuous spread footings and buried pipes, these de Doctorat, Université de Bordeaux, France (2013)
4. Vanmarcke, E.H.: Probabilistic modeling of soil profiles, America, society of civil engineers. J. Geotech. Eng. Div. **103**(11), 1227–1246 (1977)
5. Mounji, M., Lahmili, A., Ouadif, L., Baba, K., Bahi, L.: Probabilistic approach for the selection of the shallow foundation's safety factor. Int. J. Eng. Technol. (IJET) **8**(2), 7 (2016)
6. Chew, Y.M., NG, K.S., NG, S.F.: The effect of soil variability on the ultimate bearing capacity of shallow foundation. J. Eng. Sci. Technol., 1–13 (2015). Special issue on ACEE 2015 Conference
7. Boutahir, M., Belabed, L., Benyaghia, H.: Analyse fiabiliste de la capacité portante des fondations superficielles, Séminaire International, innovation et valorisation en génie civil et materiaux de construction, No 5P-092, Rabat, Maroc (2011)
8. Hahn, G.J., Shapiro, S.S.: Statistical models in engineering. Wiley Interscience Publisher (1994)

Improvement of Collapsible Soil Conditions for Industrial Floors

Nelson L. Fonte Jr.[1], David de Carvalho[2(✉)], and Roberto Kassouf[3]

[1] Head of Geoponto Engineering, Mogi das Cruzes, São Paulo, Brazil
lopesfontejr@bol.com.br
[2] University of Campinas, Campinas, São Paulo, Brazil
d33c@uol.com.br
[3] Metropolitan College of Campinas, DeVry Metrocamp, Campinas, Brazil
kassouf.engenharia@kassouf.com.br

Abstract. Unsaturated surface soils with porosities above 50% cover great extensions of areas in Midwest Brazil. Because of their large volumes of voids, these soils undergo great strain under loads. In addition, many of these soils are collapsible, i.e., when the soils are under load and in case of a significant increase in the moisture content or saturation of the soil, the structure collapses, thus producing unacceptable displacement values for the buildings. Because of these characteristics, problems often occur in industrial floors, pavements and other types of slabs on ground and shallow foundations.

To solve this problem, compaction of the topsoil is performed before starting the construction. For this study, the soil characteristics are analyzed using both geotechnical laboratory tests and field tests in order to predict the soil behavior in terms of deformability, resistance and collapsibility.

The study was conducted at the experimental site located at the State University of Campinas - Unicamp, in the municipality of Campinas, State of São Paulo, Brazil. The geotechnical properties of the subsoil were determined by collecting undisturbed samples down to 8 m in depth and deformed samples up to 9 m of depth (impenetrable). Simple recognition surveys were performed: standard penetration test (SPT) and electric static cone penetration test (CPT).

The edometric tests conducted on undisturbed samples with flooding at different levels of stress revealed collapsible characteristics of the soil. With the conduction of tri-axial tests (CU), numerical values were obtained for the angle of friction and the cohesion intercepts for each depth. Paschoalin Filho (2008) verified a significant reduction in these values with soil saturation.

The influence of soil deformability and of the type of load on industrial floors is analyzed. The results indicate huge influence of soil deformability in the case of distributed loads, and a minor influence in case of concentrated loads.

The thickness of the compacted soil layer is very important to the definition of the modulus of subgrade reaction (k) for the studied soil.

W. Frikha et al. (eds.), *Soil Testing, Soil Stability and Ground Improvement*,
Sustainable Civil Infrastructures, DOI 10.1007/978-3-319-61902-6_15

1 Introduction

Collapsible soils are non-saturated soils with a potentially unstable porous structure. Once under the action of a load, their index of voids is suddenly reduced in case of a certain increase in moisture content. For such, there must be cementation or suction to keep the soil structure stable in its natural condition. However, In case the soil is flooded with water, suction ceases to exist and the cement may dissolve destabilizing the structure of the soil.

Soil collapse may damage buildings and floors due to the magnitude of differential settlements in case of rupture of pipes, flooding, water table increase, among others (Gon 2011).

Excavating the soil down to a certain depth and filling the same material back in a controlled manner is a way to improve the supporting soil of shallow foundations, industrial floors and pavements on collapsible soils by increasing resistance, rigidity and stability against collapse, and reducing both strain and permeability (Ribeiro and Futai 2010).

This improvement can be verified in the soil analyzed in this work by the increased rigidity and resistance obtained via edometric and triaxial tests using compacted and undisturbed samples of soil. The effect of this improvement on the design of industrial floors is analyzed.

The study was conducted at the campus of the State University of Campinas - Unicamp, city of Campinas, State of São Paulo, Brazil. Several researches have already been conducted on the soil of the campus, and laboratory tests were made in non-deformed and deformed samples down to 9-m depth, as well as field tests: SPT with torque check (SPT-T), electric (CPTU) and mechanical (CPT) cone tests with Begeman sleeve, cross-hole test, vertical seismic sounding, and Menard-type pressiometric and dilatometric (DMT) tests (Paschoalin 2008).

1.1 Occurrence of Collapsible Soils in Brazil

Collapsible soils cover approximately 50% of the southwest region of Brazil. Giacheti (1991) presents a map pinpointing collapsible soils with geological/geotechnical characteristics potentially similar to the soil studied in the city of Campinas (Fig. 1). Besides this vast region, Ferreira et al. (2007) present records of collapsible soils in different states of the northeast and north of Brazil: Amazonas, Pará, Tocantins, Piauí, Paraíba, Rio Grande do Norte, Pernambuco and Bahia. Bandeira et al. (2012) also mention occurrence of such collapsible soils in the state of Ceará.

Taking into account the large areas of collapsible soils in Brazil, it is necessary to learn more about the behavior of such soils and take their characteristics into due account when designing shallow foundations, industrial floors and rigid pavements.

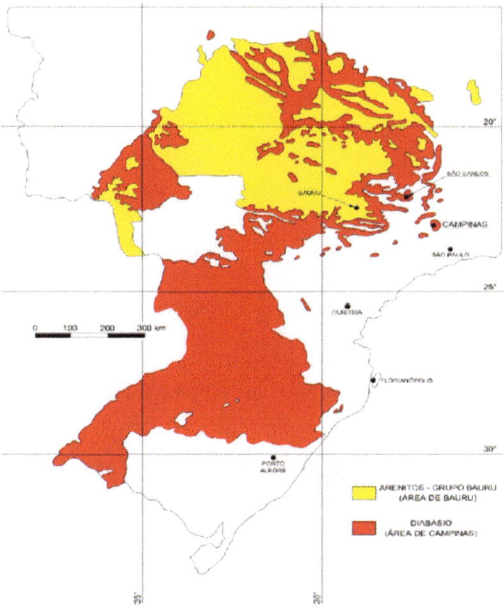

Fig. 1. Regions in Brazil where the same profile as that of Campinas may exist (Giacheti 1991 adapted by Cury Filho 2016)

1.2 Compaction as Soil Improvement

The most commonly method used worldwide to construct buildings with the use of surface foundations on porous, collapsible ground is to remove the surface layer, compact it and return it to the same place (Ribeiro and Futai 2010).

Souza and Cintra (1994) carried out four circular plate load tests and two tests in continuous footings measuring 0.70 m × 3.02 m, all of which were seated at a depth of 0.70 m. The study was carried out in porous soils in the countryside of the state of São Paulo, in the city of Ilha Solteira. Two of the plate tests were performed with soil in its natural moisture, one of which on a compacted soil layer. Two more plate tests were performed likewise but in flooded soil.

The authors concluded that compaction reduced collapse settlement by 86%. Settlements in other steps of loading were reduced by 50%. The conclusion was that compaction was a cost-effective, efficient and easy to use process to enable use of shallow foundations in this type of soil. After an analysis of the results shown in this research, the observation was that compacted soil increased the bearing capacity of foundations by approximately 100% when compared to the same moisture content. Another conclusion was that compacted soil, when flooded, resisted 36% more than soil in its natural condition.

In the cases mentioned, the soil was compacted down to a "Z" depth equal to the smallest dimension of "B" footing; the width of the compacted regions is the width of "B" footing, with an addition of B/2 on each side, as shown in Fig. 2.

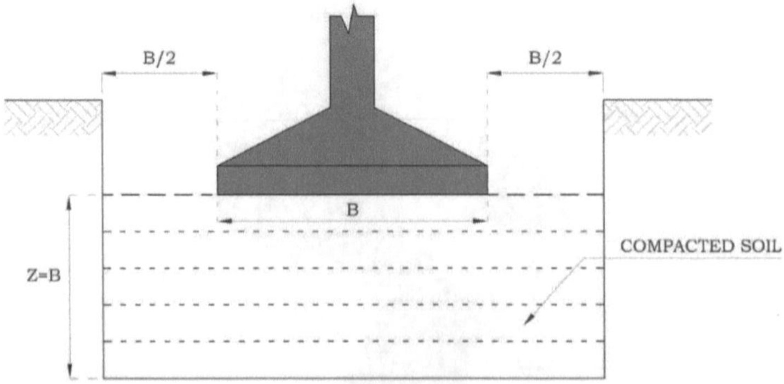

Fig. 2. Use of footings (shallow foundations) In compacted collapsible soil (Cintra et al. 2003)

Considering that stresses propagate at depth in a 1:2 distribution, compaction with the width of B plus B/2 on each side causes the added stress to occur only on the compacted soil at depth Z = B and, at this depth, this added stress is only 25% of the stress applied by the footing, thus reducing the collapse of the natural soil due to reduced stress (Cintra and Aoki 2013).

Guimarães and Ferreira (1998) conducted several laboratory tests with compacted samples and also confirmed the efficacy of controlled compaction to improve soils in the state of Pernambuco as to reduction of collapse.

1.3 Study Site

There are two experimental sites at the campus of the State University of Campinas - Unicamp, in the city of Campinas, state of São Paulo, where several researches have been conducted. One is the experimental site of the School of Agricultural Engineering (FEAGRI). The other is the experimental site of the School of Civil Engineering, Architecture and Urbanism (FEC). The distance between the sites is approximately 300 m and both are installed in an area with the same type of soil, so the results can be interpreted as representative of one large region.

The FEC experimental site has a 600-m^2 area. Its geodesic coordinates in the WGS84 geodesic system are: latitude −22°49′01″ and longitude −47°03′41″; the experimental site of FEAGRI as a 400-m^2 area, with the following coordinates: latitude −22°49′10″ and longitude −47°03′40″. Figure 3 shows the location of the experimental sites at the Unicamp campus. Paschoalin (2008) described part of the researches conducted at the experimental site of FEAGRI until 2008: Peixoto (2001), in which several SPT-T type tests were performed; Fontaine (2004) executed electric cone tests and pressiometer tests; Carvalho et al. (2000) provided geotechnical characteristics obtained by means of several field and laboratory tests; Albuquerque (2001), Nogueira (2004) and Paschoalin (2008) carried out several load tests on piles. Giachetti (1991) and Albuquerque (1996) conducted several laboratory tests for geotechnical characterization; Monacci (1995) studied the soil as to its collapsibility. Paschoalin (2002)

Fig. 3. Location of the experimental sites

studied several characteristics of this soil in its "natural" and compacted condition in different moisture contents.

Gon (2011) collected samples down to 9-m depth at the FEC experimental site for tests of characterization, permeability, edometric, triaxial and proctor tests.

Based on these researches, it can be stated that the soil in the region under analysis comprises primarily one first layer of approximately 6.0 m of highly porous sandy clay; in some spots a 0.50-m layer of clayey-silty fine and medium sand can be found. There is a second layer of residual diabase soil composed of clayey sandy silt down to the depth of 16 m. The water table was not found down to this depth. Figure 4 shows this profile.

The first layer is formed by mature soil which underwent intense weathering. The second layer is formed by young residual soil, which has preserved inherited characteristics from the original rock (Albuquerque 2001). Figure 5 shows the geological profile of the Unicamp region (Barão Geraldo District).

Figures 6, 7, 8, 9 and 10 show results of field tests carried out by Carvalho et al. (2004) and resubmitted by Cury Filho (2016). Figures 6, 7 and 8 show average, maximum and minimum data of SPT index, Tmax e Tmin resistance of SPT-T.

Figures 9 and 10 show results of static cone penetration tests (CPT) using electric cone, obtained by Carvalho et al. (2000). The average, minimum and maximum values of point resistance of the cone (qc) and lateral friction (fs) are shown.

Monacci (1995) and Gon (2011) carried out edometric compression tests in non-deformed samples, with flooding at different levels of stress to check soil collapsibility in conformance with Vargas method (1978).

According to Tables 1 and 2, most tests show that the soil under study is collapsible.

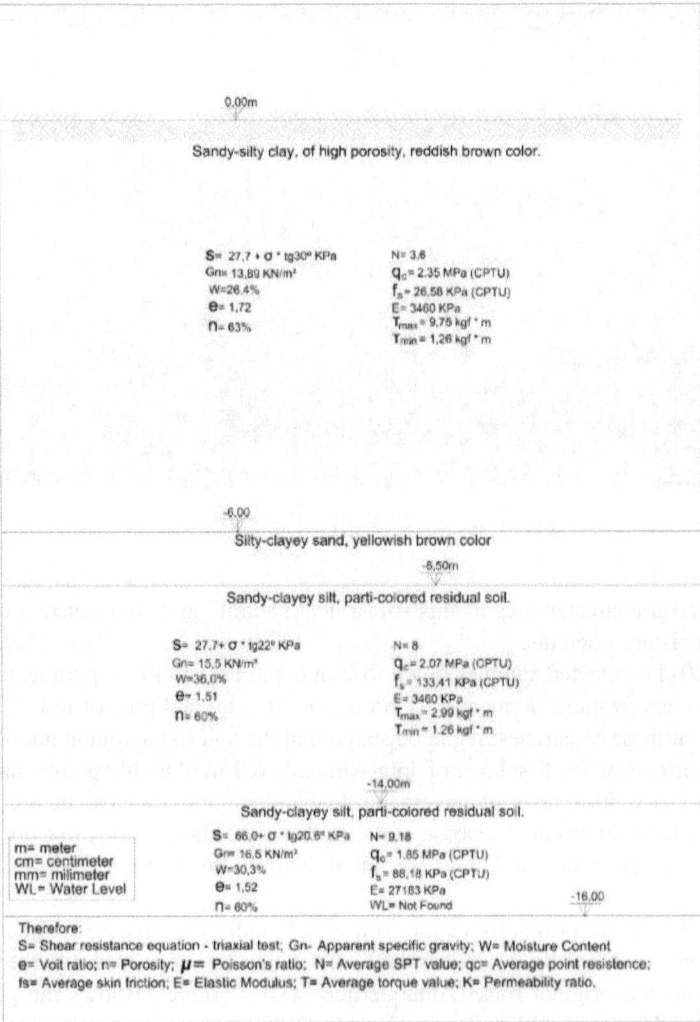

Fig. 4. Geotechnical profile - average values

In order to analyze the influence of soil deformability both in its "natural" condition and after compaction when designing slabs on soil, the initial elasticity modulus was obtained via triaxial test.

Figure 11 shows the stress - strain behavior of a non-deformed sample collected at the depth of 1.0 m, with different stresses of confinement (Gon 2011). This Figure shows the initial modulus of deformability for stress of confinement of 100 kPa, resulting in the value of 3.3 MPa.

Figure 12 shows the stress - strain behavior of a sample of compacted soil at its optimal moisture of the Normal Proctor Test. Through these curves, the data in Table 3 was obtained, which shows the tensile strength and the value obtained for the initial

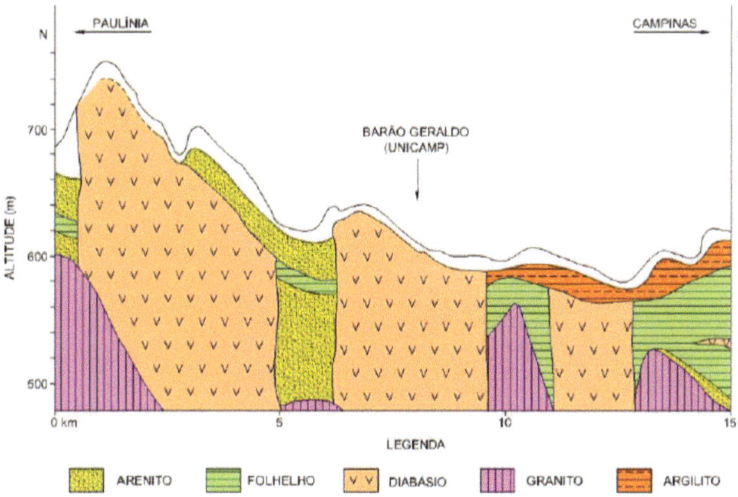

Fig. 5. Geological profile of the region of Unicamp (Dirney 2016 adapted from Zuquete 1987)

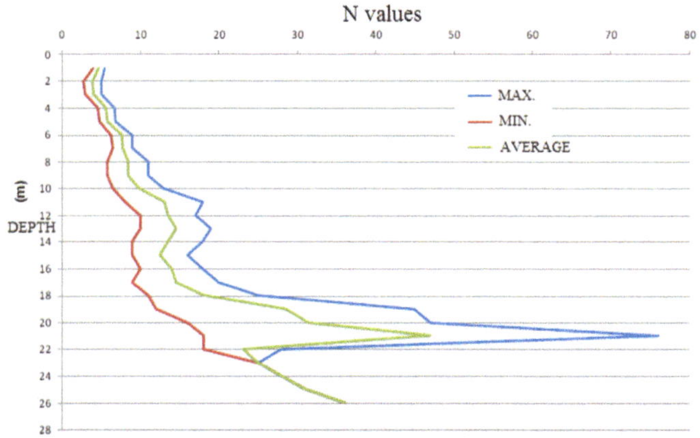

Fig. 6. Summary N resistance (Cury Filho 2016)

modulus of elasticity for each confinement stress. This table demonstrates that the average value of the initial modulus of elasticity for the compacted soil was 22 MPa.

The increase of the modulus of elasticity due to compaction is about 6,7, similar the difference between a structural steel (E = 205 GPa) and a structural concrete (E = 30 GPa).

Figures 13 and 14 show the curves obtained by Paschoalin (2002) in the edometric test both for natural and compacted soil. A large reduction in voids can be seen when the soil is compacted.

Although compaction is a common method to improve most types of soil around the world, it is especially efficient for the studied soil.

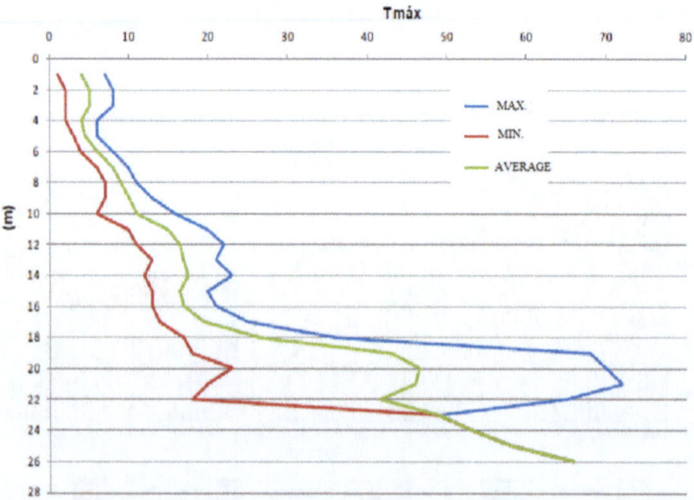

Fig. 7. Summary of Tmáx resistance (Cury Filho 2016)

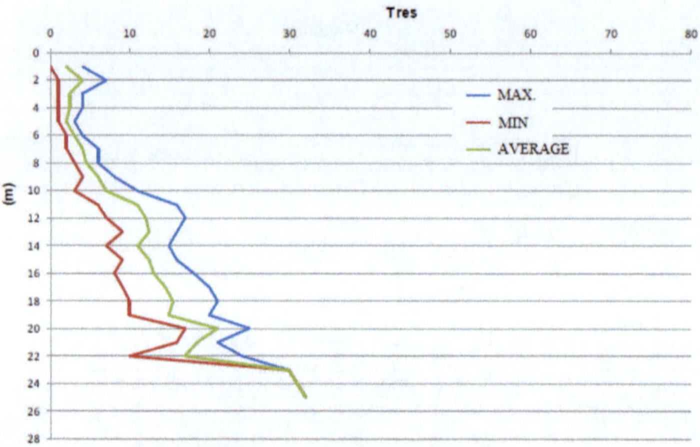

Fig. 8. Summary of Tres resistance (Cury Filho 2016)

Three CBR tests were performed with compacted soil in its optimal moisture. The results are described in Table 4. The average value is 4%.

1.4 Impact on Industrial Floors Design

When designing slabs on ground, whether for industrial floors, rigid pavements or radier-type foundations, in many methods the soil influences the dimensioning because of its deformability via consideration of the settlement coefficient k, also named vertical

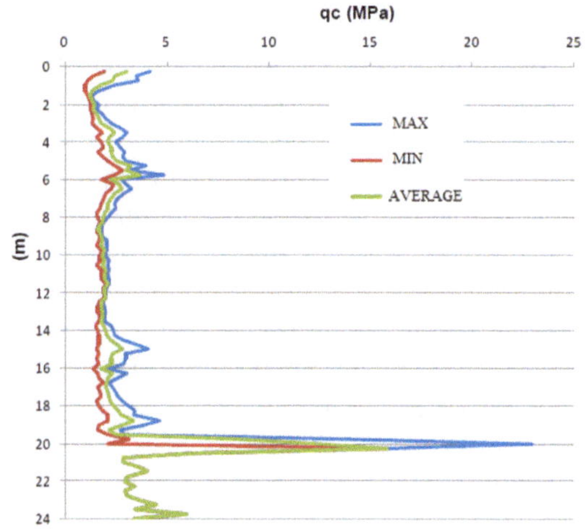

Fig. 9. Summary of qc values (Carvalho et al. 2004 apud Cury Filho 2016)

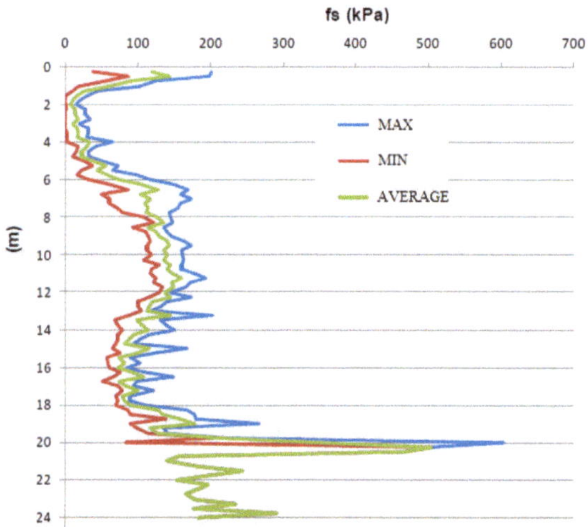

Fig. 10. Summary of fs values (Carvalho et al. 2004 apud Cury Filho 2016)

Table 1. Colapse coefficient (Monacci 1995)

Depth	0,5 a 0,75 m	4,75 a 5,00 m	7,75 a 8,00 m
σ [kPa]	*i* [%]		
5,00	4,97	–	–
9,80	11,97	2,41	–
19,40	7,40	3,76	–
29,10	9,20	–	–
38,70	9,98	7,72	3,10
77,30	23,19	15,51	5,26

Table 2. Colapse coefficient (Gon 2011)

Depth	CC 100 kPa	CC 200 kPa	CC 400 kPa
1	7,80%	27,00%	6,00%
2	1,30%	0,65%	23,00%
3	0,45%	0,04%	19,84%
4	9,90%	7,50%	26,80%
5	0,28%	7,03%	7,24%
6	1,10%	3,40%	9,20%
7	1,40%	5,60%	10,74%
8	5,20%	6,60%	0,09%

*CC > 2 are collapsible soils.

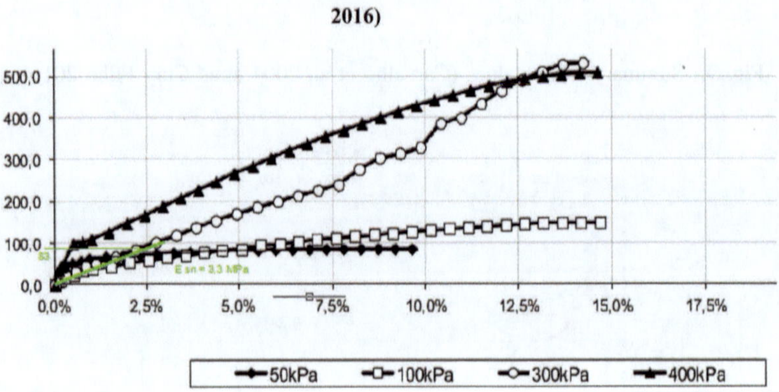

Fig. 11. Stress-strain behavior of natural soil for the 1st meter (Adapted from Gon 2011)

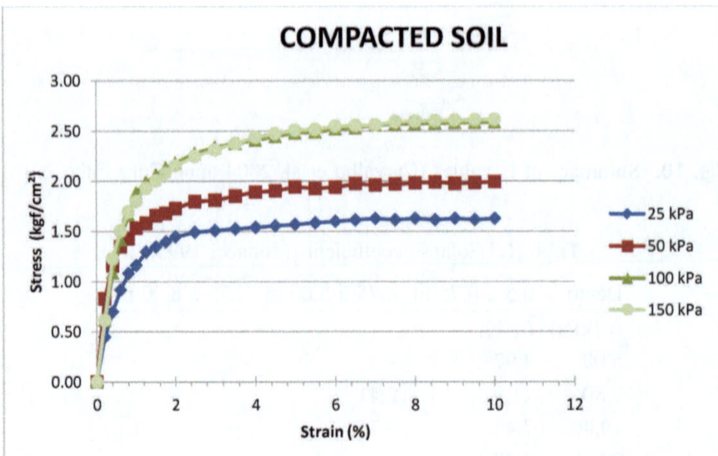

Fig. 12. Stress-strain behavior of compacted soil

Table 3. Summary of compacted soil stress-strain behavior – triaxial tests

Sample	Confinant stress (kPa)	Rupture stress (kPa)	Elastic modulus (MPa)
CP1	25	160	13
CP2	50	200	30
CP3	100	260	20
CP4	150	260	23
Average	–	–	22

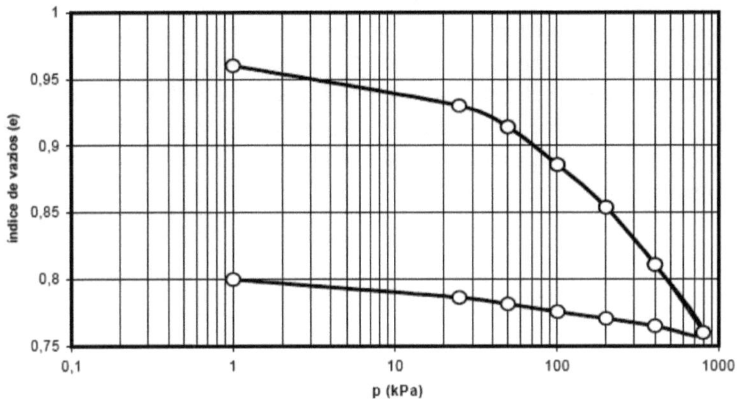

Fig. 13. Variation in the void index with addition of pressure for compacted soil (Paschoalin 2002)

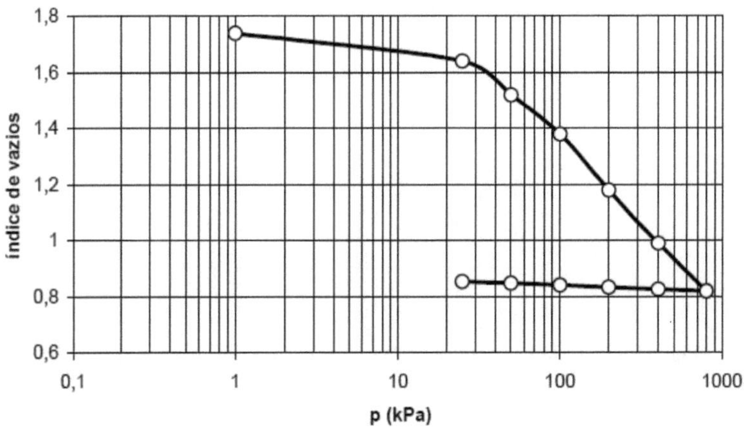

Fig. 14. Variation in the void index with addition of pressure for undisturbed soil (Paschoalin 2002)

Table 4. CBR tests

Corpo de Prova	CBR (%)	Expansão (%)
CP1 – S	3	0,015
CP2 – S	3	0,075
CP3 – S	6	0,20
Média	4	0,097

reaction modulus. In an elastic model, k, is the ratio between the stress applied on a point on the soil and the corresponding displacement.

One of the ways to get the modulus of vertical reaction of the soil (subgrade) is through correlation with the CBR index (ACI 360R-06). Figure 15 shows a curve relating the CBR index of the soil with k (Rodrigues et al. 2006). With this curve, it was possible to get the value of k of 33 MPa/m for the soil under study, considering the CBR index of 4%. Using this method is not possible to take into account the thickness of the compacted soil layer.

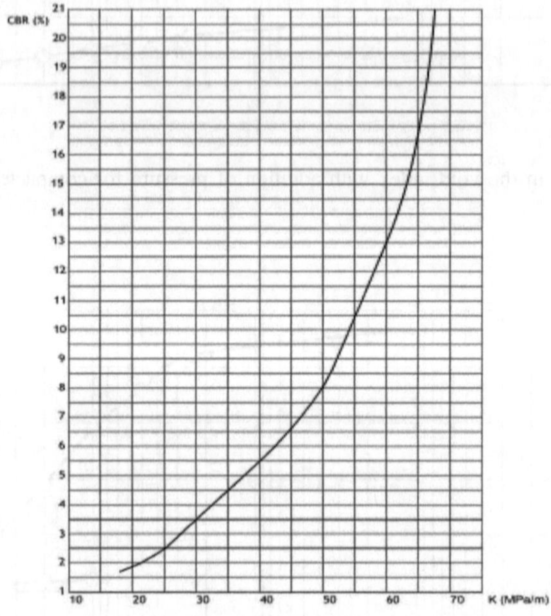

Fig. 15. Correlation between CBR × k (Vaz Rodrigues et al. 2006)

Another form to get k is by means of a plate load test. There are standards that set a form to adopt k. For the specific purpose of designing rigid pavements, there is ASTM D 1196. In Brazil, there is standard 055-2004 ME of DNIT. Both standards determine a load test on a plate with 76 cm diameter and adopt as settlement coefficient the ratio of stress that causes settlement of 1.27 mm by the value of this settlement (Fig. 16).

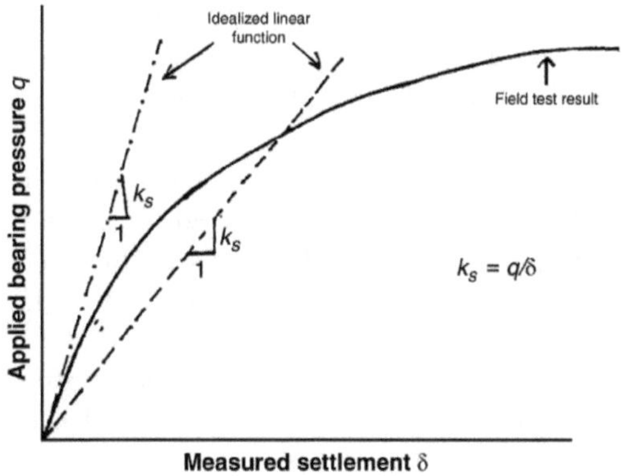

Fig. 16. Plate load-deformation diagram (ACI 360R-06)

Through the theory of elasticity, one can simulate the load test, if the modulus of elasticity of the soil is known, and get K for different conditions of the soil profile. Using the modulus of elasticity of the natural and compacted soil obtained via triaxial test (En = 3.3 MPa and Ec = 22 MPa), it was possible to get k for different thicknesses of compacted soil from zero to the 8-m layer. To do so, a soil profile with 8-m thickness on a non-deformed layer was adopted. Using the method of Janbu et al. (1956, apud Cintra 1998) k was obtained for different situations, including variation of Poisson's coefficient of 0.20, 0.35 and 0.50. The results are shown in Table 5, which shows that k ranged from 5.6 MPa/m to 48.3 MPa/m.

Table 5. k for different thickness of compacted soil layer

Compacted soil thickness (m)	Ens = 3,3 MPa e Ecs = 22 MPa		
	k (MPa/m) p/ν = 0,5	k (MPa/m) p/ν = 0,35	k (MPa/m) p/ν = 0,20
0	7,2	6,2	5,6
0,2	9,5	8,1	7,4
0,25	10,4	8,9	8,1
0,3	10,7	9,1	8,4
0,4	12,4	10,6	9,7
0,5	13,8	11,8	10,8
0,8	17,5	15,0	13,7
1	19,4	16,6	15,2
1,5	25	21,4	19,5
2	31	26,5	24,2
8	48,3	41,3	37,7

Two situations of loading were used to assess the influence of variation of k in designing an industrial floor. The first situation is the case of a concentrated load of

50 kN applied by a tire with calibration pressure of 700 kPa. The second situation is the case of a distributed load of 40 kN/m^2 with aisles without loading.

For the first case, the method of Meyerhof (1962) is used for floors with distributed reinforcement on the top, considering the load applied in the interior region of the floor (away from the edges of the plate). The bending moment acting in the concrete slab is obtained by means of Eq. 1, which, in turn, uses Eq. 2, which uses the modulus of elasticity of concrete of 26,840 MPa and Poisson's coefficient of concrete of 0.20, where h is the thickness of the concrete of the industrial floor. In Eq. 1, 'a' is the radius of the area of contact between the tire and the industrial floor. Once the acting moment is obtained, the dimensioning is made by limiting the tensile stress in concrete by the admissible tensile stress in bending. For concrete of fck 30 MPa, an admissible stress of 2.07 MPa is adopted.

$$M_i = \frac{P}{6[1 + (2a/l)]} \tag{1}$$

$$l = \sqrt[4]{\frac{E \times h^3}{12x(1 - \mu^2) \times K}} \tag{2}$$

For the case of distributed load, the method of Packard (1996) is used. This method uses Eq. 3. C is the admissible distributed load, σ_{adm} is the admissible tensile stress of concrete (adopted 2.07 MPa), h is the thickness of concrete in cm and k is the coefficient of settlement in MPa/m.

$$C = 1,03.\sigma_{adm}.\sqrt{h.k} \tag{3}$$

Figure 17 shows the variation in thickness of a floor submitted to concentrated load with variation in the value of k. It can be noted that, even with such a large variation of k ranging from 5 to 50 MPa/m, the thickness of concrete varied approximately 1.20 cm, i.e., approximately 10%.

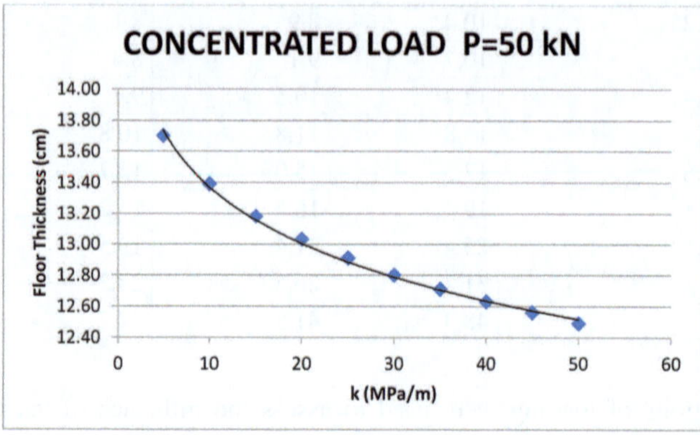

Fig. 17. Influence of k on floor thickness for cases of concentrated load

Figure 18 shows the same variation for distributed load, where the large influence of the value of k can be noted. By increasing 10 times its value, the thickness of the floor varied from 7 to 70 cm, which is directly proportional to the variation of k.

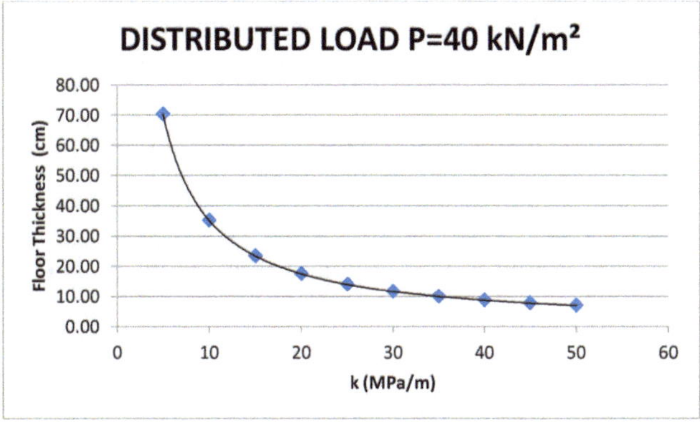

Fig. 18. Influence of k on floor thickness for cases of distributed load

Considering two different support condition for an industrial floor: one without any compacted soil layer, the slab directly on the natural soil, and other with 1 m of compacted soil. Using the results presented in Table 5 ($\nu = 0{,}35$) the k value for the first condition is 6,2 MPa/m, and 16,6 MPa for the second condition. With these K values, Fig. 17 shows that for concentrated load the slab thickness in a design would be 14 cm for both support conditions. Figure 18, for distributed load, shows for the first condition the slab thickness would be about 70 cm, and about 21 cm for the second support condition. The influence of the compacted soil in this case is huge.

2 Conclusions

Based on the results obtained, it can be stated that, for the soil under study, compaction improved the performance of the soil as to the following aspects:

- Increased bearing capacity.
- Increased rigidity and consequent reduction in strain due to stress.
- Eliminated the collapsible characteristic of natural soil.

The soil under study is unsaturated, highly porous and collapsible. Its compaction is easy to be executed and cost-effective.

Compaction of collapsible surface soil is a solution that can make it possible to use shallow foundations in soils similar to the soil of the region under study, within certain limits of stress.

The layer of compacted soil significantly influences the value of the modulus of subgrade reaction for the soil under analysis.

For concentrated loads, the value of k has little influence on the industrial floors design, thus justifying the use of simpler methods to obtain it, such as the correlation with the CBR index. In these cases, the recommendation is a minimum layer of 20–25-cm thick compacted soil to even out the support and to get stabilization against collapse.

For distributed loads, the value of k has large influence on the industrial floors design, thus justifying the use of more reliable methods to obtain it, such as plate load tests.

The compacted soil layer affects strongly the slab support deformability for the studied soil.

Acknowledgements. We wish to thank Vander Ferreira Alves, Iago Leandro dos Santos, Paulo Gustavo Krejci Nunes and Juliana Silva for their help in developing the tests.

References

Albuquerque, P.J.R.: Analysis of the behavior of instrumented driven piles of small diameters in diabasic residual soil of Campinas region. M.Sc. dissertation of Agriculture Engineering College, State University of Campinas (1996)

Albuquerque, P.J.R.: Bored piles, CFA and omega piles: study of the behavior subjected to compression in diabasic residual soil, through instrumented load test. Ph.D. thesis of Polytechnic School of São Paulo University (2001)

Bandeira, A.P.N., et al.: Characterization of collapsible soils in municipality of Juazeiro do Norte, interior of Ceará. Brazilian Congress of Soil Mechanics and Geotechnical Engineering, COBRAMSEG, ABMS, Porto de Galinhas (2012)

Carvalho, D., et al.: Experimental site for studies of soil mechanics and foundation in Campinas – SP. In: Seminar of Specials Foundation Engineering, vol. 4, São Paulo (2000)

Carvalho, J.M., Viana da Fonseca, A., Almeida, F., Hermosilha,H.: ISC'2 experimental site investigation and char acterization: conventional and to mographic P and S waves refraction seismics vs. electrical resistivity. In: Viana da Fonseca, A., Mayne, P.W. (eds.) Geotechnical and Geophysical Site Characterization. Millpress, Rotterdam (2004)

Cintra, J.C.A.: Foundation on Collapsible Soils, 1st edn. Graphical Service of Engineering School of São Carlos, University of São Paulo, São Carlos (1998)

Cintra, J.C.A., Aoki, N.: Foundation Design on Collapsible Soils, 1st edn. Graphical Service of Engineering School of São Carlos, University of São Paulo, São Carlos (2013)

Cury Filho, D.: Load capacity estimation for steel piles subjected to compression in diabasic soil of Campinas region. M.Sc. dissertation of Civil Engineering College, State University of Campinas (2016)

Ferreira, R.S.M.: Research report soil volume variation behavior due to flooding. Civil Engineering Department, UFP (2007)

Fontaine, E.B.: Utilization of specials field tests (electric cone and pressiometer test) for estimation of geotechnical properties and pile load capacity in some tropical soils. Ph.D. thesis of Agriculture Engineering College, State University of Campinas (2004)

Giachetti, H.L.: Experimental study of dynamic parameters of some São Paulo state tropical soils. Ph.D. thesis of Engineering School of São Carlos, University of São Paulo, São Carlos (1991)

Gon, F.S.: Geotechnical characterization through laboratory tests of a diabasic soil of Campinas region. M.Sc. dissertation of State University of Campinas (2011)

Guimarães, N., Ferreira, S.R.M.: Collapse due to flooding in compacted soils. Brazilian Congress of Soil Mechanics and Geotechnical Engineering, COBRAMSEG, ABMS, Brasília (1998)

Meyerhof, G.G.: Load-carrying capacity of concrete pavements. In: ASCE Proceedings (1950)

Monacci, M.D.: Study of soil collapsibility in experimental site of the Agriculture Engineering College – Unicamp. M.Sc. dissertation of the Agriculture Engineering College, State University of Campinas (1995)

Nogueira, R.C.R.: Behavior of root piles, instrumented, subjected ti axial compression, in diabasic soil. M.Sc. dissertation of Civil Engineering College, State University of Campinas (2004)

Packard, R.G.: Slab Thickness Design for Industrial Concrete Floors on Grade. Portland Cement Association, Concrete Information, Skokie (1996)

Paschoalin, J.A.: Study of different piles subjected to axial tensile in diabasic soil of Campinas region. Ph.D. thesis of State University of Campinas – Unicamp (2008)

Paschoalin, J.A.: Utilization of diabasic residual soil as foundation element and construction material of compacted embankments for small dams. M.Sc. dissertation of State University of Campinas – Unicamp (2002)

Peixoto, A.S.P.: Study of SPT-T test and its application in foundation engineering practice. Ph.D. thesis of Agriculture Engineering College, State University of Campinas (2001)

Junior, I.R., Futai, M.M.: Study case of collapsible soil improvement with compaction. In: Brazilian Congress of Soil Mechanics and Geotechnical Engineering, COBRAMSEG, ABMS, Gramado (2010)

Rodrigues, P.P.F., et al.: Gerdau Manual of Industrial Floors, 1st edn. Ed. Pini, São Paulo (2006)

Souza, A., Cintra, J.C.A.: Shallow foundation on collapsible soil of Ilha Solteira – SP. In: 10th Brazilian Congress of Soil Mechanics and Foundation Engineering, Foz do Iguaçu, vol. 1 (1994)

Vaz Rodrigues R., Muttoni A., Burdet O.: Large scale tests on bridge slabs cantilevers subjected to traffic loads. In: Proceedings of the 2nd fib Congress, vol. 1, p. 232, Naples (2006)

Ground Improvement of Tank Foundations in the Middle East

Babak Hamidi[1(✉)] and Serge Varaksin[2]

[1] ISSMGE TC-211 Member, Perth, Australia
babakhamidi@hotmail.com
[2] Apageo, ISSMGE TC-211 Co-chairman, Paris, France

Abstract. Although tank construction is well advanced in the Middle East, not all tanks are located on suitable grounds, and many require the implementation of specific foundation measures to ensure that design specifications are satisfied. While ground improvement of tank foundations in the region has a proven track record that dates back to the 1980s, and can potentially be the most appropriate solution for many tanks, it is the authors' experience that occasionally the projects' settlement requirements appear to have been influenced by building specifications that have not been developed for the purpose. Consequently and unnecessarily, more expensive solutions with longer construction schedules may have to be adopted that will not bring any additional benefits to the foundation solution. This paper initially reviews settlement criteria of a number of internationally recognized standards and then presents a number of case studies of tank foundation improvement by dynamic compaction and quality control by Menard pressuremeter tests.

1 Introduction

Industrial scale tanks are usually large and heavily loaded structures that are used to store fluids and must remain in uninterrupted service due to their commercial value and social importance. For example, any disruption in the supply of potable water or fuel due to a tank becoming temporarily non-operational will immediately attract public attention and can lead to social anxiety or unrest if the matter is not addressed and resolved rapidly.

With consideration that tanks are relatively thin structures, unforeseen and unaccounted ground deformations can impede the performance of the tanks, and ultimately lead to tank failure. Hence, implementation of specific foundation techniques has become common practice when in-situ ground conditions are not able to provide foundations with the required bearing or deformation limitations.

Ground improvement in general and dynamic compaction, in particular, have been used extensively for tank foundation improvement since the early 1970s. Dynamic compaction was first used for the improvement of a Petro France tank in Villeneuve La Garenne in 1972. The same technique was also used for the first time respectively for oil tanks in Africa (Congo) in 1973, in Asia (Japan) in 1975, in North America (Los Angles) in 1977, and in South America (La Plata, Argentina) in 1980. However, it was only not until 1988 that dynamic compaction and replacement were first used in the Middle East

W. Frikha et al. (eds.), *Soil Testing, Soil Stability and Ground Improvement*,
Sustainable Civil Infrastructures, DOI 10.1007/978-3-319-61902-6_16

for the treatment of Iraqi Sate Company for Oil Project's tank farm in Yanbu, Saudi Arabia. Among other structures and infrastructures, this project included 10 large crude oil storage tanks, each with diameters and heights that were respectively 110.8 m and 20 m. Two smaller fuel storage tanks were also treated in the same manner.

Today, hundreds of tank foundations have been treated using ground improvement throughout the world, many of which have been documented through conference proceedings or journal publications. Some interesting publications include Cognon et al. (1983); Hendy and Muir (1997); Ihm and Masse (2001); Hamidi and Jullienne (2007); Buschmeier et al. (2012); Debats et al. (2013) and Hamidi et al. (2016).

2 Tank Settlements

Unlike buildings that are basically hollow structures, tanks are heavily loaded thin-shelled structures that exert large loads to large foundations. Hence, tank owners are generally and rightfully very sensitive to foundation movements as any disturbance in the performance of tanks can have significant commercial or social impacts. However, it is the authors' experience that when decision-making engineers have a civil or building background or are not sufficiently familiar with existing research and current tank codes of practices, they tend to stipulate settlement criteria requirements that are unrealistically and unnecessarily onerous and more appropriate and applicable to buildings rather than tanks. The authors have even been involved in discussions where the settlement requirement was zero, which is physically meaningless.

In reality, the impact of ground movement on tanks has been well researched since years ago (Marr et al. 1982; D'Orazio and Duncan 1987), and contrary to most structures whose global performances due to full loadings are only assessable after completion of construction and during utilization phase, tanks are typically subjected to a full-scale hydro test loading before becoming operational.

Internationally recognized standards such as American Petroleum Institute (API), American Concrete Institute (ACI), and Engineering Equipment and Materials Users Association (EEMUA) provide guidelines and minimum requirements for tank foundation deformations. Based on internal knowledge and know-how, other organizations such as Mobil (1990) have also developed their own specifications.

2.1 Types of Tank Settlement

Tank settlements can be envisaged to be uniform, rigid body tilting or planar tilting, and out-of-plane settlement or differential settlement. These types of movements can result in various forms of failure, of which the ones that are most concerning are the distortion of the shell, rupture of the shell or bottom plate or shell-bottom plate connection (Marr et al. 1982).

Uniform Settlement

The uniform settlement of a tank is shown in Fig. 1. Most standards do not appear to be overly sensitive to this type of settlement, but occasionally draw attention to its effect on the piping, which can be resolved by using flexible connections or periodically repositioning the pipe supports.

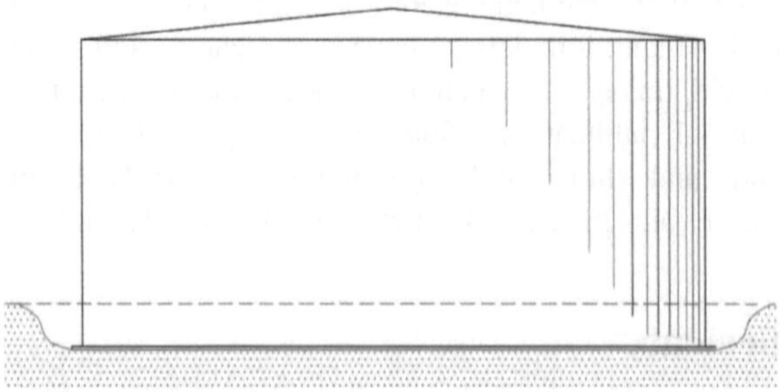

Fig. 1. Uniform settlement

API 620 (American Petroleum Institute 2002) that is applicable to the design and construction of large, welded, low-pressure storage tanks does not stipulate limits on settlement, but notes that uniformity of support and avoidance of excessive settlement are much more important for tanks that have formed bottom plates than they are in the case of flat-bottom, vertical storage tanks.

API 625 (American Petroleum Institute 2010) that is used for refrigerated liquefied gas storage tanks systems simply notes that the amount of acceptable uniform settlement depends on piping and structural connections between the tank system and adjacent structures.

API 650 (American Petroleum Institute 2007) that is utilized for welded tanks also does not limit this type of settlement but states that total settlement must not strain connecting piping or produce gauging inaccuracies, and that settlement should not continue to a point at which the tank bottom is below the surrounding ground surface. This standard further specifies that estimated settlement should be within the acceptable tolerances for the tank shell and bottom, and adds that if a large settlement is expected, then the tank bottom elevation should be raised so that the final elevation above grade will be a minimum of 150 mm after settlement.

API 653 (American Petroleum Institute 2003) that is applicable to tank inspection, repair, alteration, and reconstruction notes that excessive foundation settlement of storage tanks can affect the integrity of tank shells and bottoms, and adds that this type of settlement could affect tank nozzles that have piping attached to them. While this standard does not limit total settlements, it states that for existing tanks with histories of successful services, it may be possible to accept greater settlements and distortions of the foundation from a true plane than new tank construction standards allow.

ACI 376 (American Concrete Institute 2010) that has been developed for the design and construction of concrete structures for the containment of refrigerated liquefied gasses permits uniform settlement of concrete shallow foundations provided that the other provisions of the standard are met, and the connecting piping system accommodates the settlement.

Rigid Body Tilting

Figure 2 shows rigid body tilting or planar tilting. This form of settlement of the shell reduces freeboard, alters the shape of the fluid surface, and places additional stresses in the shell (Marr et al. 1982). Marr et al. show that a planar tilt that is 5% of the tank diameter will increase the hoop stress at the bottom of the shell by 10%. While they assess that failure of the shell from planar tilt alone seems unlikely, and note that this type of tilting appears to have no detrimental consequence to the structural integrity of the bottom plate, Marr et al. propose limits to avoid spilling oil from floating roofs or exceeding the tolerance of the floating roof by stressing and deforming the shell.

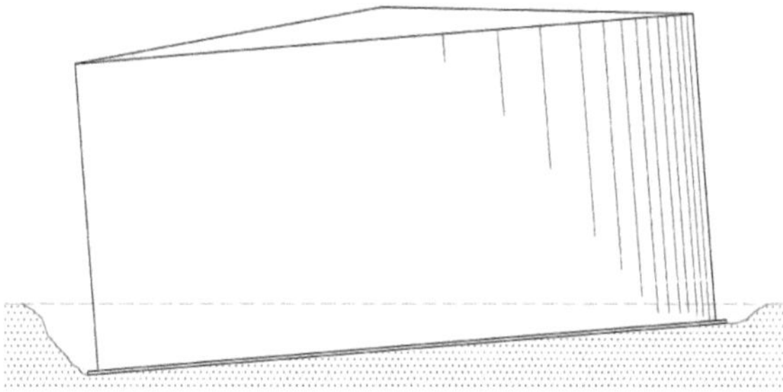

Fig. 2. Rigid body or planar tilting

While API 620 does not stipulate limits to planar tilting of steel tanks, API 625 allows variations from the settlement limits that it specifies provided that they are accounted for in the design of the tank system and interconnecting components, which in the authors' opinion, is a very rational approach that should be considered as good practice. API comments that while large tanks may be able to accommodate significant tilting without damage, other components usually require lower values of tilt, and thus limits its predicted value to Eq. 1:

$$GT < 25.4\alpha \frac{D}{H} \tag{1}$$

GT = global (rigid body) tilt settlement, in mm

α = 5, but often restricted to lesser values as a result of the same issues that are of concern for uniform settlement.

D = tank diameter

H = tank height.

ACI 376 limits differential settlement or planar tilting to 1/500.

Out of Plane Settlement

Dishing of the tank bottom, which is shown in Fig. 3 is the most commonly envisaged form of out-of-plane settlement. However, non-planar settlements do not necessarily

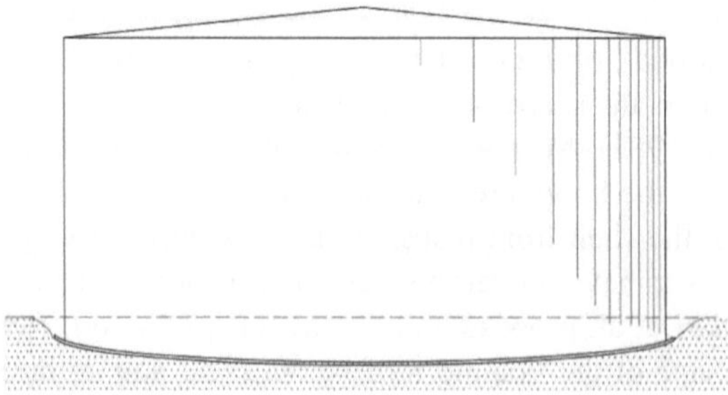

Fig. 3. Out of plane settlement in the form of tank bottom dishing

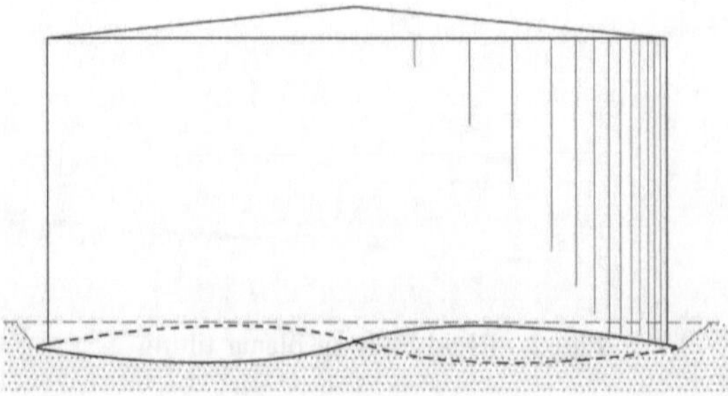

Fig. 4. Out of plane settlement in the form of warping

have to be from the center of the tank to the shell and can be localized at the connection of the shell to the bottom plate in the form of a localized bulge or warping, which is shown in Fig. 4.

Out-of-plane, non-planar or differential settlement may radially distort or overstress the shell, which can result in the malfunction of a floating roof. Overstressing can also rupture the shell. Dish-shaped settlements and localized depressions in the bottom plates of tanks will overstress the plate and welds. Alternatively, the shell-bottom plate connection may rupture when the shell bridges over a soft spot, but the bottom plate settles with it (Marr et al. 1982).

API 620 does specify any limits for non-planar settlement of steel tanks. However, API 625 limits predicted differential settlements between the edge of the tank and the center (*DS*), and around the periphery of the tank (*TS*) respectively to Eqs. 2 and 3.

$$DS < R/240 \tag{2}$$

R = tank radius

$$TS < 1/1000 \tag{3}$$

API 653 seems to be more tolerant on differential settlement of existing tanks. Based on the publication of Marr et al. (1982) this standard uses Eq. 4 to calculate the maximum permissible out of plane deflection of the shell:

$$|S| \leq \frac{11L^2Y}{2EH} \tag{4}$$

S = deflection
L = arc length between two points
Y = yield strength
E = Young's modulus.

The units of L, S, and Y are the same as are the units of E and Y.
API 653 limits the permissible bulge or depression of the tank bottom to:

$$B_B \leq 30.8R \tag{5}$$

B_B = maximum height of bulge or depth of local depression, in mm
R = radius of inscribed circle in bulged area or local depression, in m.

ACI 376 limits dishing along a radial line from the outer perimeter to the tank center to 10 mm drop of 1/300, and notes that restricting the dishing settlement to 1/300 maintains the bending curvatures within acceptable limits so that insulation materials are not damaged. Specialists have commented on the 10 mm criterion as being unclear and too conservative. ACI Committee's response is that it is agreed that 10 mm criterion should be deleted (ACI Committee 376 2010).

ACI also limits footing settlement around the perimeter of the tank to the lesser of 1/500 and the maximum settlement limit calculated for the uniform (planar) tilting of the tank.

3 Case Histories of Dynamic Compaction Application to Tanks in the Middle East

3.1 LNG Tanks in Ras Laffan, Qatar

Introduction and Ground Conditions
LNG Tanks T-4, T5, and T-6 are located within the reclaimed port area of Ras Laffan in Qatar, and the foundations of all three have been treated using dynamic compaction. Although there were some differences in the specifications and construction procedures of these tanks, as the ground improvement methodology of all three were very similar, only the process implemented during the ground improvement works of Tank T-6 (Hamidi and Jullienne 2007) will be reviewed.

Tank T-6's capacity is 140,000 m^3, and its inner steel tank diameter and height are respectively 74.3 m and 35 m. The outer concrete shell has an internal diameter of 76.3 m. The LNG storage level is 34 m, and the maximum hydrostatic test level was specified to be 19.98 m.

The tank's site was fairly flat. Based on boreholes that were as deep as 80 m, cross-hole tests to depths of 45 m, cone penetration tests (CPT) within the soil layers, Menard Pressuremeter Tests (PMT), and laboratory tests on recovered samples, the ground at the tank's location was understood to consist of a superficial well graded fine to coarse calcareous sand layer that was mixed with angular gravel. This layer's fines content was less than 30%. At localized areas, the sand was slightly cemented at the base of the layer. Weathered limestone, approximately 3.5 m thick, was encountered at about 4 m below ground level, followed by 3 m of upper limestone, 12 m of lower limestone, 30 m of calcareous siltstone, and at least 90 m of calcareous sandstone down to end of boring. Cavities were detected in the siltstone and sandstone layers, and groundwater was encountered approximately 2 m below ground level.

Acceptance criteria for the project were defined to be:

- Maximum settlement at the tank's perimeter: 80 mm
- Maximum deviation from any planar tilt of the foundation between any two points: 1/700 (25 mm differential settlement over any 17.5 m horizontal distance).

Using the specification's differential settlement criterion over a distance of 40 m between the center and edge of the tank measured from the concreting of the joint at the wall to the mat, it can be calculated that the center of the tank was able to settle an additional 57 mm or a total of 137 mm.

Hamidi and Jullienne (2007) have tabulated the design values for the thicknesses, densities, Young moduli, Poisson ratios, cohesion, and internal friction angle of each layer. Initial calculations suggested that the ground would not be able to satisfy the specifications, and ground improvement was considered as a viable solution of interest.

Foundation Solution: Dynamic Compaction with a Shear Ring Trench

The ground improvement works for satisfying the design basis were awarded to a specialist contractor who had proposed the application of dynamic compaction. As shown in Fig. 5, in this scheme a shear ring trench was constructed by excavating a

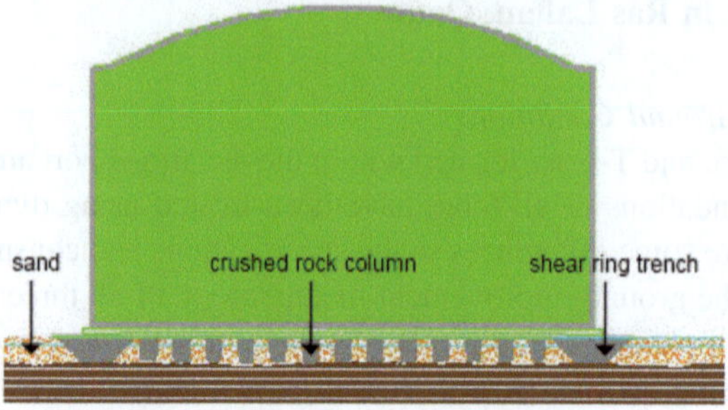

Fig. 5. Schematic cross section of tank and ground treatment scheme

3.5 m deep trench around the tank and simultaneously backfilling it with crushed stone. The trench had a trapezoidal cross-section, with the top and bottom bases being respectively 12 m and 5 m.

The crushed stone used was up to 200 mm in diameter, the Los Angles abrasion of aggregate coefficient was less than 35, and fines content was less than 5%.

Similar to dynamic replacement that is typically performed by backfilling the impact craters with crushed stone and applying additional blows to enlarge, deepen and compact the granular inclusions, in this project the prints were initially excavated and backfilled with crushed stone to provide larger granular columns with better mechanical properties than compacted sand. The trench was then dynamically compacted to increase local bearing capacity, resistance against local shear failure under the tank's shell and to reduce edge settlements.

Testing and Verification

Analysis of settlements was undertaken using finite element analysis software with consideration of the tank's construction methodology; i.e. the construction of the perimeter ring beam and tank wall to approximately 38 m height, construction of the central portion of the mat, construction of the joint between the ring beam and the mat, construction of the steel cupola, construction of the inner tank, construction of the outer tank wall haunch and roof, hydro testing, and operation of the tank.

It was calculated that the maximum settlement of the tank would occur in its center during the hydro test (57.5 mm), but the maximum differential settlement (0.83/700) would take place after construction of the ring beam and tank wall and prior to the placement of concrete at the joint.

Field tests were performed during calibration of dynamic compaction parameters and after treatment. PMTs were carried out at 25 locations, of which 3 were during the calibration phase and the remaining were after the treatment. The tests were performed at every one meter before reaching bedrock. 14 plate load tests were also carried out on circular plates with diameters of 1 m, of which 2 were performed during the calibration phase.

Testing and interpretation of the results indicated that the average moduli of deformation in the dynamic replacement columns within the tank, compacted sand, and ring trench were respectively 75.2 MPa, 36.9 MPa, and 70.1 MPa.

While maximum vertical settlement at the end of construction was calculated to be 31.7 mm under the tank wall, surveying of 18 plates installed on the wall indicated that measured total settlement during the hydro test were from 4 to 10 mm, and differential settlement between any two adjacent points was in the range of null to 3 mm.

3.2 STP Tanks in Palm Jumeira, Dubai

Introduction and Ground Conditions

A sewerage treatment plant (STP) tank has been built on each of the two outer crescent islands of Palm Jumeira in Dubai, UAE. Due to the similarity of the ground improvement technique applied at these two sites, only the first plant (Lot A-A) will be reviewed in this paper; however, Hamidi et al. (2011) and Hamidi and Varaksin (2015) have described the ground improvement works of both tanks (Lots A-A and G-G) in detail.

Palm Jumeira is a reclaimed group of man-made islands that consists of a trunk, a crown with 17 fronds, three surrounding crescent islands that form a breakwater and two smaller islands that are in the shape of the logo of project's developer. The main body of the development consists of calcareous sand that was dredged from the Persian Gulf. When possible, sand was discharged from the bottom of the hulls of trailing suction hopper dredgers, but when the water was shallow, the dredgers sprayed the sand and water mixture onto the reclamation by rainbow discharge.

Due to the low strength and high compressibility of the soil, ground improvement by vibro compaction was carried out almost throughout the reclamation. Additionally, piles were also installed to support the heavily loaded structures. The two STP tanks are the only heavily loaded structures of the development that are only found on ground that has been improved by dynamic compaction and dynamic surcharging.

The STP lots are located on the tips of the outer crescents. Each plant includes one reinforced concrete tank with a diameter of 35.1 m, which was envisaged to be subjected to a total uniform load of 120 kPa.

As part of the general geotechnical investigation of the reclamation, two SPT (standard penetration test) boreholes and two CPTs (cone penetration test) were performed near the location of Tank A-A. The boreholes indicated that the upper crust of the soil was generally very dense with SPT blow counts, N, reaching 28. The deeper layers of soil were less dense, with minimum N in the upper 8 m of soil being as low as 5. The soil then appeared to become denser with a minimum N of 18 and exceeding 50 at the depth of 13 m. The fines content of the soil in these two boreholes were from 2 to 10% in the upper 13 m of soil but increased to 22% at the depth of 14.5 m. Ground level was at +4 m RL (reduced level) and groundwater was at the depth of about 3 m. CPT cone resistance, q_c, in the upper 2 m of sand was as high as 25 MPa. The soil then became loose with q_c dropping to as low as 3 to 4 MPa to the depth of about 13 m where refusal was achieved. Although these tests suggested that the soil was clean sand, fines content as high as 30% was observed in a number of boreholes that were not very far from the project.

At later stages, four additional SPT boreholes were drilled in Lot A-A's tank location. These boreholes indicated that the upper 3 m of sand was indeed very dense, but the soil then became very loose to medium dense at water level. N values at depths of 3 to 8 m ranged from 4 to 14, and then shifted to values in between 11 to 20 to depths of approximately 12 to 13 m. N then exceeded 50. The average fines content of the 38 samples that were extracted from the four boreholes generally ranged from 16 to 21% and up to 30%, which was significantly more than the average 5% that was indicated by the preliminary geotechnical investigation.

Two PMTs were also carried out in Lot A-A. Testing commenced below sea level. Limit pressure, P_{LM}, was from less than 100 to about 700 kPa. Pressuremeter Moduli, E_M, varied from less than 1 MPa to approximately 6 MPa.

Preliminary studies suggested that mat foundations could not be used for the tanks without the implementation of specific foundation measures.

Foundation Solution: Dynamic Compaction and Dynamic Surcharging
Piling and ground improvement were both deemed as possible solutions; however, the former solution was discarded due to the high costs associated with it.

Several ground improvement techniques were considered. Although vibro compaction was the commonly practiced method of ground improvement on Palm Jumeira, the possible presence of silty sand made the applicability of this technique dubious. Stone columns were feasible but appeared to be very costly. Dynamic compaction was considered as both economical and reliable, and the ground improvement works were awarded to a specialist contractor who had proposed this technique.

Soil improvement acceptance criteria for a tank foundation level at level +2.5 m RL were developed as:

- Allowable bearing capacity: 160 kPa with a safety factor of 3
- Differential settlement: 1/750 for a uniformly distributed load of 120 kPa.

A study on the available rigs in the region at the time indicated that the maximum pounder lift capacity was limited to 15 tons; thus, in addition to classical dynamic compaction, the specialist contractor decided to implement dynamic surcharging to improve the achievable results at depth.

Dynamic surcharging is the combined effect of static loading and high energy impacts. Its purpose was to create acceleration in the soil under the static loading in such a way to generate a shearing process around the surcharge fill to reduce the load spreading zone of the high strength superficial layer. Furthermore, vibrations and increase of the pore pressure under the tank were to reduce the friction in between the granular soil particles and ultimately result in the collapse and densification of the foundation soil under the influence of dynamic surcharging.

In this technique a surcharge is initially placed over the treatment area, then dynamic compaction is performed. Although granular materials settle under static loads, as dynamic shear modulus has been found to decrease significantly with increasing values of shear strain amplitude (Silver and Seed 1971), it can be expected that introducing vibration will increase the amount of settlement under the surcharge. Furthermore, the rate of consolidation of fine soils is most when the pore water pressure is high, and it is possible to increase the rate of consolidation back to previously high values by inducing pore water pressure through vibration.

Four settlement monitoring plates were installed on the ground surface. One plate was installed in the center of the tank and the others were installed at 120° angles on a ring whose radius was 17.5 m.

4 m of surcharge was then placed at the tank location in the shape of a truncated cone whose top and bottom diameters were respectively 32.2 m and 44.2 m. It can be calculated that the surcharge volume was approximately 4,700 m^3. Assuming that the in-situ unit weight of surcharge was 17 kN/m^3, it can be further estimated that its total weight was 80 MN. As tank foundation level was at +2.5 m RL, an additional overburden weight of 25 MN can be added to total surcharge weight. 4 m of surcharge and 1.5 m of overburden generates 93.5 kPa of pressure within the tank's layout, which was equivalent to 78% of the tank's total load; however, the total surcharge load of 105 MN approximately equated to 90% of the tank's total load.

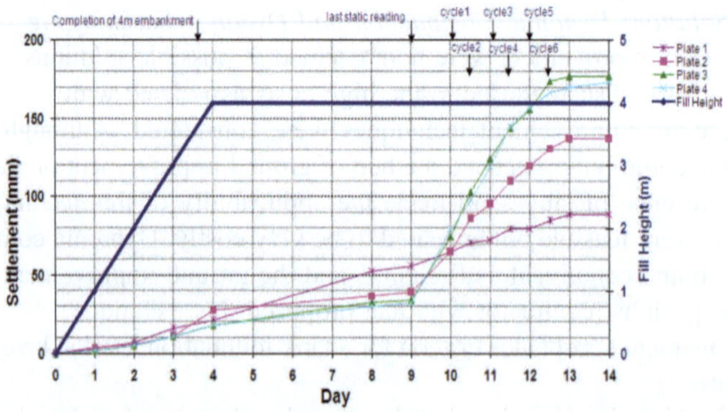

Fig. 6. Ground settlement in tank A-A during static and dynamic surcharging (Hamidi and Varaksin 2015)

Tank A-A settlements during placement of surcharge, static loading and dynamic surcharging are shown in Fig. 6. It can be seen that the plates settled almost prorate with surcharge height. Once placement of surcharge was completed, the preload was left in place for five additional days. By then the settlement rate considerably decreased, and it can be extrapolated that the ground could have additionally settled up to 5 mm in the long term.

To increase the depth of treatment influence, 26 prints were excavated to the depth of approximately 1 m on a periphery ring with a diameter of 49.2 m before applying dynamic surcharging by dropping a 15-ton pounder from 20 m a total of 30 times in five cycles at the location of each print.

It can be observed in Fig. 6 that dynamic surcharging increased the settlements of Tank A-A by 1.6 to 5.2 times the values of the static settlements. The maximum effect was on the tank's periphery where impact wave amplitudes were greatest. Also, although the maximum differential settlement of the outer monitoring plates was only 7 mm at the end of static surcharging, maximum differential settlement during dynamic surcharging increased by 4 times to 40 mm, which suggests the possibility of large differential settlements under seismic and vibratory loads in untreated areas of the same ground.

The surcharge was removed and the ground was excavated within a diameter of 41.2 m to working platform level at +2.8 m RL upon completion of dynamic surcharging. The excavation sides sloped outwards in such a way that the top of excavation had a diameter of 46 m.

Dynamic compaction was performed on prints located in the center of the tank and on 4 concentric rings around the central print. Each print was additionally excavated by approximately 1 m to facilitate pounder penetration and to increase the depth of influence. 150 m^3 of crushed rock and cobbles were also added to the total of 58 prints in each tank. This amounts to about 2.6 m^3 of added rock per print or an equivalent of approximately 0.13 m of rock thickness per every meter of ground within the treatment zone. This amount of stone was insufficient to efficiently increase the ground strength as was intended and described in Ras Laffan Tank T-6 but was rather used to increase soil permeability. Pounder drop height during the first and second phases of compaction was 20 m. A lesser impact energy was used during the ironing phase.

Pounder penetrations and top of crater diameters were measured during the first two phases of dynamic compaction for each print. The average values of these two in phase 1 were respectively 1.7 m and 5 m. These average figures respectively reduced to 0.4 m and 2.3 m in the second phase. It is noted that the pounder's dimension was 1.7 m.

The ground level dropped to +2.25 m RL at the end of dynamic compaction. With consideration of the added stone volume, it can be calculated that the ground had settled 0.68 m in addition to the settlements induced by dynamic surcharging. Although the magnitude of this settlement is significantly greater than what was measured during dynamic surcharging, it is noted that the purpose of dynamic surcharging was to reduce the settlements of the deep layer that were less impacted by the energy of the 15-ton pounder.

Testing and Verification

Upon completion of dynamic compaction and leveling of the site, 4 PMTs were carried out in Tank A-A. P_{LM} and E_M before and after ground improvement are shown in Fig. 7. It can be observed that most of the improvement has occurred to depths of approximately 8 to 9 m; however, due to the combination of dynamic surcharging and pre-excavated dynamic compaction, P_{LM} increased by 380% and 70% respectively at depths of 5 m and 10 m. Furthermore, minimum P_{LM} after improvement was observed to be greater than 600 kPa, which demonstrates that the young hydraulic fill was no longer subject to creep due to self-weight (Menard 1975).

Bearing capacity can be calculated using the equation proposed by Menard (1975). In Tank A-A, the geometric mean of the average of the 4 post ground improvement PMT was 1,684 kPa. Conservatively assuming that the deeper layers also have the same value and that the foundation was on ground level, the allowable bearing capacity can be calculated to be 449 kPa, which is considerably greater than the required 160 kPa.

Tank settlements were calculated by taking the tank-soil interaction into account and using finite element analysis and three-dimensional modeling. The modeled tank was supported by a 0.5 m thick concrete raft, which was underlain by a very dense upper layer and a lesser dense bottom layer. The ground in the model was made stiffer on one side by varying the thicknesses of two soil layers to assess differential settlement effects. In this procedure, the thicknesses of the upper and lower sand layers on the left side of the tank were respectively assumed to be 7.5 m and 3.5 m. On the right side, the upper and lower sand layers were each assumed to be 6.5 m. Young's moduli, E_y, for the sand layers were calculated from E_M (Menard 1975), and assumed to be 57.9 and 22.5 MPa respectively for the upper and lower sand layers.

Finite element analysis suggested that maximum settlement at the center of the tank and minimum settlement at the shell were respectively 21.35 mm and 10.91 mm. Thus, the differential settlement over the radius length of 17.55 m was 10.44 mm or less than 1/1,681, which is substantially less than the allowed value of 1/750. Differential settlement from one side to the other side of the tank was calculated to be 3.13 mm or less than 1/11,200.

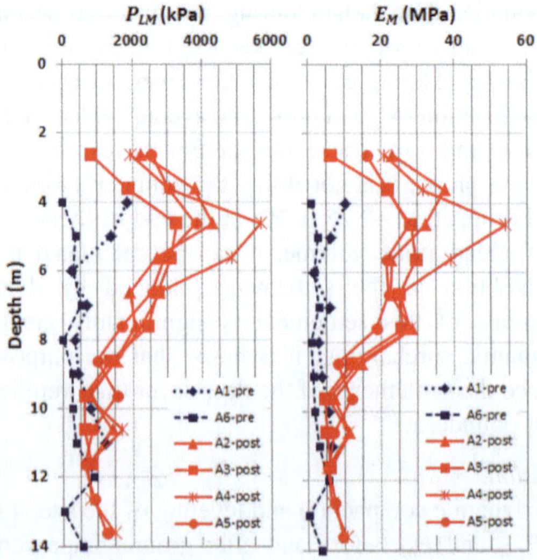

Fig. 7. P_{LM} and E_M values before and after ground improvement (Hamidi and Varaksin 2015)

3.3 Fuel Tanks in Sohar, Oman

Introduction and Ground Conditions

Sohar Aluminum Factory is located in Sohar, Oman, and includes a power plant that contains 4 gas turbines, 2 steam turbines, heavy recovery generators, associated buildings, and tanks. Phase one of the power plant includes 2 fuel tanks, and a third tank has been envisaged for the second phase of the plant. The diameters of these steel tanks were approximately 31.25 m, and the required bearing capacity was 200 kPa.

Initial geotechnical studies indicated that the ground at the location of the tanks was composed of loose to dense silty sand from the natural ground level at +5.35 m CD (Chart Datum) to −13 m CD, where very dense silty sand or sandstone was encountered. Groundwater level was identified at +1.2 m CD.

Supplementary PMTs that were later carried out indicated that the upper crust of the ground surface was very dense with P_{LM} of 5 MPa. The limit pressure then varied in the range of approximately 1.5 to 2.1 MPa.

It was also assessed that the in-situ ground conditions did not satisfy the designer's requirements, and implementation of specific foundation measures was investigated.

Foundation Solution: Dynamic Compaction

Acceptance criterion that was specified required the allowable bearing capacity of the tank foundations to be 200 kPa with a safety factor of 3. Tank base levels were defined at +4.5 m CD. It is noted that it is the authors' experience that due to their foundation sizes, tanks practically do not fail in bearing and that local shear failure beneath the shell, disruption of tank serviceability or structural damage due to ground deformations are more probable problems that need to be considered.

Prior to commencement of treatment the working platform level was cut down to elevation +4.5 m CD.

A 14.5-ton pounder was used to dynamically compact the ground. Pounder drop height in the first and second phases of ground improvement was 20 m but was reduced to 10 m for the ironing phase. 10 and 8 blows were applied per print respectively in phase 1 and phase 2, but only a single blow was applied to each ironing print.

Testing and Verification
Pressuremeter test results before and after dynamic compaction are shown in Fig. 8. The reference level for depth is at +4.5 m CD. It can be observed that even though the strength and deformation parameters at the tank locations was on the denser side of materials, nevertheless P_{LM} and E_M respectively improved by approximately 1.6 and 1.9 times in the 4 m of soil beneath the hard upper crust.

Calculation of allowable bearing capacity using the method of Menard (1975) indicates that the project's criterion has been well satisfied. Also, it can be calculated that settlements within the treated layers of the ground will have reduced to approximately half of the untreated values.

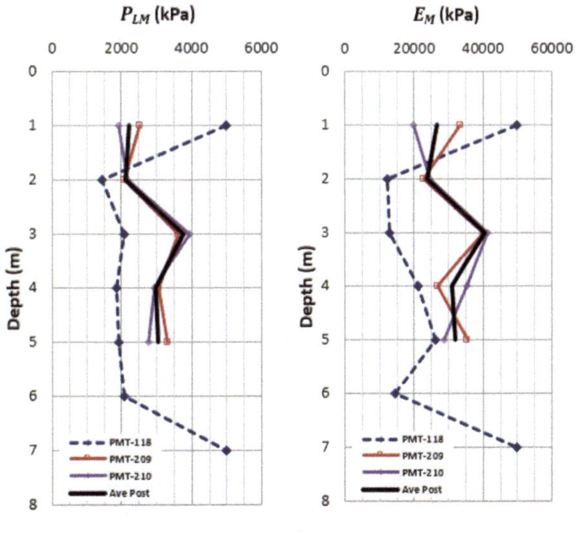

Fig. 8. P_{LM} and E_M values before and after ground improvement

4 Conclusion

Tanks undergo a combination of uniform settlements, planar tilting and non-planar differential settlements that can undermine the performance and serviceability of tanks, result in damages or ultimately lead to failure. While they are relatively thin, heavily loaded and sensitive structures, international specifications appear to be relatively tolerant to uniform settlements but restrict planar tilting and non-planar deformations more stringently.

Ground improvement techniques have been used successfully in numerous tank projects throughout the world in general and in the Middle East in particular. Several successful cases of ground improvement by application of dynamic compaction and verification by Menard pressuremeter test for various types of tanks in the region were presented in this paper.

Acknowledgments. The authors would like to express their appreciation to Menard for providing the information used in this paper.

References

ACI Committee 376: Agenda for Webinars on November 30, December 2 and 16, 2010, p. 23 (2010)

American Concrete Institute: Code Requirements for Design and Construction of Concrete Structures for the Containment of Refrigerated Liquefied Gases, p. 170. ACI, Farmington Hills (2010)

American Petroleum Institute: API Standard 653: Tank Inspection, Repair, Alteration, and Reconstruction, 3rd edn., Washington, D.C., p. 112 (2003)

American Petroleum Institute: API Standard 625: Tank Systems for Refrigerated Liquefied Gas Storage, 1st edn., Washington, D.C., p. 72 (2010)

American Petroleum Institute: API Standard 650: Welded Tanks for Oil Storage, 11th edn., Washington, D.C., p. 445 (2011)

Buschmeier, B., Masse, F., Swift, S., Walker, M.: Full scale instrumented load test for support of oil tanks on deep soft clay deposits in Louisiana using controlled modulus columns. In: International Symposium on Ground Improvement (IS-GI) Brussels 2012, Brussels, pp. 359–372 (2012)

Cognon, J.M., Liausu, P., Vialard, R.: Combination of the drains and surcharge method with dynamic compaction. In: 8th European Conference on Soil Mechanics and Foundation Engineering, Helsinki, pp. 219–224. A.A. Balkema, Rotterdam, Netherlands (1983)

Debats, J.M., Scharff, G., Balderas, J., Melentijevic, S.: Ground improvement efficiency and back-analysis of settlements. Ground Improv. **166**(3), 138–154 (2013)

D'Orazio, T.B., Duncan, J.M.: Differential settlement in steel tanks. J. Geotech. Eng. ASCE **113**(9), 967–983 (1987)

Hamidi, B., Jullienne, D.: The construction and performance of foundation using dynamic compaction and dynamic replacement for an LNG tank in Qatar. In: 16th South East Asian Geotechnical Conference, pp. 523–527 (2007)

Hamidi, B., Masse, F., Racinais, J., Varaksin, S.: The boundary between deep foundations and ground improvement. Geotech. Eng. **169**(GE2), 201–213 (2016)

Hamidi, B., Varaksin, S., Nikraz, H.: Application of dynamic surcharging for construction of tanks on reclaimed ground. In: International Conference on Advances in Geotechnical Engineering (ICAGE), Perth, pp. 873–878 (2011)

Hamidi, B., Varaksin, S.: Dynamic compaction and dynamic surcharging at Dubai's Palm Jumeira STP. In: Ground Improvement: Case Histories and New Directions, pp. 309–335. Elsevier, Oxford (2015)

Hendy, M.S., Muir, I.C.: Experience of dynamic replacement on a 40 m deep reclamation in Hong Kong. In: Third International Conference on Ground Improvement Geosystems: Ground Improvement Geosystems - Densification and Reinforcement, pp. 76–80. Thomas Telford, London (1997)

Ihm, C.W., Masse, F.: Successful application of Menard vacuum consolidation method to Nakdong river soft clay in Kimhae, South Korea. In: 15th International Conference on Soil Mechanics and Foundation Engineering (ICSMGE), Istanbul (2001)

Marr, W.A., Ramos, J.A., Lambe, T.W.: Criteria for settlement of tanks. J. Geotech. Geoenviron. Eng. **108**(GT8), 1017–1039 (1982). ASCE

Menard, L.: The Menard Pressuremeter, Sols Soils No. 26 (1975)

Mobil: Chapter 6 - Foundation Design. Mobil Engineering Guide: EGS 262-90. Mobil (1990)

Silver, M.L., Seed, H.B.: Deformation characteristics of sands under cyclic loading. J. Soil Mech. Found. Div. **9**(SM8), 1081–1098 (1971). ASCE

Default k-Values for Estimating Resilient Modulus of Coarse-Grained Nigerian Subgrade Soils

Abdulfatai Adinoyi Murana(✉)

Department of Civil Engineering, Ahmadu Bello University, Zaria, Kaduna, Nigeria
fatinoyi2007@gmail.com

Abstract. In this paper subgrade materials from different locations in Nigeria were characterized for use in the Mechanistic-Empirical Pavement Analysis and Design. The engineering properties of the coarse-grained Nigerian subgrade soils materials were obtained in the laboratory. Seven selected resilient modulus constitutive equation for estimating the resilient modulus of coarse-grained subgrade soils were used to estimate the default values using the repeated load triaxial test result conducted on coarse-grained Nigerian subgrade soils. These default resilient modulus parameters developed can be used to estimate the resilient modulus of the compacted subgrade soils with reasonable accuracy and utilized as level 3 resilient modulus input for Mechanistic-Empirical Pavement Analysis and Design.

Keywords: Coarse-grained soils · Resilient modulus · Mechanistic-Empirical · Flexible pavement · Subgrade soils

1 Introduction

A conventional flexible pavement consists of a prepared subgrade or foundation and layers of sub-base, base and surface courses (AASHTO 1993).

For the roadbed soils, the seasonal variation of resilient moduli is considered and used directly to determine the design or effective roadbed soil resilient modulus. However, seasonal variation of the resilient moduli for pavement materials is not used or considered in the design process, even though the resilient modulus of pavement materials can vary substantially throughout the year (Von Quintus and Killingsworth 1997).

The design and evaluation of pavement structures on base and subgrade soils requires a significant amount of supporting data such as traffic loading characteristics, base, subbase and subgrade material properties, environmental conditions and construction procedures. Characterization of pavement materials is a key requirement for the pavement design process. The characterization task involves obtaining material properties that identify the material response to external stimuli of traffic loading and environmental conditions. Characterization of subgrade materials using resilient modulus involves obtaining material properties (index properties, physical and compaction properties) that identify the material response to external stimuli of traffic loading and environmental

W. Frikha et al. (eds.), *Soil Testing, Soil Stability and Ground Improvement*,
Sustainable Civil Infrastructures, DOI 10.1007/978-3-319-61902-6_17

conditions. In its 2002 design guide, the AASHTO advocated the use of the resilient modulus parameter for describing granular material behaviour. Laboratory techniques provide a means for directly measuring the resilient modulus parameter, wherein the process usually involves elaborate and extensive testing at various stress levels and physical conditions to completely map the range of the resilient modulus parameter for any material under consideration. To be able to adopt the Mechanistic-Empirical pavement design method for use in Nigeria, calibration of the subgrade material models to reflect Nigerian conditions need to be carried out.

A more widely used recent test method on which pavement designs are based is the resilient modulus value. It is defined as the ratio between repeated deviator stress and resilient strain. The laboratory testing procedures for determining the resilient modulus values is time consuming and needs expensive equipment and highly trained personnel (Vogrig et al. 2003).

The resilient modulus is a fundamental engineering material property that describes the non-linear stress-strain behaviour of pavement materials under repeated loading. It is defined as the ratio of the maximum cyclic stress to the recoverable resilient (elastic) strain in a repeated dynamic loading (Mohammad et al. 2007).

It is a measure or estimate of the elastic modulus of the material at a given stress or temperature. Mathematically it is expressed as the ratio of applied deviator stress to recoverable strain (George 2004).

In Level 1 design/analysis, the MEPDG requires input of the regression constants of the stress-dependent constitutive equation for resilient modulus of a particular unbound material (subgrade soil or base aggregate). This ensures a more accurate assessment of the modulus during the analysis over the design period including seasonal variation and varying stress conditions. Constitutive equation coefficients (k-values) are usually obtained from the regression analysis of resilient modulus test data for an actual soil/aggregate sample (Hossain 2010).

Some agencies consider the cost, time, complication, and sampling resolution required for meaningful resilient modulus testing to be too cumbersome for its application in less critical projects. Regardless of project size, it is often difficult to predict and consequently reproduce the in-situ conditions, usually with respect to the state of stress, further complicating the use of resilient modulus testing. Because of this, correlations are desired for estimating resilient modulus, especially for use (or verification of default values) associated with MEPDG Level 2 design/analysis. A common method to predict a resilient modulus value is to use the stress-dependent constitutive equation with the k-values estimated from soil index properties through further regression equations. MEPDG Level 3 design/analysis also requires a specific resilient modulus value as input (Hossain 2010).

2 Resilient Modulus Constitutive Models

Mathematical models are generally used to express the resilient modulus of subgrade soils such as the bulk stress model and the deviatoric stress model. These models were utilized to correlate resilient modulus with stresses and fundamental soil properties.

A valid resilient modulus model should represent and address most factors that affect the resilient modulus of subgrade soils (Titi et al. 2006).

Several other models were reported in the literature, which use both stresses (either confining and deviatoric stresses or bulk or octahedral stresses) that are functions of confining and deviatoric stresses. The most general form of a three-parameter model is as shown in Eq. 1 (Ooi et al. 2006; NCHRP 2008):

$$M_R = k_1 P_a [f(c)]^{k_2} [g(s)]^{k_3} \tag{1}$$

where $f(c)$ is a function of confinement; $g(s)$ is a function of shear and k_1, k_2, and k_3 are constants.

The effects of confinement in these models can be expressed in terms of the minor principal stress (σ_3), bulk stress (θ), or octahedral stress ($\sigma_{oct} = \theta/3$), while the parameter options for modelling the effects of shear include the deviatoric stress or octahedral shear stress (τ_{oct}). The three-parameter models represented by the Eq. 1 are more versatile and apply to all soils (NCHRP 2008).

Uzan (1985) studied and discussed different existing models for estimating resilient modulus. The Uzan equation was developed as a combination of the bulk and deviator stress models in an effort to improve the predicted response of M_r test results by including both axial and shear effects. The model defined the resilient modulus as shown in Eq. 2 (Uzan 1985; NCHRP 2008):

$$M_R = k_1 P_a \left(\frac{\theta}{P_a}\right)^{k_2} \left(\frac{\sigma_d}{P_a}\right)^{k_3} \tag{2}$$

where k_1, k_2, and k_3 are material constants; θ = bulk stress; σ_d = deviatoric stress; P_a is the atmospheric pressure

An equation similar to Uzan's model using the octahedral shear stress instead of the deviator stress was developed by Witczak and Uzan as shown in Eq. 3 (Witczak and Uzan 1988; NCHRP 2008):

$$M_R = k_1 P_a \left(\frac{\theta}{P_a}\right)^{k_2} \left(\frac{\tau_{oct}}{P_a}\right)^{k_3} \tag{3}$$

where θ = bulk stress ($\sigma_1 + \sigma_2 + \sigma_3$); σ_1 = major principal stress

σ_2 = intermediate principal stress = σ_3 for M_R test on cylindrical specimen.

σ_3 = minor principal stress/confining pressure; τ_{oct} = Octahedral shear stress.

$$\tau_{oct} = \frac{1}{3}\left((\sigma_1 - \sigma_2)^2 (\sigma_2 - \sigma_3)^2 (\sigma_3 - \sigma_1)^2\right)^{1/2}$$

An equation similar to Uzan's model using the confining pressure instead of the bulk stress was recommended by Pezo as shown in Eq. 4 (Pezo 1993; NCHRP 2008):

$$M_R = k_1 P_a \left(\frac{\sigma_3}{P_a}\right)^{k_2} \left(\frac{\sigma_d}{P_a}\right)^{k_3} \tag{4}$$

An equation similar to Pezo's model using the confining pressure and deviator stress in a three-parameter formulation was recommended by Ni et al. as shown in Eq. 5 (Ni et al. 2002; NCHRP 2008):

$$M_R = k_1 P_a \left(1 + \frac{\sigma_3}{P_a}\right)^{k_2} \left(1 + \frac{\sigma_d}{P_a}\right)^{k_3} \tag{5}$$

Ooi et al. slightly modified the equation recommended by Ni et al. using the bulk stress, octahedral shear stress and deviator stress in a three-parameter formulation into two models as shown in Eqs. 6 and 7 (Ooi et al. 2004; NCHRP 2008):

$$M_R = k_1 P_a \left(1 + \frac{\theta}{P_a}\right)^{k_2} \left(1 + \frac{\sigma_d}{P_a}\right)^{k_3} \tag{6}$$

$$M_R = k_1 P_a \left(1 + \frac{\theta}{P_a}\right)^{k_2} \left(1 + \frac{\tau_{oct}}{P_a}\right)^{k_3} \tag{7}$$

An equation similar to Ooi et al.'s model using the octahedral shear stress and bulk stress was recommended by the NCHRP project 1-28 A as shown in Eq. 8 (NCHRP 2008):

$$M_R = k_1 P_a \left(\frac{\theta}{P_a}\right)^{k_2} \left(1 + \frac{\tau_{oct}}{P_a}\right)^{k_3} \tag{8}$$

3 Evaluation of the Resilient Modulus Model Parameters for Coarse-Grained Soils

The resilient modulus of coarse-grained soils obtained in the laboratory were statistically analysed. These values were used in evaluating the M_r parameters of the coarse-grained soils using the seven resilient modulus equations presented in literature. Figures 1a, 1b, 1c, 2a, 2b, 2c, 3a, 3b, 3c, 4a, 4b, 4c, 5a, 5b, 5c, 6a, 6b, 6c, 7a, 7b and 7c presents the histogram of the resilient modulus parameters (k_i) values of coarse-grained soils obtained from the resilient modulus equations evaluated.

Figure 1a, 1b and 1c shows the histogram of resilient modulus parameters (k_i) values obtained from Uzan model.

Figures 2a, 2b and 2c present the histogram of the resilient modulus model

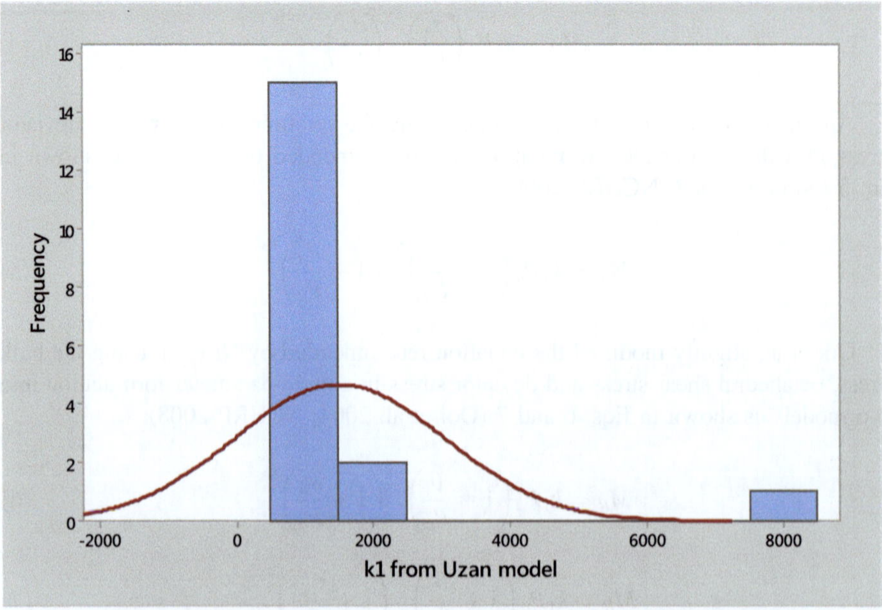

Fig. 1a. Histograms of Uzan resilient modulus model parameters k_1

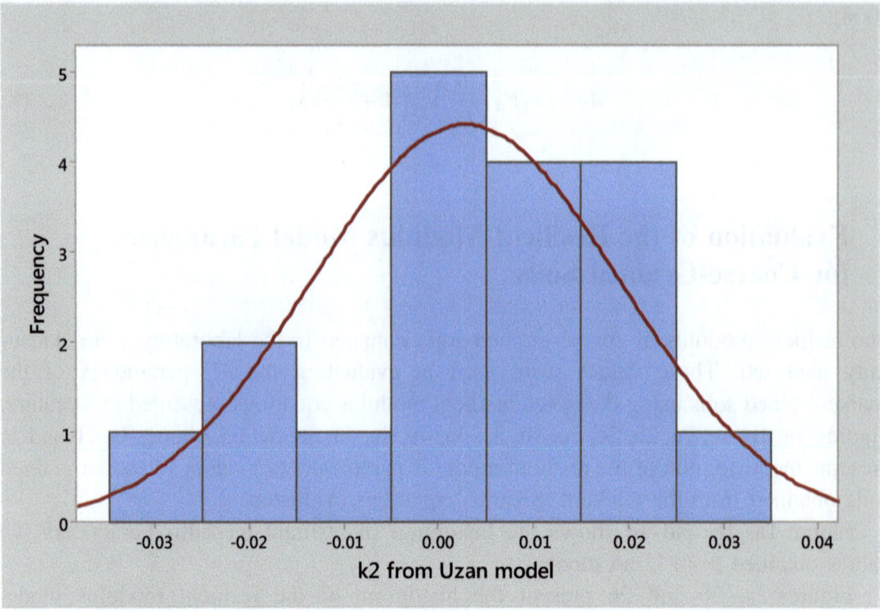

Fig. 1b. Histograms of Uzan resilient modulus model parameters k_2

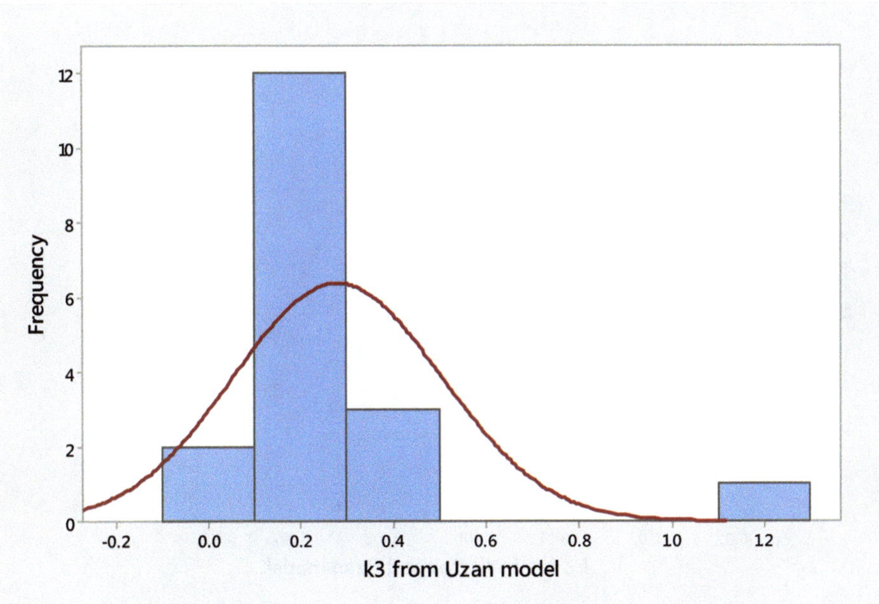

Fig. 1c. Histograms of Uzan's resilient modulus model parameters k_3

parameters k_i obtained using the Witczak and Uzan resilient modulus model.

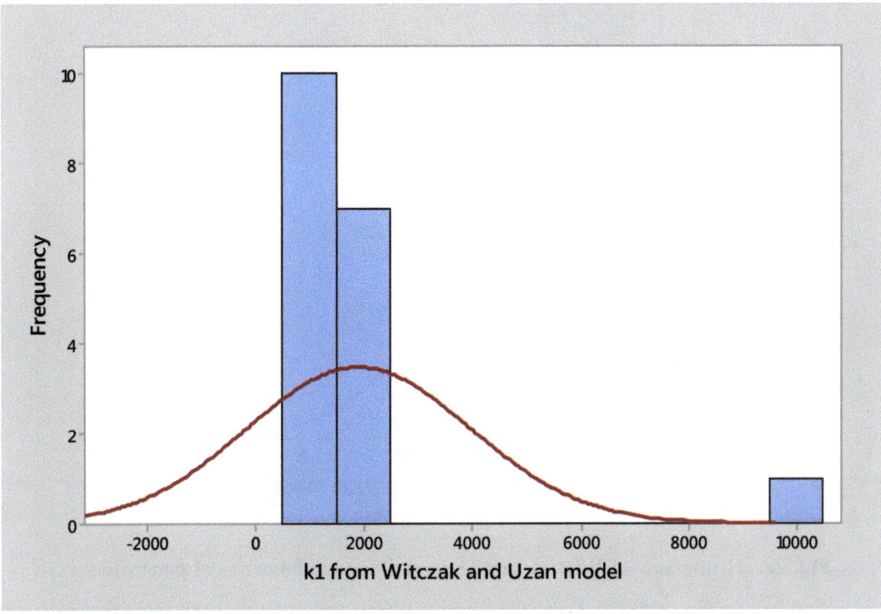

Fig. 2a. Histograms of Witczak and Uzan resilient modulus model parameters k_1

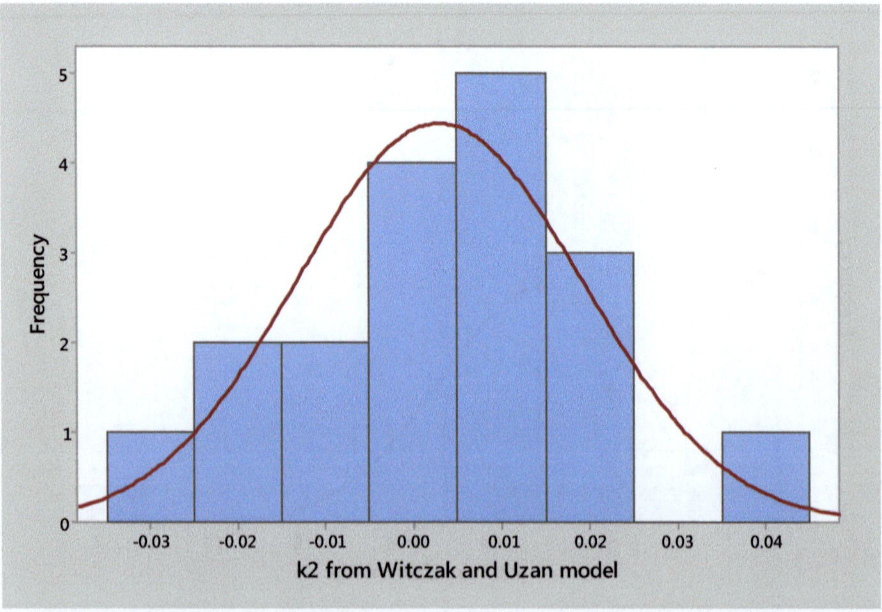

Fig. 2b. Histograms of Witczak and Uzan resilient modulus model parameters k_2

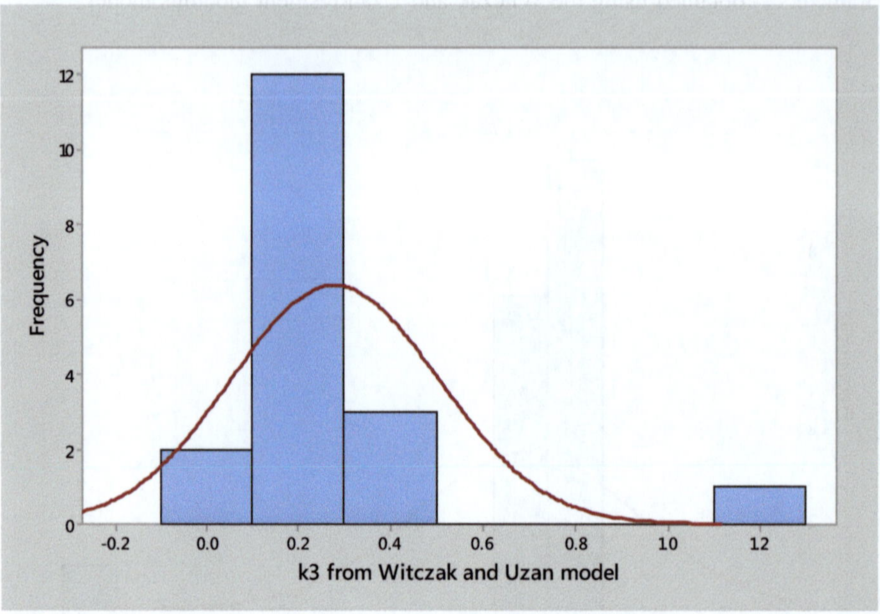

Fig. 2c. Histograms of Witczak and Uzan resilient modulus model parameters k_3

Figures 3a, 3b and 3c present the histogram of the resilient modulus model parameters k_i obtained using the Pezo resilient modulus model.

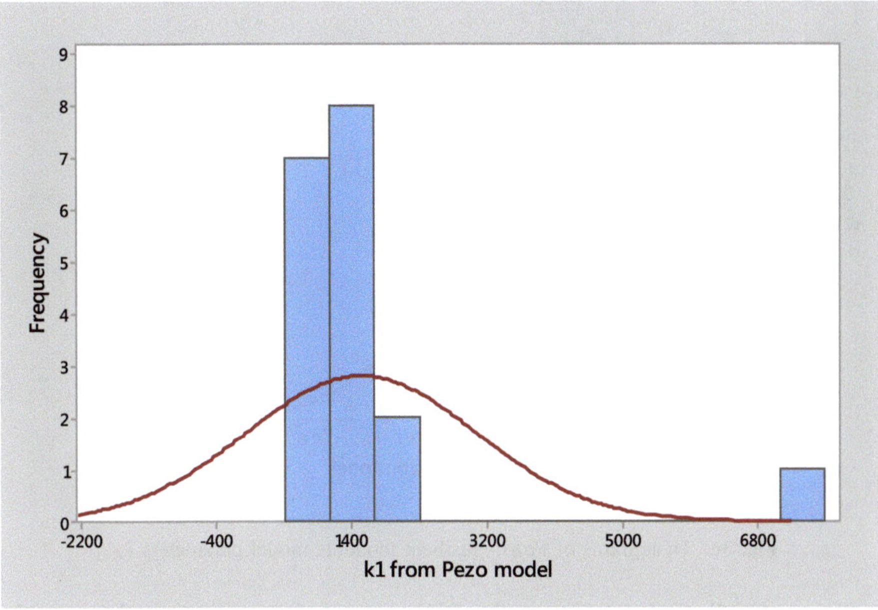

Fig. 3a. Histograms of Pezo's resilient modulus model parameters k_1

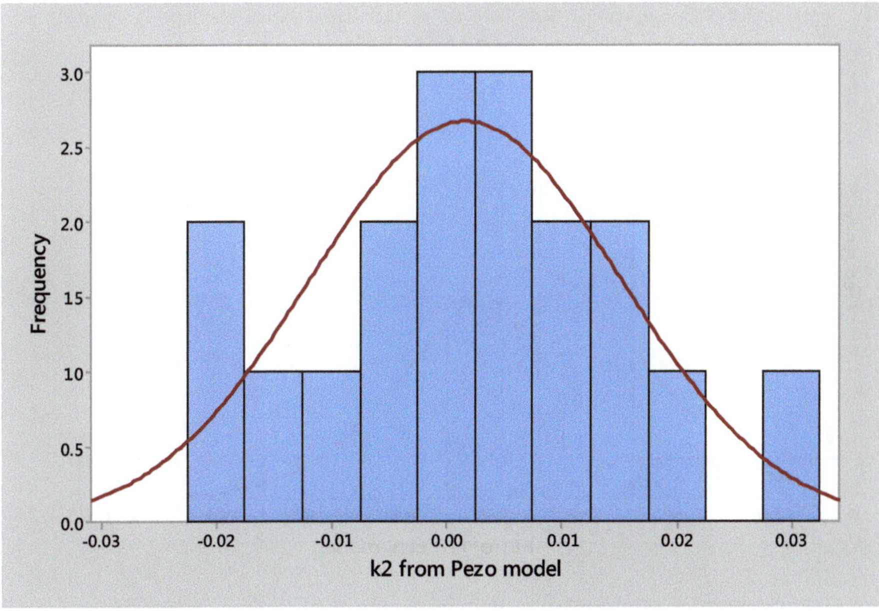

Fig. 3b. Histograms of Pezo's resilient modulus model parameters k_2

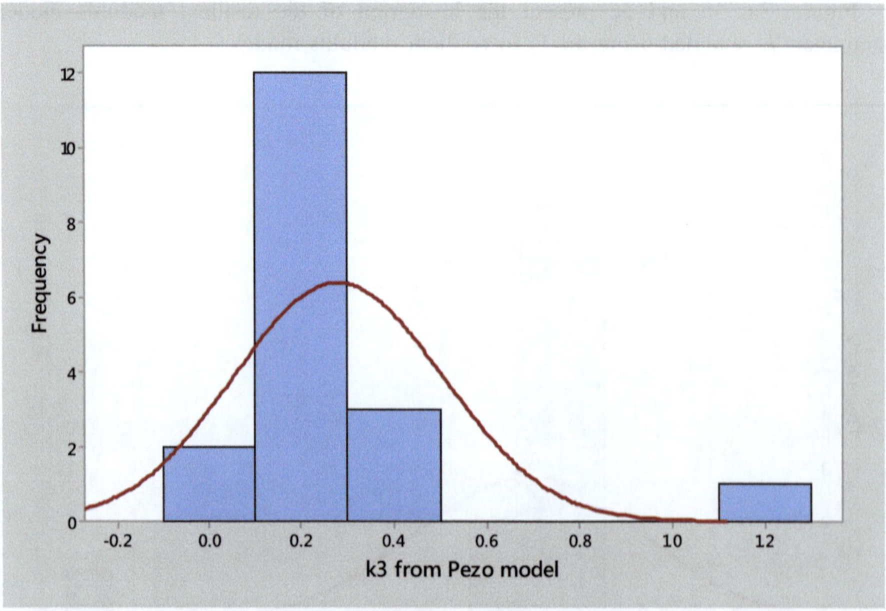

Fig. 3c. Histograms of Pezo's resilient modulus model parameters k_3

Figures 4a, 4b and 4c present the histogram of the resilient modulus model parameters k_i obtained using the Ni et al. resilient modulus model.

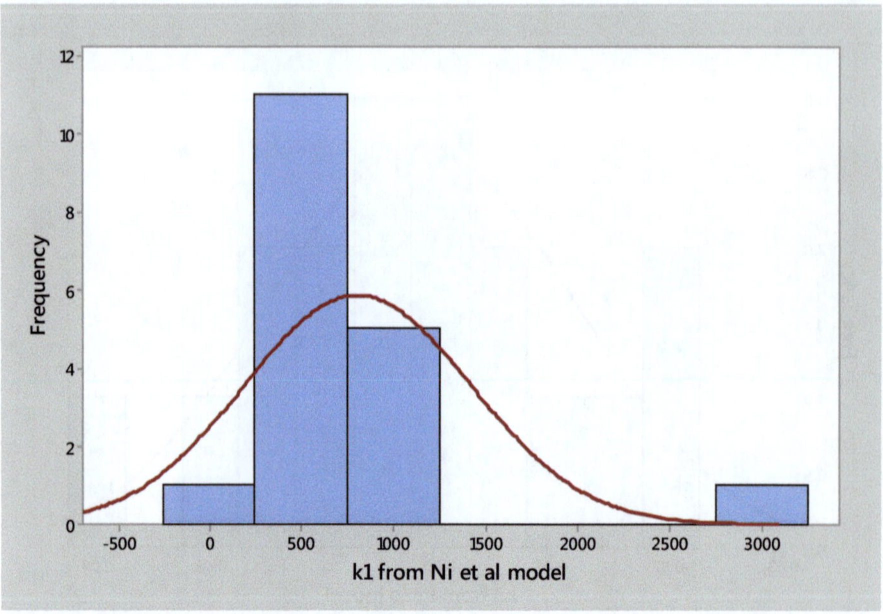

Fig. 4a. Histograms of Ni et al.'s resilient modulus model parameters k_1

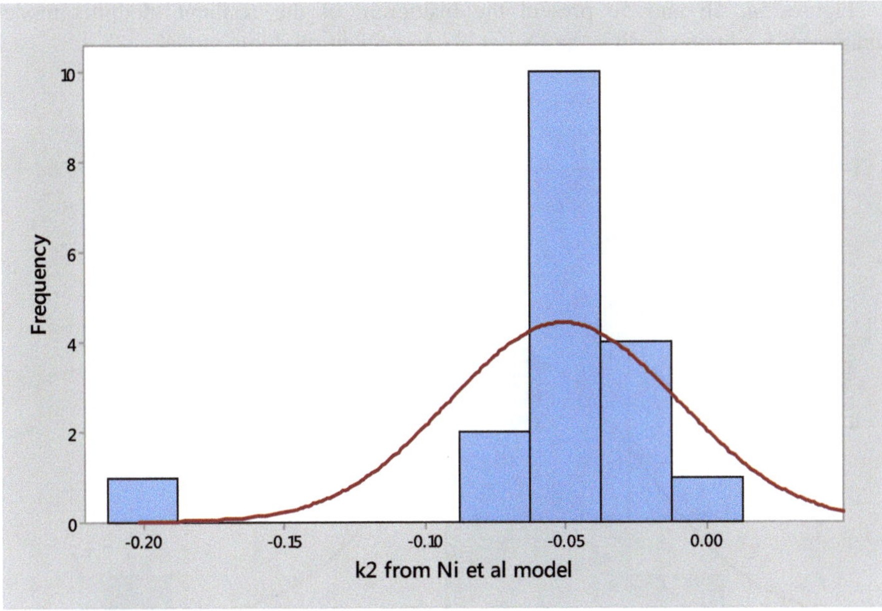

Fig. 4b. Histograms of Ni et al.'s resilient modulus model parameters k_2

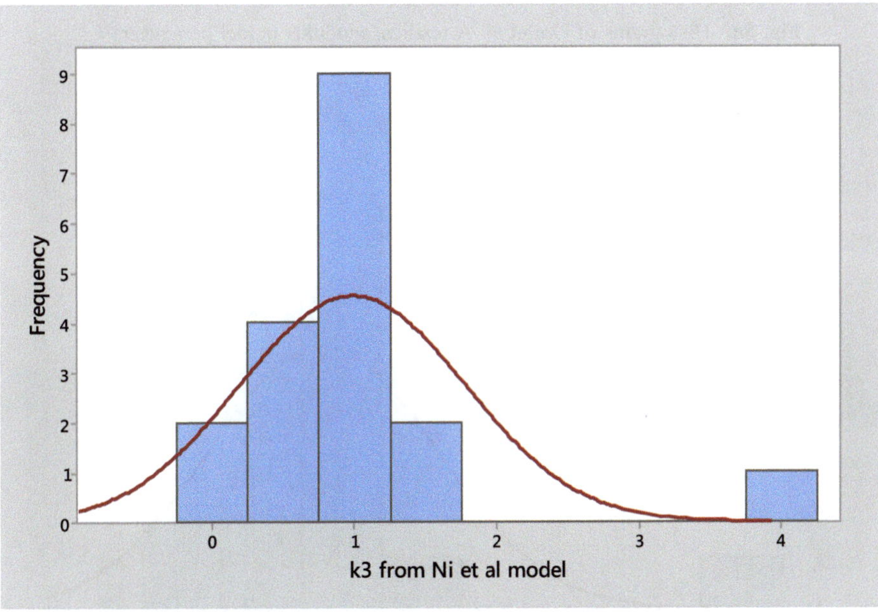

Fig. 4c. Histograms of Ni et al.'s resilient modulus model parameters k_3

Figures 5a, 5b and 5c present the histogram of the resilient modulus model parameters k_i obtained using the Ooi et al. A resilient modulus model.

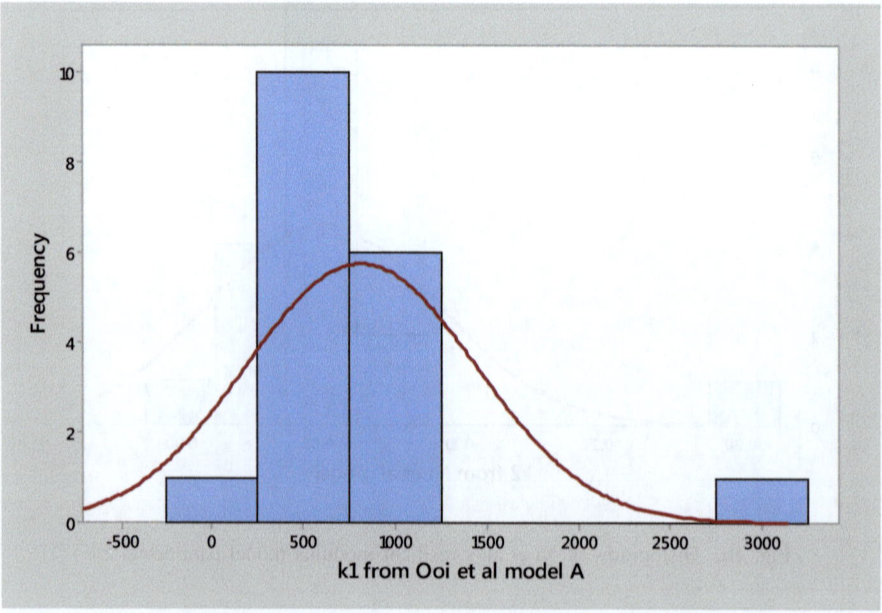

Fig. 5a. Histograms of Ooi et al. A resilient modulus model parameters k_1

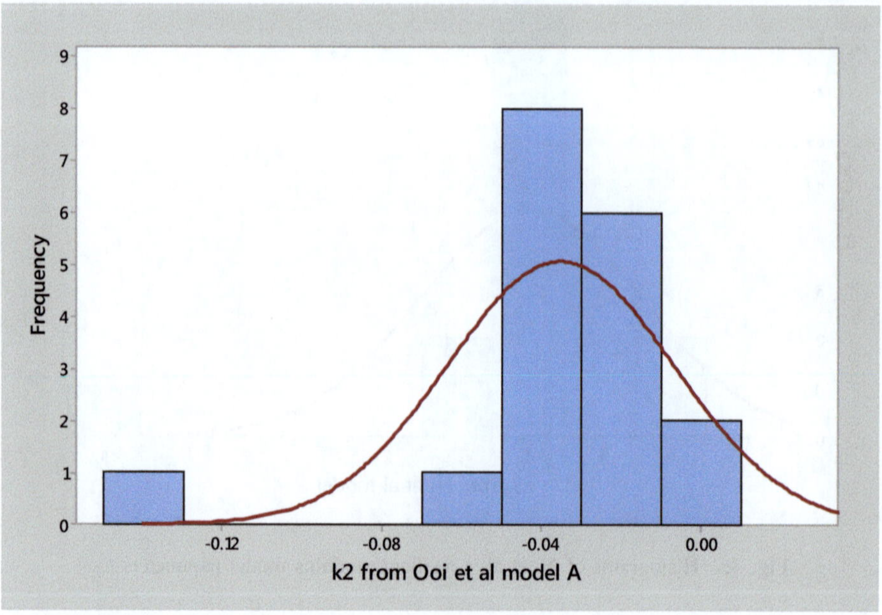

Fig. 5b. Histograms of Ooi et al. A resilient modulus model parameters k_2

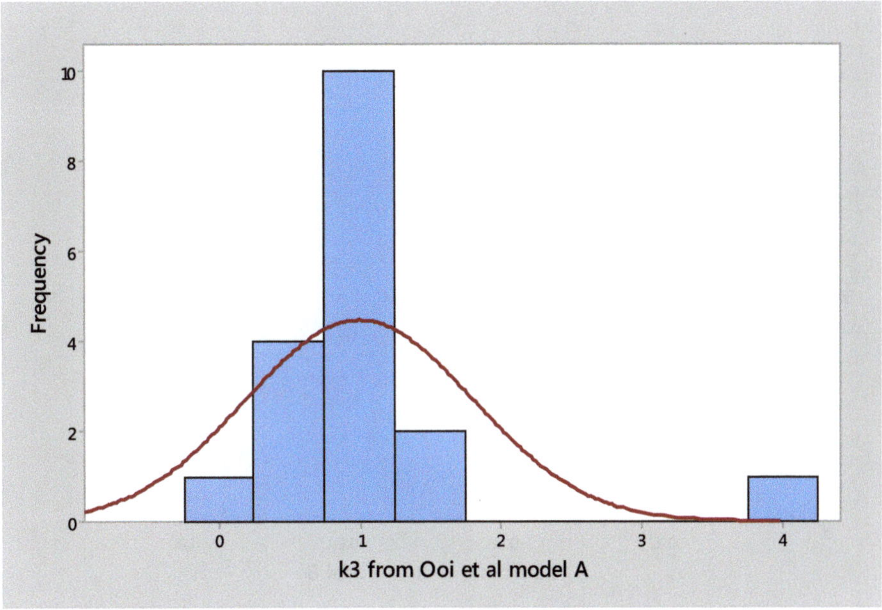

Fig. 5c. Histograms of Ooi et al. A resilient modulus model parameters k_3

Figures 6a, 6b and 6c present the histogram of the resilient modulus model parameters k_i obtained using the Ooi et al. B resilient modulus model.

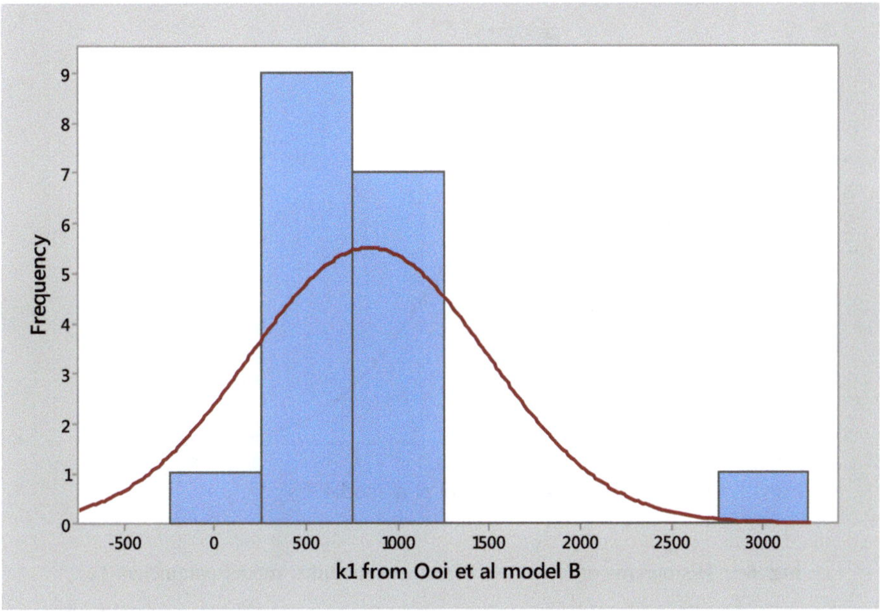

Fig. 6a. Histograms of Ooi et al. B resilient modulus model parameters k_1

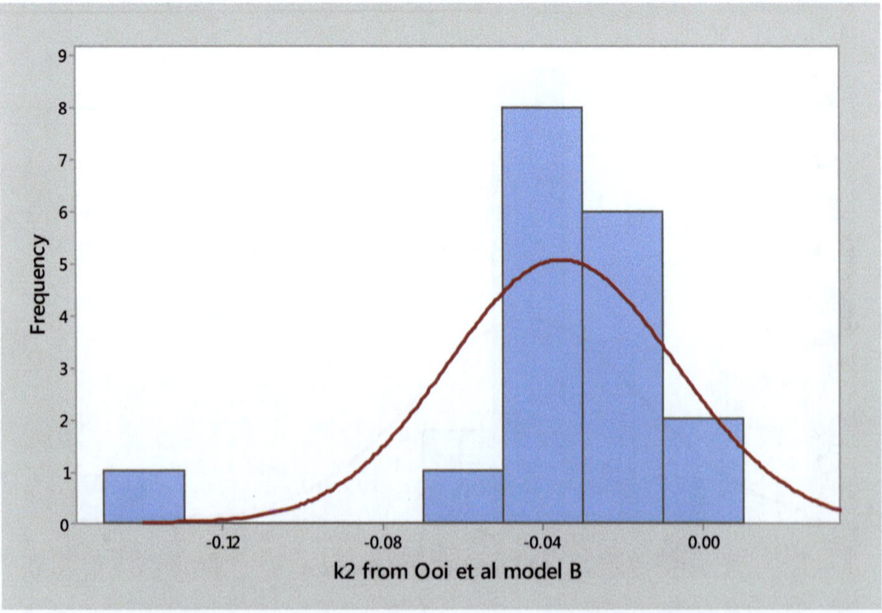

Fig. 6b. Histograms of Ooi et al. B resilient modulus model parameters k_2

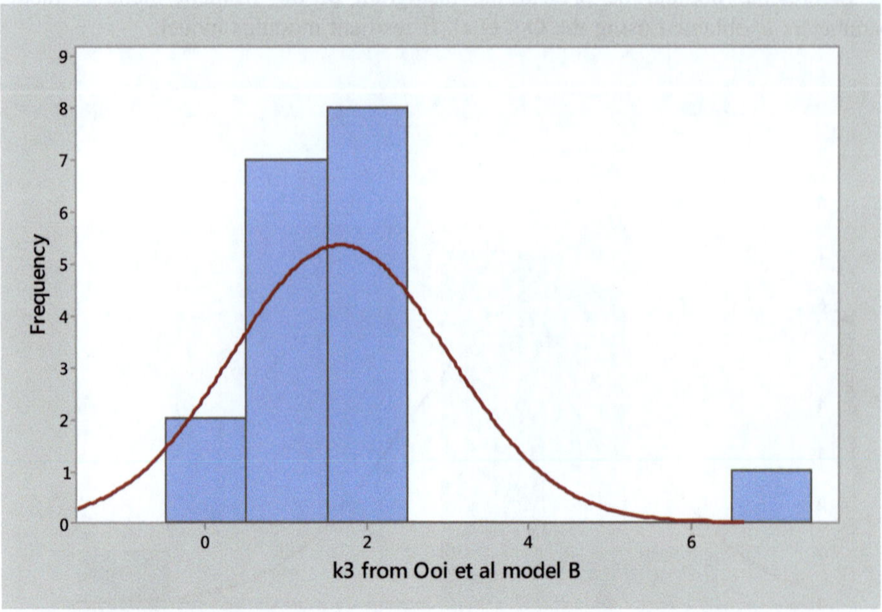

Fig. 6c. Histograms of Ooi et al. B resilient modulus model parameters k_3

Figures 7a, 7b and 7c present the histogram of the resilient modulus model parameters k_i obtained using the NCHRP's resilient modulus model.

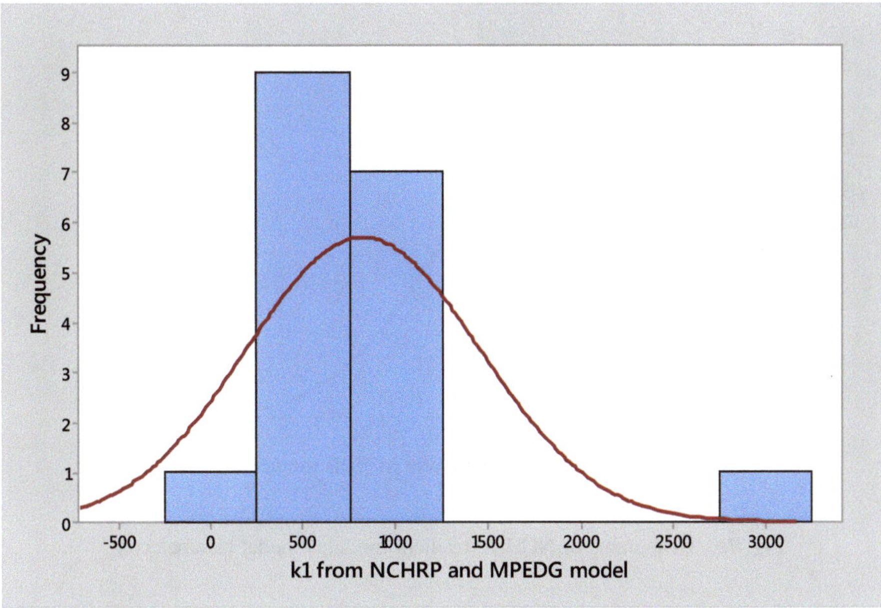

Fig. 7a. Histograms of NCHRP resilient modulus model parameters k_1

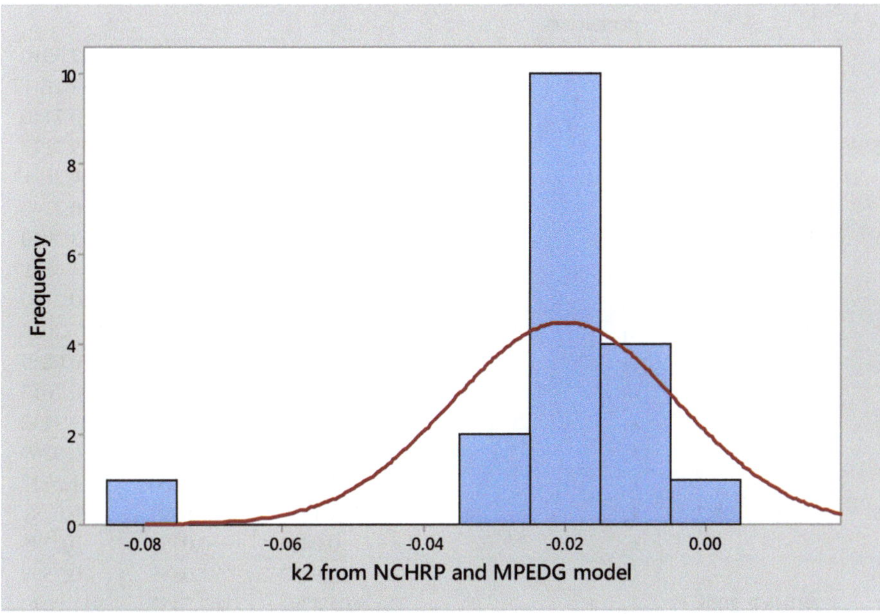

Fig. 7b. Histograms of NCHRP resilient modulus model parameters k_2

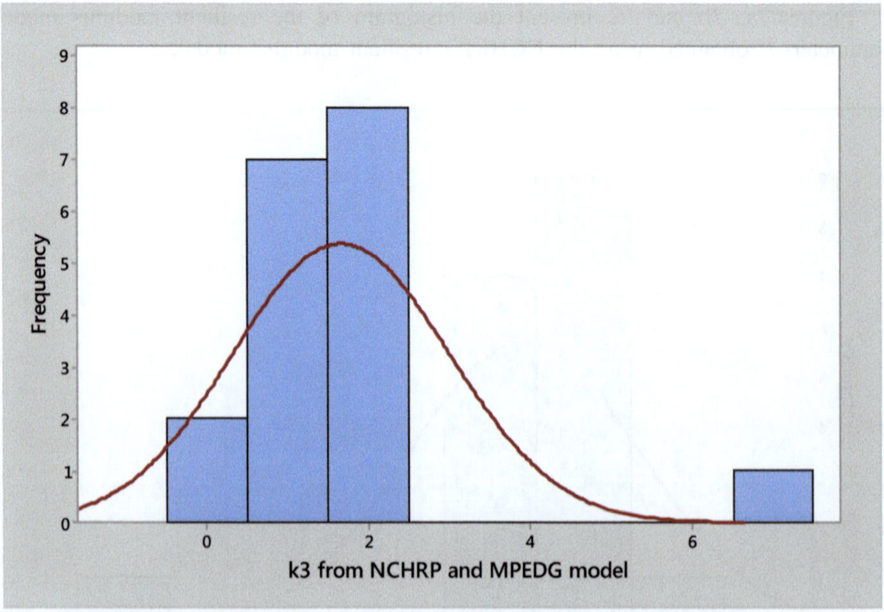

Fig. 7c. Histograms of NCHRP resilient modulus model parameters k_3

Table 1. Statistical data of k_i obtained from the test results of the coarse-grained soils using resilient modulus models.

S/No.	Model	Resilient modulus model parameters	A-1-b	A-2-4	A-2-7
1	Uzan 1985	k_1	1128.2647	1567.4997	1493.3910
		k_2	7.98E-17	5.31E-17	−4.1E-17
		k_3	0.2695	0.3094	0.2140
2	Witczak and Uzan 1988	k_1	1381.7457	2011.3930	1769.3172
		k_2	8.88E-17	5.45E-17	−4.2E-17
		k_3	0.2695	0.3094	0.2140
3	Pezo 1993	k_1	1128.2647	1567.4997	1493.3910
		k_2	8.67E-17	3.2E-17	−3.4E-17
		k_3	0.2695	0.3094	0.2140
4	Ooi et al. 2004	k_1	601.3008	797.2932	907.2052
		k_2	−0.0339	−0.0389	−0.0269
		k_3	0.9587	1.1006	0.7613
5	Ooi et al. 2004	k_1	620.7723	823.0154	929.3132
		k_2	−0.0339	−0.0389	−0.0269
		k_3	1.6028	1.8400	1.2727
6	Ni et al. 2002	k_1	594.1272	787.8599	899.0481
		k_2	−0.0485	−0.0557	−0.0385
		k_3	0.9457	1.0856	0.7509
7	NCHRP 2008	k_1	605.4392	802.7457	911.9080
		k_2	−0.0192	−0.0220	−0.0152
		k_3	1.5964	1.8326	1.2676

From the evaluation, as presented in Figs. 1a, 1b, 1c, 2a, 2b, 2c, 3a, 3b, 3c, 4a, 4b, 4c, 5a, 5b, 5c, 6a, 6b, 6c, 7a, 7b and 7c, the resultant resilient modulus parameters of coarse-grained soils with the following classifications (A-1-b, A-2-4 and A-2-7) using the resilient modulus models are as presented in Table 1.

Based on the evaluation of the resilient modulus equations for coarse-grained soils, it was observed from Table 1 for level 3 analysis that the resilient modulus equation adopted by NCHRP was the best in determining resilient modulus of coarse-grained soils.

4 Conclusion

Based on the results of this research, the following conclusions are reached:

1. Resilient modulus constitutive equation adopted by NCHRP and MEPDG was adopted for estimating resilient modulus of coarse-grained soils.
2. Default values of resilient modulus parameters was determined for coarse-grained soils as level 3 resilient modulus input.

References

AASHTO: Guide for Design of Pavement Structures. American Association of State Highway and Transportation Officials, Washington, DC (1993)

George, K.P.: Prediction of Resilient Modulus from Soil Index Properties, FHWA/MS-DOT-RD-04-172, Department of Civil Engineering, The University of Mississippi, Mississippi (2004). http://www.mdot.state.ms.us/research/pdf/ResMod.pdf

Hossain, S.M.: Characterization of Unbound Pavement Materials From Virginia Sources for Use in the New Mechanistic-Empirical Pavement Design Procedure. Virginia Transportation Research Council, Commonwealth of Virginia, Virginia (2010)

Mohammad, L.N., Kevin, G.P., Ananda, H.P., Munir, D.N.: Comparative Evaluation of Subgrade Resilient Modulus from Non-destructive, In-situ, and Laboratory Methods. Louisiana Transportation Research Center, Louisiana Department of Transportation and Development. National Technical Information Service, Springfield (2007)

NCHRP: Synthesis 382: Estimating Stiffness of Subgrade and Unbound Materials for Pavement Design. National Cooperative Highway Research Program, Transportation Research Board, Washington, DC (2008)

Ni, B., Hopkins, T.C., Sun, L., Beckham, T.L.: Modeling the resilient modulus of soils. In: Proceedings of the 6th International Conference on the Bearing Capacity of Roads, Railways, and Airfields, vol. 2, pp. 1131–1142. Balkema Publishers, Rotterdam, the Netherlands (2002)

Ooi, P.S., Archilla, A.R., Sandefur, K.G.: Resilient modulus models for compacted cohesive soils. Transportation Research Record No. 1874 Transportation Research Board, National Research Council, pp. 115–124 (2004)

Ooi, P.S., Sandefur, K.G., Archilla, A.R.: Correlation of resilient modulus of fine-grained soils with common soil parameters for use in design of flexible pavements. Honolulu: Report No. HWY-L-2000-06, Hawaii Department of Transportation (2006)

Pezo, R.F.: A general method of reporting resilient modulus tests of soils: a pavement engineer's point of view. In: 72nd Annual Meeting of the Transportation Research Board. Transportation Research Board, Washington, DC (1993)

Titi, H.H., Mohammed, B.E., Sam, H.: Determination of Typical Resilient Modulus Values for Selected Soils in Wisconsin. University of Wisconsin, Department of Civil Engineering and Mechanics. National Technical Information Service 5285 Port Royal Road, Springfield, Milwaukee (2006)

Uzan, J.: Characterization of granular material. In: Transportation Research Record 1022, Transportation Research Board, National Research Council, pp. 52–59 (1985)

Vogrig, M., McDonald, A., Vanapalli, S.K., Siekmeier, J., Roberson, R., Garven, E.: A laboratory technique for estimating the resilient modulus of unsaturated soil specimens from CBR and unconfined compression tests. In: 56th Canadian Geotechnical Conference, 4th Joint IAH-CNC/CGS Conference, 2003 NAGS Conference, Canada (2003)

Von Quintus, H., Killingsworth, B.: Design Pamphlet for the Determination of Design Subgrade in Support of the AASHTO Guide for the Design of Pavement Structures. Office of Engineering R&D, Federal Highway Administration, National Technical Information Services, Springfield (1997)

Witczak, M.W., Uzan, J.: The Universal Airport Pavement Design System, Report 1 of 4, Granular Material Characterization. University of Maryland, College Park (1988)

Finite Element Analysis of Rock Slope Stability Using Shear Strength Reduction Method

Greg You[1(✉)], Maged Al Mandalawi[1], Ahmed Soliman[2], Kim Dowling[1], and Peter Dahlhaus[1]

[1] Faculty of Science and Technology, Federation University Australia, Ballarat, Australia
g.you@federation.edu.au
[2] Glencore, Mount Isa Mines, Zinc Assets, Mount Isa, Australia

Abstract. Finite element analysis incorporating the shear strength reduction method was applied to study the west slope stability of an open cut mine in Australia using Mohr–Coulomb and generalized Hoek–Brown criteria. The pit of the mine had multiphase excavations and reached 180 m in depth. The study investigated three slope configurations, namely, Stage 1 inter ramp slope 43°, Stage 2 inter ramp slope 49° and optimized Stage 2 slope 54°. When implementing the generalized Hoek–Brown failure criterion, the equivalent factor of safety was 1.96, 1.87 and 1.40 under dry slope for the three configurations, respectively. However, under partly saturated conditions, the optimised slope would have a factor of safety 1.16. Furthermore, the generalised Hoek–Brown criterion generated lower factors of safety than the Mohr–Coulomb failure criterion. The difference is related to an overestimation of the shear strength parameters by the linear Mohr–Coulomb criterion under low confining stresses compared with the non-linear Hoek–Brown.

1 Introduction

Slopes need be designed and cut with a margin of safety for open pit mines. Economics could be improved by steepening the slope thereby reducing the amount of waste excavation; however, excessive steepening of slope could result in failure leading to loss of life and damage to property (Singh et al. 1989; Singh and Singh 1992). The factors, which mainly influence the stability of a typical open-pit slope, are the shear strength parameters of slope forming material, the presence and characteristics of discontinuities in the slope mass and the groundwater conditions (Singh and Monjezi 2000; Singh et al. 2008).

The shear strength reduction (SSR) method was used for soil slope stability analysis in 1975 (Zienkiewicz et al. 1975). Duncan (1996) defined the factor of safety of soil slopes as the ratio of actual shear strength to the minimum shear strength required to avoid failure, or the factor by which shear strength must be reduced to bring a slope to failure. This method was applied for rock masses (Shangyi et al. 2003). Yingren and Shangyi (2004) and Hammah et al. (2007) demonstrated the efficiency of SSR method for slopes of soil and rock masses. This method was also adopted in several other studies, such as Yeo and Chan (1993), Dowson et al. (1999), Hammah et al. (2004a, b),

W. Frikha et al. (eds.), *Soil Testing, Soil Stability and Ground Improvement*,
Sustainable Civil Infrastructures, DOI 10.1007/978-3-319-61902-6_18

Zheng et al. (2009) and Gupta et al. (2016). The advantages of the SSR method over limit equilibrium method (LEM) are: (1) No assumption is required of the interslice shear force distribution; (2) the critical failure surface in the slope can be found from the shear strain; and (3) this method is suitable for complex slope conditions in order to interpret details of displacements, stresses and water pressures. LEM may give lower estimates of failure volumes than SSR method in numerical modelling (Chiwaye 2010).

The SSR approach includes the search for a stress reduction factor (SRF) value that brings the slope to fail. The shear strength reduced by a factor of safety F can be determined using a series of trials to adjust the friction angle (ϕ'), and the cohesion (c′) of slope rock mass. For example of the Mohr-Coulomb criterion (Eq. 1), the process of shear stress reduction can be expressed in Eq. 1a.

$$\tau = c' + \sigma' \tan \varphi' \tag{1}$$

$$\frac{\tau}{F} = \frac{c'}{F} + \frac{\sigma' \tan \varphi'}{F} \tag{1a}$$

Where, the reduced value of cohesion $c' = \frac{c'}{F}$ and the reduced value of internal friction angle $\varphi' = arctan\left(\frac{tan\varphi'}{F}\right)$.

This paper presents a finite element (FE) analysis of slope stability using SSR method for three inter-ramp slopes at the Handlebar open pit mine, namely Stage 1 (43°), Stage 2 (49°) and an optimised Stage 2 slope (54°), under both dry and partly saturated conditions that implements the Mohr–Coulomb and the generalised Hoek–Brown failure criteria.

2 Site Geology

Handlebar Hill open - pit mine is located at Mt. Isa, north Queensland, Australia (Fig. 1). In the Mount Isa Valley, the rock formation is represented by Magazine Shale, Spears-Kennedy Siltstone, Urquhart Shale, Native Bee Siltstone, and Eastern Creek Volcanics (Figs. 1 and 2). The ground is oxidized and leached of sulphides. The leaching can extend to great depths aided by faults (Fig. 2). The leached zone in the rock slope is regarded as an unconfined aquifer (Rosengren and Associates 2007). Pre-mining groundwater level data indicated that groundwater levels coincided with the base of the oxidized zone at a depth of about 50 m below ground level.

Mining started in 2008 through multiple stages. Stage 1 was excavated to a depth of 77 m and Stage 2 to 180 m. There were six benches at each stage of varying bench height from 10 m to 17 m (Fig. 2). The inter-ramp angle (IRA) was 43° and 49° for Stage 1 and Stage 2, respectively, on the west wall at the 4440 N cross-section.

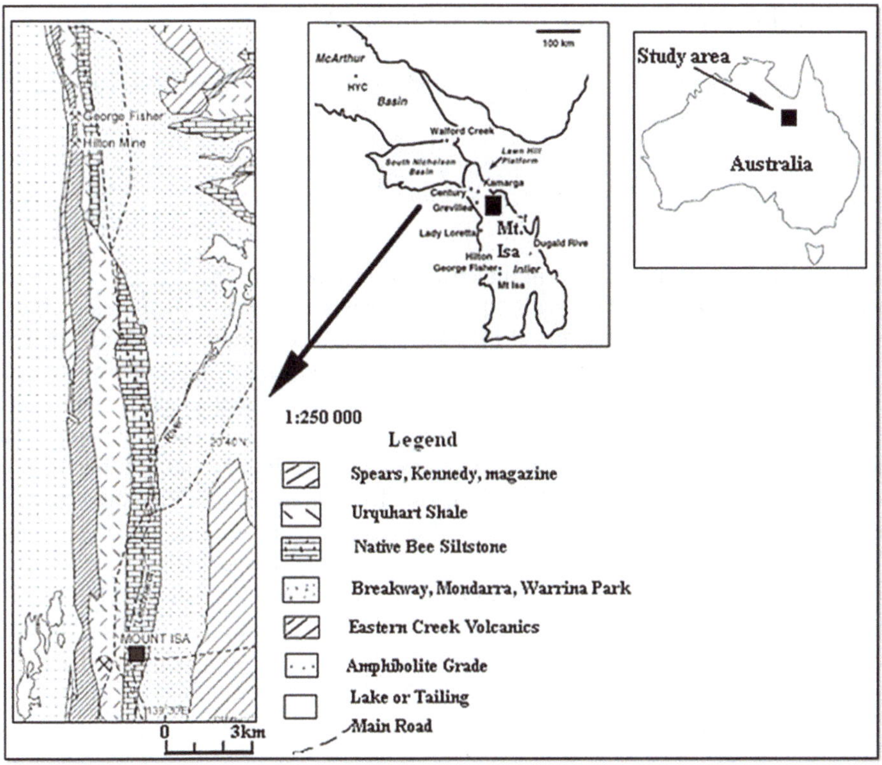

Fig. 1. Location and geological map of the study area (after Conaghan et al. 2003)

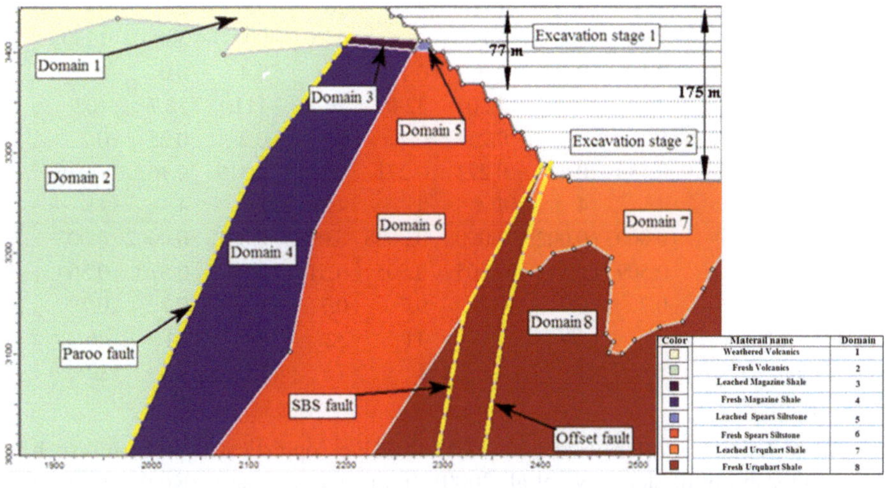

Fig. 2. Rock domains of the west slope at N4440 cross-section

3 Data Preparation and Finite Element Analysis

Both the Mohr–Coulomb criterion (Eq. 1) and the generalized Hoek–Brown criterion (Eq. 2) are used in this analysis.

$$\sigma_1' = \sigma_3' + \sigma_{ci}\left(m_b \frac{\sigma_3'}{\sigma_{ci}} + s\right)^a \tag{2}$$

Some of the parameters used in the Hoek–Brown criterion (e.g., s, m_b, and a) are available from rock mass data, while other parameters used (m_i, v, σ_{ci}) were derived using laboratory rock tests (Seville 1981; Tarrant and Lee 1984). From June 2006 to June 2007, a total of 20,189.6 m of diamond core was drilled and 3,080 m of rock chips were completed (Rosengren and Associates 2008). The geological strength index GSI was determined from the diamond core logging and in-situ surface mapping. The mean GSI value of the overall slope was reduced by 20% in this study (Table 1). For the weathered Magazine Shale, a GSI value of 24 was used because this represented a poorer rock mass quality below the base of complete oxidization in the simulations. The rock mass modulus (E_{rm}) of all domains was estimated according to Hoek and Diederichs (2006). A disturbance factor of D = 0.7 was used for small scale blasting in civil engineering slopes that results in modest rock mass damage if controlled blasting is used (Hoek et al. 2002). The parameters used in the study are detailed in Table 1.

Table 1. Input data used in the slope stability analysis

Parameter	Unit	Domain							
		1	2	3	4	5	6	7	8
γ	kN/m^3	27.4	28.3	26.5	26.5	23.5	27.2	26.9	31.1
E_{rm}	MPa	3000	10000	4500	4500	1500	8500	2000	12000
φ	(°)	23	32	13	25	20	35	22	44
c	kPa	350	800	50	410	240	990	300	1250
σ_t	MPa	0.1	0.3	0.0	0.1	0.0	0.4	0.0	0.2
σ_{ci}	MPa	36.4	55.5	32.0	32.0	26.3	111.1	32.7	108
v	–	0.3	0.2	0.25	0.2	0.25	0.2	0.25	0.2
m_b	–	0.184	0.42	0.271	0.271	0.148	0.319	0.165	1.036
m_i	–	4	4	4	4	4	4	4	11
s	–	0.0003	0.0026	0.0008	0.0008	0.0002	0.0013	0.0002	0.002
a	–	0.509	0.503	0.505	0.505	0.511	0.504	0.510	0.504
D	–	0.7	0.7	0.7	0.7	0.7	0.7	0.7	0.7
80% GSI	–	36	48	24	41	32	43	34	46
Dilation	(°)	5	8	4	5	7	12	4	12

SSR technique is often used with FEM to solve sophisticated problems such as estimating stability of slope (Ng et al. 2000). In this study, Phase^2 (RocScience 2014) FE program is used for the slope stability simulations under different scenarios. The model comprises an area of 750 m in width and 450 m in depth (Fig. 2).

The pit bottom is located at an elevation of 3,368 m (175 m deep), and the top of the west slope is at an elevation of 3,543 m (ground–surface level). The ramp crosses the west slope section at an elevation of 3,463 m (77 m deep).

Phase[2] is a powerful FE program, which can be used for a wide range of engineering projects including slope stability, groundwater seepage and probabilistic analysis. It can simulate and analyze a complex multi-stage model. Progressive failures and explicit modelling of discontinuities can be simulated to gain further insight into the rock mass behavior of the slope. When setting up the project, the multi-stage model is available to simulate the stresses resulting from different excavations. The program was designed to calculate the last excavation stage for the critical SRF, or factor of safety. Therefore, the strength reduction factor analysis has to be run many times for the two stages of excavations. The FE simulations were carried out to further improve the overall slope stability analysis, namely to optimize the Stage 2 IRA.

Figure 3 shows maximum shear strains simulated for IRA 43° and 49° using the Mohr–Coulomb criterion and the generalized Hoek–Brown criterion. At the upper IRA 43° (Stage 1, excavation depth 77 m), there is a localized high shear strain zone in

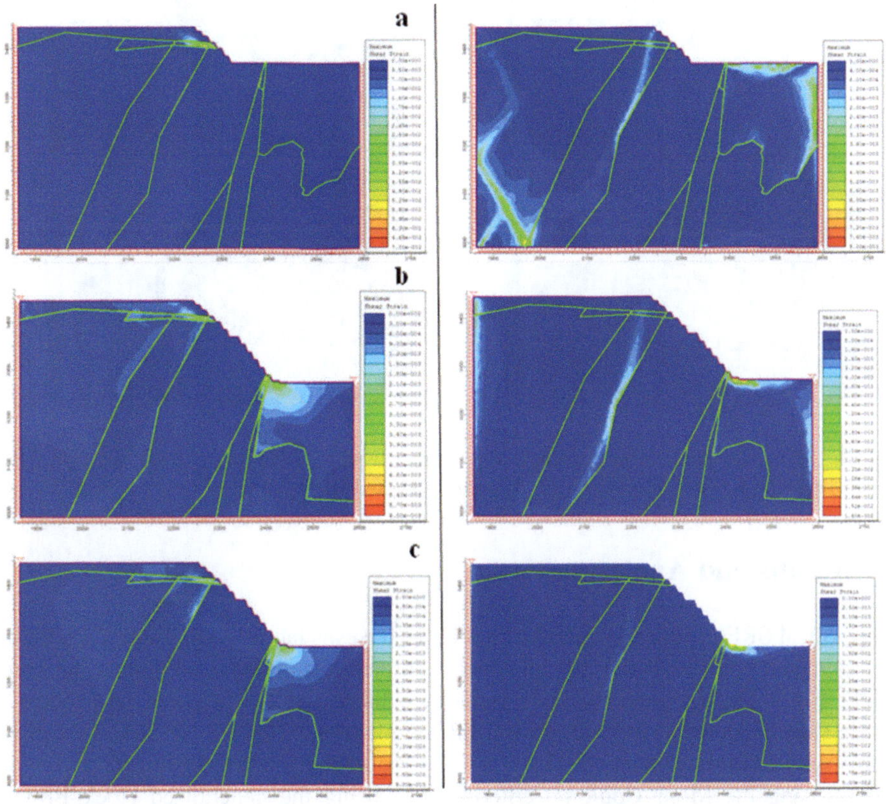

Fig. 3. Maximum shear strain contours, (a) Stage 1, IRA 43°, (b) Stage 2, IRA 49° and (c) IRA 49° with enhanced slope cohesion strength, dry slope conditions. Right: Generalized Hoek–Brown criterion, left: Mohr–Coulomb criterion

Domains 3 and 5, which is structurally controlled leached Magazine shale and Spears siltstone (Fig. 3a, left). At the lower IRA 49° (Stage 2, excavation depth 180 m), the shear strain concentration at toe is prominent (Fig. 3b, c). In comparison, there is less deformation in the slope using the Mohr–Coulomb criterion than using the generalized Hoek–Brown failure criterion (Fig. 3). These results indicate that an increase in maximum shear stress occurs at the final excavation of stage 2.

In the third case of the study, the Stage 2 IRA was increased 5° to optimize the open–pit design to be as steep as possible. Figure 4 shows the maximum shear strain contours for the IRA 54° under dry and partly saturated conditions. As the effective normal stress decreases under partly saturated slope conditions, larger maximum shear strains develop. There is a clear increase in shear strain in Domains 3 and 5 and at the toe of the slope (Fig. 4b, c).

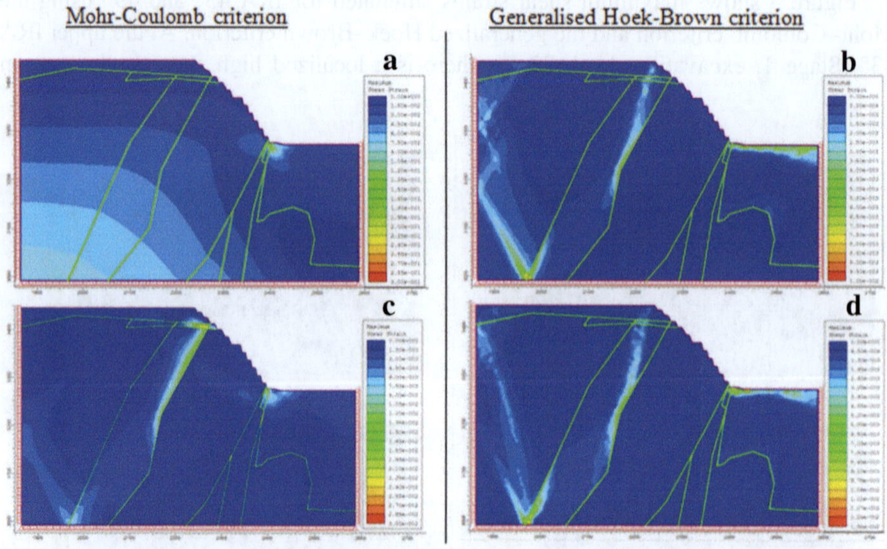

Fig. 4. Maximum shear strain contours for optimized slope after two excavation stages, for lower IRA of 54°, (a, b) dry slope, (c, d) partly saturated slope

4 Results and Analyses

The critical SRF or factor of safety of the FE analyses is tabulated in Table 2 for IRAs 43°, 49° and optimized IRA 54° under dry and partly saturated conditions using both Mohr–Coulomb and generalized Hoek–Brown failure criteria. As can be expected, rising slope angle leads to lower critical SRF. Furthermore, the critical SRF is lower in partly saturated slope than the dry one. The modelling predicted a higher factor of safety using the Mohr–Coulomb criterion than using the generalized Hoek–Brown criterion (Table 2). This may attribute to the linear failure envelope in the Mohr–Coulomb criterion (Fig. 5). Figure 5 is plotted based on the shear strength analysis of Urquhart Shale rock using RocLab 1.0 (RocScience 2014). Rock mass strength in the

Table 2. Critical SRF from FE analyses for different slopes

IRA	Dry slope		Partly saturated slope	
	Mohr–Coulomb	Generalized Hoek–Brown	Mohr–Coulomb	Generalized Hoek–Brown
43°	2.76	1.96	2.70	1.71
49°	2.00	1.87	1.86	1.59
54°	1.98	1.40	1.85	1.16

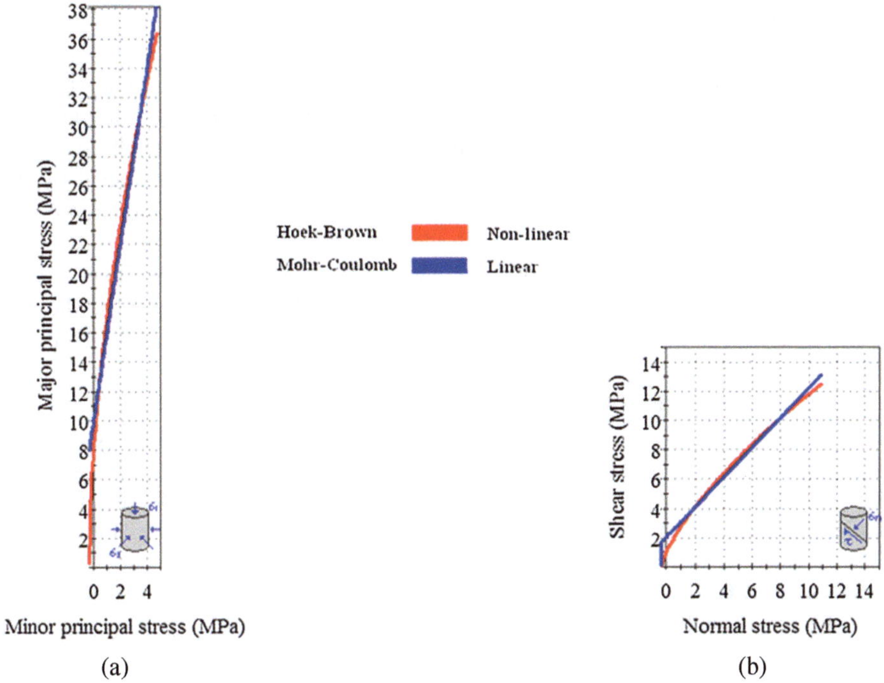

Fig. 5. Comparison between the Mohr–Coulomb criterion and the generalized Hoek–Brown failure criterion, (a) in term of major and minor stresses, (b) in term of shear and normal stresses

area is a non-linear stress function. The linear Mohr–Coulomb failure criterion may not agree well with the rock mass failure envelope where the low confining stresses in an open-pit slope make the non-linearity of rock behaviors more possible (Shen 2013). On the other hand, the Mohr–Coulomb strength parameters may overestimate the shear strength of a rock mass at a high stress level because it was derived using a straight line fitted over the Hoek–Brown curve (Lin et al. 2014). Hammah et al. (2004a, b) stated that the generalized Hoek–Brown criterion is the most suitable strength model for predicting the failure of rock masses, especially in low normal stress ranges.

If the Stage 2 slope increases 5° to IRA 54°, the critical SRF is 1.16 under partly saturated condition using the generalized Hoek–Brown criterion, which is less than 1.3. This will have an influence on the stability of the west slope if the shear strength failure mechanisms of the rock masses are considered to dominate the slope stability in wet conditions. The results for slopes under the other conditions show a stable slope with SRF > 1.3 using both failure criteria. For the long term operations and after heavy rain-fall, bench scale instabilities can be expected. Given that a GSI reduction of 20% was assumed and the E_{rm} was estimated as a worst case scenario for the rock mass strength parameters, the slope is expected to be stable. The groundwater has influence on the slope stability, in particular, in the case of steeper slope with an IRA of 54°.

5 Conclusions

Slope stability analysis of the west slope of Handlebar Hill open-pit mine was conducted using non-directional strength of rock mass properties through finite element analyses implemented the shear strength reduction technique. The generalized Hoek–Brown criterion did not generate any SRF greater than the results predicted by the Mohr–Coulomb failure criterion. The difference is related to an overestimation of the shear strength parameters by linear fit under low confining stresses compared with the non-linear Hoek–Brown. The stability of the west slope would be overestimated when directly using the shear strength parameters for the analysis. Consequently, the disturbance factor value (D), E_{rm} and the GSI must be significantly calibrated and considered as main inputs to model the slope.

Presuming that a minimum factor of safety 1.3 is adopted for long term excavations, the Stage 2 IRA could be optimized and increased to 54°, in which SRF was 1.4. However, this IRA would have a SRF of 1.16 in case of partly saturated slope conditions. Inter-ramp stability can be controlled by both geological structure and rock mass strength, however, the combined failure mechanism is not the scope of this study.

Acknowledgments. The authors would like to express their appreciation to Glencore zinc for providing permission to carry out this research and publish this paper.

References

Chiwaye, H.: A Comparison of the Limit Equilibrium and Numerical Modelling Approaches to Risk Analysis for Open-Pit Mine Slopes. University of the Witwatersrand, Johannesburg (2010)

Conaghan, L.E., Hannan, W.K., Tolman, J.: Mount Isa Cu and Pb-Zn-Ag deposits, NW Queensland, Australia. In: Advances in Regolith: Proceedings of the Cooperative Research Centre for Landscape Environments and Mineral Exploration (CRC LEME). Regional Regolith Symposia, Queensland, Mt. Isa (2003)

Dawson, E.M., Roth, W.H., Drescher, A.: Slope stability analysis by strength reduction. Géotechnique **49**(6), 835–840 (1999)

Duncan, J.M.: State of the art: limit equilibrium and finite-element analysis of slopes. J. Geotech. Eng. **122**(7), 577–596 (1996)

Gupta, V., Bhasin, R.K., Kaynia, A.M., Kumar, V., Saini, A.S., Tandon, R.S., Pabst, T.: Finite element analysis of failed slope by shear strength reduction technique: a case study for Surabhi Resort Landslide, Mussoorie township, Garhwal Himalaya. Geomatics Nat. Hazards Risk **7**(5), 1677–1690 (2016). doi:10.1080/19475705.2015.1102778

Hammah, R.E., Curran, J.H., Yacoub, T. E., Corkum, B.: Stability analysis of rock slopes using the finite element method. In: Proceedings of the ISRM Regional Symposium EUROCK (2004a)

Hammah, R.E., Curran J.H., Corkum, B., Yacoub T.E.: Stability analysis of rock slopes using the finite element method. In: Euoreka 2004 and 53rd Geomechanics Colloquium. Schubert (2004b)

Hammah, R.E., Yacoub, T.E., Corkum, B., Wibowo, F., Curran, J.H.: Analysis of blocky rock slopes with finite element shear strength reduction analysis. In: Proceedings of the 1st Canada-U.S. Rock Mechanics Symposium, Vancouver, Canada, pp. 329–334 (2007)

Hoek, E., Diederichs, M.S.: Empirical estimation of rock mass modulus. Int. J. Rock Mech. Min. Soc. **43**, 203–215 (2006). Canada

Hoek, E., Carranza-Torres, C., Corkum, B.: Hoek-Brown failure criterion-2002 edition. In: Proceedings of 5th North American Rock Mechanics Symposium, pp. 267–273. University of Toronto Press, Toronto (2002)

Lin, H., Zhong, W., Xiong, W., Tang, W.: Slope stability analysis using limit equilibrium method in nonlinear criterion. Sci. World J. (2014). Article ID 206062, 7 pages. Hindawi Publishing Corporation

Yeo, K., Chen, A.: Finite element slope stability analysis by shear strength reduction technique. Jpn. Geotech. Soc. **33**(2), 194–196 (1993)

Ng, C.W.W., Zhang, L.M., Ho, K.K.S., Choy, C.K.: Influence of laterally loaded sleeved piles on slope stability. In: GeoEng 2000, Melbourne, Australia (2000)

Phase^2 V9.0/RocScience: A Two-Dimensional Finite Element Analysis of Soil and Rock for Underground and Excavated Slopes, User's Manual, RocScience Inc., Toronto (2014)

RocLab 1.0/RocScience: Program for Determining Rock Mass Strength Parameters, User's Manual, RocScience Inc., Toronto (2014)

Rosengren, K., and Associates: Proposed Handlebar Hill Open Cut geotechnical review (Technical Report No. 27010). Xstrata, Mt. Isa (2007)

Rosengren, K., and Associates: Handlebar Hill Open Cut Pit. Internal Technical report, in Xstrata report in 2011. Xstrata, Mt. Isa (2008)

Seville, R.: Review of Hilton Mine Rock Property Data. Mount Isa Mines Limited (1981)

Shangyi, Z., Yingren, Z., Weidong, D.: Stability analysis on jointed rock slope by strength reduction FEM. Chin. J. Rock Mech. Eng. **2**, 020 (2003)

Shen, J.: Analytical and numerical analyses for rock slope stability using the generalized Hoek-Brown criterion. Doctoral Thesis, Adelaide University, Australia (2013)

Singh, T.N., Ulabhaje, A., Singh, D.P.: Planning of Slope Stability in an Opencast Mine- A Model Approach, National Seminar on Surface Mining, Dhanbad edited by the Ind. Min. and Eng. Jr., pp. 5.2.1–5.2.10 (1989)

Singh, T.N., Singh, D.P.: Slope stability study in an opencast mine over previously worked seam. In: International Symposium on Rock Slope, New Delhi, pp. 467–477 (1992)

Singh, T.N., Gulati, A., Dontha, L., Bhardwaj, V.: Evaluating cut slope failure by numerical analysis – a case study. Nat. Hazards **47**, 263–279 (2008)

Singh, T.N., Monjezi, M.: Slope stability study in jointed rock mass - a numerical approach. Min. Eng. J. **1**(10), 12–13 (2000)

Tarrant, G., Lee, M.: Mount Isa rock properties (Technical No. RES MIN 60). Mount Isa Mines Limited (1984)

Yingren, Z., Shangyi, Z.: Application of strength reduction FEM in soil and rock slopes. Chin. J. Rock Mech. Eng. **23**(19), 3381–3388 (2004)

Zheng, Y., Tang, X., Zhao, S., Deng, C., Lei, W.: Strength reduction and step-loading finite element approaches in geotechnical engineering. J. Rock Mech. Geotech. Eng. **1**(1), 21–30 (2009)

Zienkiewicz, O.C., Humpheson, C., Lewis, R.W.: Associated and nonassociated. Visco-plasticity and plasticity in soil mechanics. Geotechnique **25**(4), 671–689 (1975)

Study of the Behavior of Floating Stone Columns in Soft Clay Formations Using Numerical Modeling

Ayman L. Fayed[1], Tamer M. Sorour[1], and Hany F. Shehata[2(✉)]

[1] Structural Engineering Department, Ain-Shams University, Abbasseya, Cairo, Egypt
{ayman_fayed, tamer.sorour}@eng.asu.edu.eg
[2] Housing and Building National Research Center (HBRC), Dokki, Giza, Egypt
hf.shehata@yahoo.ca

Abstract. Stone columns have been increasingly acknowledged as one of the viable foundation techniques in soft ground conditions. The concept behind this option is to enhance soft soils by insertion of special elements stiffer than the original soil. The stiffness of the improved ground depends on the geometry of the stone columns in addition to the relative stiffness between the original soft soil and the installed elements. Back analysis of case histories is always a successful and reliable approach in studying the in situ behavior of different ground improvement methods. In this paper, a well instrumented case study of floating stone columns installed in the Bothkennar soft clay formation is studied using numerical modeling utilizing the finite elements method. Different soil constitutive models are investigated to define the most appropriate one in simulating the actual Bothkennar soft clay behavior. Results of the comparison between the monitored field performance and the estimated behavior from the numerical model using different constitutive models are presented and discussed, while recommendations for the most suitable stress strain model of the studied clay are concluded.

1 Introduction

Vibro replacement is a popular form of ground improvement commonly used to enhance the settlement and bearing capacity characteristics of soft soils. Vertical columns of compact stones are formed in the ground using either the top or bottom feed systems. Vibro replacement process typically involves replacing 10–35% of the in situ soil with crushed stones. The high stiffness properties of the crushed stones reduce the overall and differential settlements of the structures founded on the treated soil. The consolidation time is also reduced due to the high permeability of the crushed stones. Stone columns can be used to provide support for a variety of loading scenarios ranging from small footings (i.e. pads/strip footings) to large area loadings (i.e. embankments and large dimensions slabs).

Design of foundations on soft soils is usually governed by settlement criteria rather than bearing capacity due to the high compressibility characteristics of soft soils. The majority of analytical design methods developed to date contain many simplifying

W. Frikha et al. (eds.), *Soil Testing, Soil Stability and Ground Improvement*,
Sustainable Civil Infrastructures, DOI 10.1007/978-3-319-61902-6_19

assumptions, such as the unit cell concept, which assumes an infinite grid of columns supporting an infinitely wide load area. Therefore, they do not account for the loss of lateral confinement associated with groups of columns supporting small footings. In addition, the reduction in vertical stresses with depth beneath small footings is much sharper than that beneath wide area loadings and, therefore, offers the possibility of partial depth treatment. While some correction factors exist to account for the loss of lateral confinement, current design methods do not consider the loss of lateral confinement for small groups of floating stone columns. While the bearing capacity of small groups of stone columns has been well researched, a lack of information exists regarding the settlement performance of small groups of columns. This is highlighted by McCabe et al. (2009) whose settlement database of over 20 case studies comprised only three cases related to small groups of columns.

2 Numerical Modelling: Procedures and Discussions

Many numerical studies conducted to study the behavior of the stone columns are axisymmetric analyses of large groups of columns. The shortage of information regarding the settlement performance of small groups of stone columns was identified by Black (2006) who conducted some high quality laboratory research; however, it is difficult to extrapolate the findings due to scale effects associated with laboratory tests and also as some of the area ratios considered are at the high end of typical values used in practice.

This research aims of studying the behavior of floating small groups of columns using finite element modelling by PLAXIS 3D 2016 for full-scale trial models in Bothkennar soft clay. The effect of the used constitutive laws on the performance of stone columns and the associated ground deformations are also investigated for both treated and untreated soft clay profiles. Lastly, some recommendations to properly simulate the small group of stone columns by numerical modelling in Bothkennar soil profile are provided.

2.1 Constitutive Models

Three common constitutive models that have already been implemented in most of the geotechnical finite element codes are used in this research; Mohr-Coulomb Model (MCM), Hardening Soil Model (HSM) and Soft Soil Model (SSM). A brief description of these models is presented below.

2.1.1 Mohr Coulomb Model (MCM)

Mohr-Coulomb model as shown in Fig. 1a is an elastic-perfectly plastic model that is often used to model soil behavior and serves as a first-order model. In general stress state, the model's stress-strain behaves linearly in the elastic range, with two defining parameters from Hooke's law (Young's modulus, E and Poisson's ratio, ν). There are two parameters define the failure criteria (the friction angle, ϕ and cohesion, c) and also a parameter to describe the flow rule (dilatancy angle, ψ which comes from the use of non-associated flow rule which is used to model a realistic irreversible change in

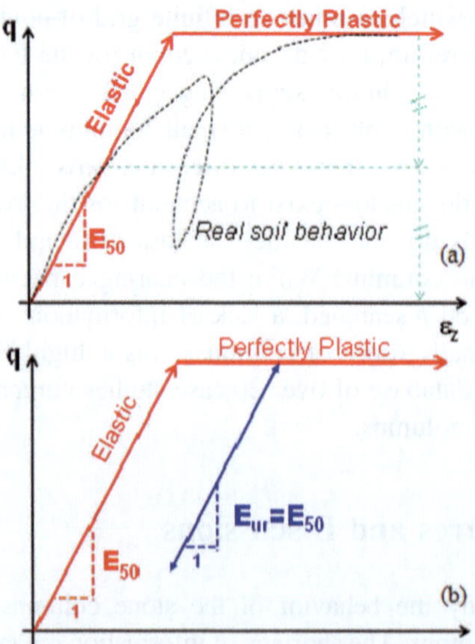

(a) Elastic-perfectly plastic assumption of Mohr-Coulomb constitutive model.

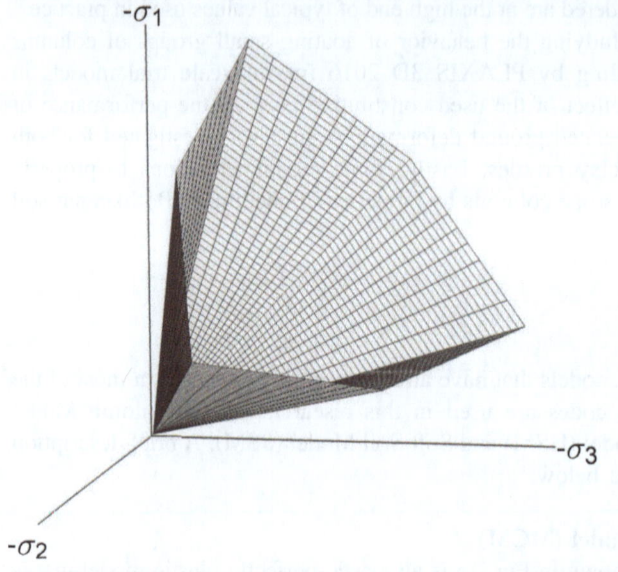

(b) The Mohr-Coulomb yield surface in principal stress space (c = 0)

Fig. 1. Mohr-Coulomb constitutive model.

volume due to shearing). Mohr-Coulomb model is a simple and applicable to three-dimensional stress space model, as shown in Fig. 1b, with only two strength parameters to describe the plastic behavior.

2.1.2 Hardening Soil Model (HSM)

The Hardening Soil model (Brinkgreve and Vermeer 1997; Schanz 1998) is a true second order model for soils in general, for any type of application (Brinkgreve 2005). The model involves friction hardening to model the plastic shear strain in deviatoric loading, and cap hardening to model the plastic volumetric strain in primary compression. Distinction can be made between two main types of hardening, namely shear hardening and compression hardening. Shear hardening is used to model irreversible strains due to primary deviatoric loading. Compression hardening is used to model irreversible plastic strains due to primary compression in oedometer loading and

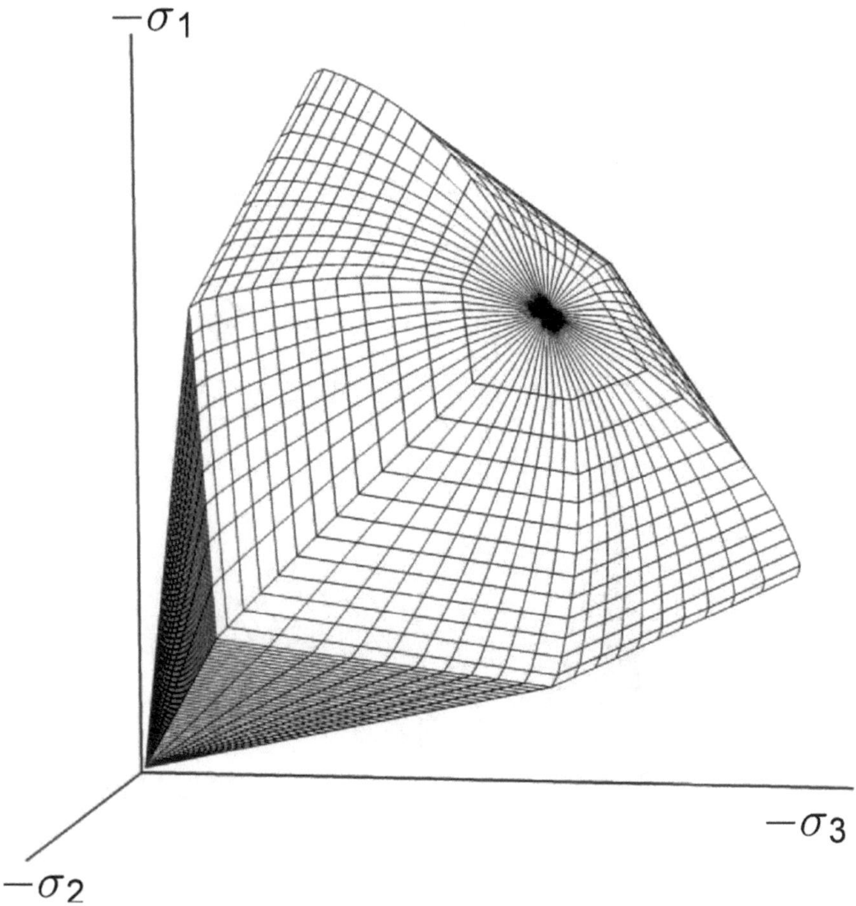

Fig. 2. Total yield contour of the hardening-soil constitutive model in principal stress space for cohessionless soil.

isotropic loading. Both types of hardening are contained in this model. Yield contour of the model in three-dimensional space is shown in Fig. 2. Failure is defined by means of Mohr-Coulomb failure criterion. Some basic characteristics of the model are Stress dependent stiffness according to a power law (m), plastic straining due to primary deviatoric loading (E_{50}^{ref}), plastic straining due to primary compression (E_{oed}^{ref}), elastic unloading/reloading input parameters (E_{ur}^{ref}, ν_{ur}) and failure criterion according to the Mohr-Coulomb model (c, φ and ψ).

2.1.3 Soft Soil Model (SSM)

The soft soils could be considered like near-normally consolidated clays, clayey silts and peat. A special feature of such materials is their high degree of compressibility. This is best demonstrated by oedometer test data as reported for instance by Janbu in his Rankine lecture (1985). Considering tangent stiffness moduli at a reference oedometer pressure of 100 kPa, he reported for normally consolidated clays E_{oed} = 1 to 4 MPa, depending on the particular type of clay considered. The differences between these values and stiffnesses for NC-sands are considerable as here the values are in the range of 10 to 50 MPa, at least for non-cemented laboratory samples. Hence, in oedometer testing normally consolidated clays behave ten times softer than normally

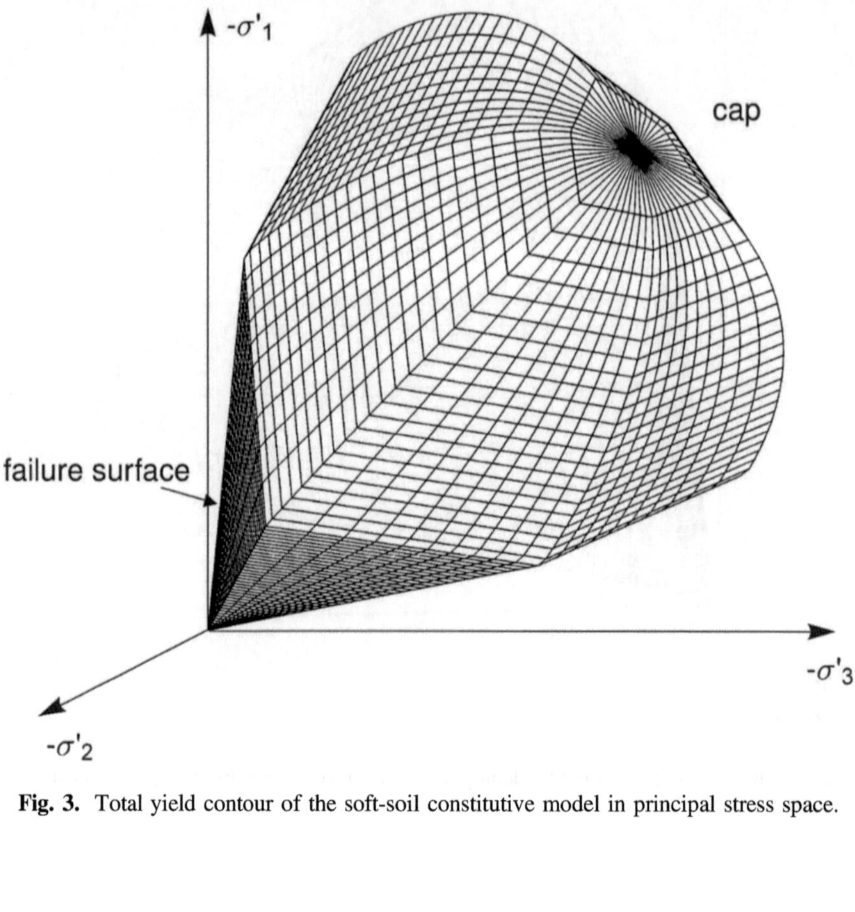

Fig. 3. Total yield contour of the soft-soil constitutive model in principal stress space.

consolidated sands. This illustrates the extreme compressibility of soft soils. The Soft Soil model is a Cam-Clay type model especially meant for primary compression of near normally consolidated clay-type soils. The total yield contour in principal stress space is shown in Fig. 3. The parameters of the Soft Soil model include compression (C_c) and swelling (C_s) indicies, which are typical for soft soils, as well as the Mohr-Coulomb model failure parameters.

2.2 Subsurface Conditions of the Case Studies Site

The site selected for the field trials at Bothkennar was a facility for large or full scale soft clay researches, with the site having been used extensively for research into in-situ testing and also 'undisturbed' sampling techniques. A comprehensive 'state-of-the-art' ground investigation and geotechnical data existed for the site. A wealth of researches have been previously undertaken, including Hight et al. (1992), Paul et al. (1992) and Nash et al. (1992a, b) and was detailed in the Institution of Civil Engineers Géotechnique Symposium (Vol. 42, No. 2: 1992). The detailed geotechnical profile of the site under study is shown in Fig. 4.

The Bothkennar clay profile shown in Fig. 4, is a result of a well-documented and well-established geotechnical investigation through many research projects. Hight et al. (1992) and Nash et al. (1992a) have provided rigorous investigation works, which had been carried out to provide data characterizing the clay for the benefit of future research projects. The clay properties used in the soil model are presented in Table 1 and are divided into crust, upper Carse clay and lower Carse clay.

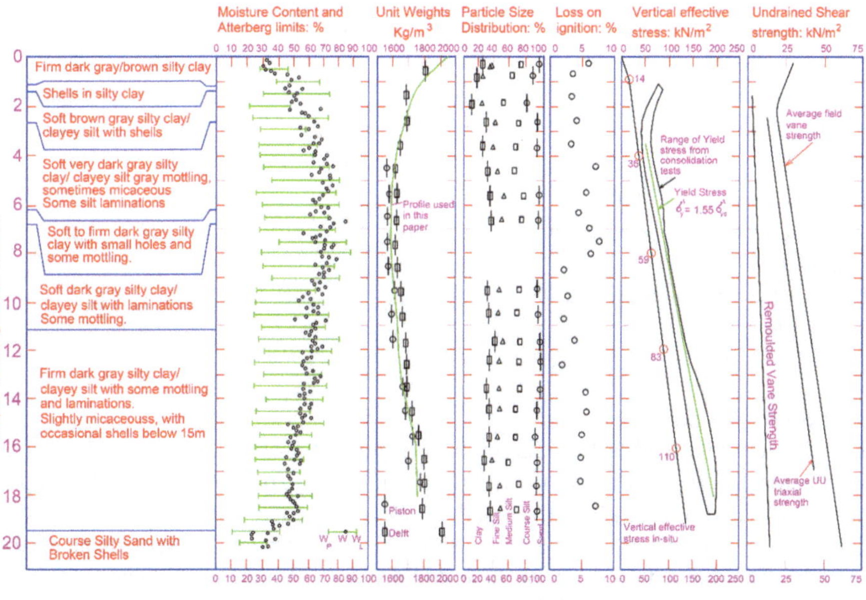

Fig. 4. Key geotechnical profile for the Bothkennar soft clay research site (after Nash et al. (1992a, b))

A critical state friction angle (φ') of 34° is used for the Carse clay. A slightly cohesion value of 3 kPa was used for the crust layers. E_{50}^{ref} is assumed equal to E_{oed}^{ref} according to Elshazly et al. (2009).

In choosing the friction angle of the stone backfill, the value of $\varphi' = 45°$ shown in Table 1, in which the stone column properties have been included, should be readily achievable, as noticed by McCabe et al. (2009). Default values of $E_{50} = 70$ MPa and $E_{ur} = 210$ MPa were adopted. The exponent, m, for the Hardening Soil model was chosen as m = 0.3. These stiffness parameters (i.e. E_{50}, E_{ur} and m) are in accordance with the values adopted by Gäb et al. (2008).

Table 1. Summary of the geotechnical material parameters adopted in the analysis of the Bothkennar test site

Soil parameter	Crust	Upper carse clay	Lower carse clay	Stone columns
Depth (m)	0.0–1.2:1.5	1.2:1.5–2.5	2.5–14.5	As needed
Bulk unit weight, γ (kN/m^3)	18.0	16.5	16.5	19.0
Over-consolidation ratio (OCR)	1.0	1.0	1.5	–
Pre-overburden pressure (kPa)	15	15	0	–
Coefficient of lateral earth pressure, K_0	1.5	1.0	0.75	0.3
Effective cohesion, c′ (kPa)	3	1	1	1
Angle of internal friction, ϕ' (degrees)	34	34	34	45
Dilatancy angle, ψ (degrees)	0	0	0	15
Initial voids ratio, e_0	1.0	1.2	2.0	–
Compression index, C_C (SSM)	0.07	0.25	1.12	–
Swelling index, C_S (SSM)	0.01	0.03	0.16	–
E_{50}^{ref}, kPa (HSM)	1068	506	231	70000
E_{ur}^{ref}, kPa (HSM)	5382	3036	1164	210000
E, kPa (MCM)	4485	2530	970.3	70000
m (HSM)	1.0	1.0	1.0	0.3
Reference pressure, p^{ref} (kPa)	13	20	30	100
Vertical coefficient of permeability, k_{vert} (m/day)	6.9×10^{-5}	6.9×10^{-5}	6.9×10^{-5}	1.2
Horizontal coefficient of permeability, k_{horz} (m/day)	1.0×10^{-4}	1.0×10^{-4}	1.0×10^{-4}	1.2

Case Study No (1)

Instrumented field load test on unreinforced rigid pad footings at the Bothkennar test site, documented by Jardine et al. (1995), is numerically modelled during the scope of this paper. The footing was loaded to failure using kentledge in less than five days, with pauses in loading occurring overnight and whenever the rate of settlement exceeds 8 mm/hr. The foundation depth of the tested pad is 0.80 m below ground level, while the square pad side length is 2.20 m.

2.2.1 Finite Element Modelling

The load test is simulated using the software PLAXIS 3D 2016 as an undrained loading (with effective stress material parameters) due to the short duration of the load test. However, the crust in the adopted soil profile is modelled as a drained material as the lower part of the crust contains a significant proportion of shelly fragments (Nash et al. 1992b) and the upper part of the crust is above the groundwater level. A fine mesh has been generated to avoid any effects from the meshing size on the results. Wide boundary conditions have been chosen in order not to affect the settlement response. Figure 5 shows the discretization of the problem.

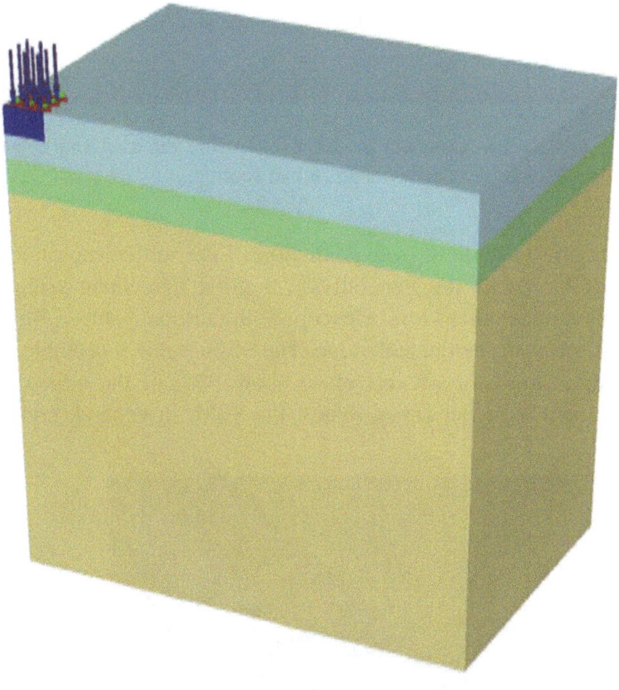

Fig. 5. Finite element model in case study no (1).

2.2.2 Results and Discussions

The relations between the load and settlement for both the field measurements and the results of the numerical model using different soil constitutive models are presented in Fig. 6. It can be recognized that the numerical model, in general, using the different constitutive models; (MCM, HSM and SSM), captures the trend of settlement versus applied pressure as compared to the monitored correlation from the field loading test. For the HSM, the settlement predicted value at 140.0 kPa, which is the maximum tested stress and represents almost the ultimate bearing pressure as can be understood from the field stress-settlement relation, is almost the same as the field measurement,

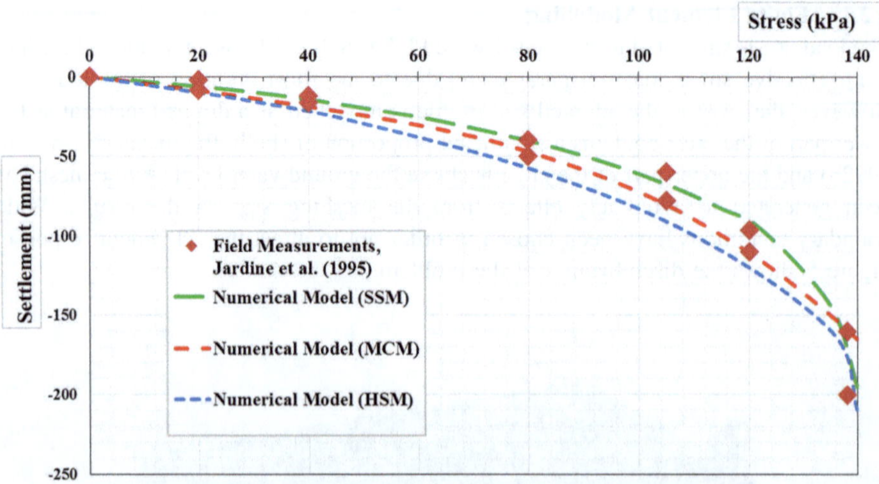

Fig. 6. Comparison between numerical model results using different soil constitutive models and the actual load–displacement behavior for a pad footing.

however there is always a slight over-prediction of the settlement values at any other stress levels. MCM results are generally in a good agreement with the measured settlement values at most stress levels, except at the ultimate stress, 140 kPa, where it slightly underestimates the settlement value. The SSM yields acceptable estimate of the settlement values at low stress levels up to about 30% of the ultimate stress value. Above that level and at higher stress values, the SSM, in general, underestimates the

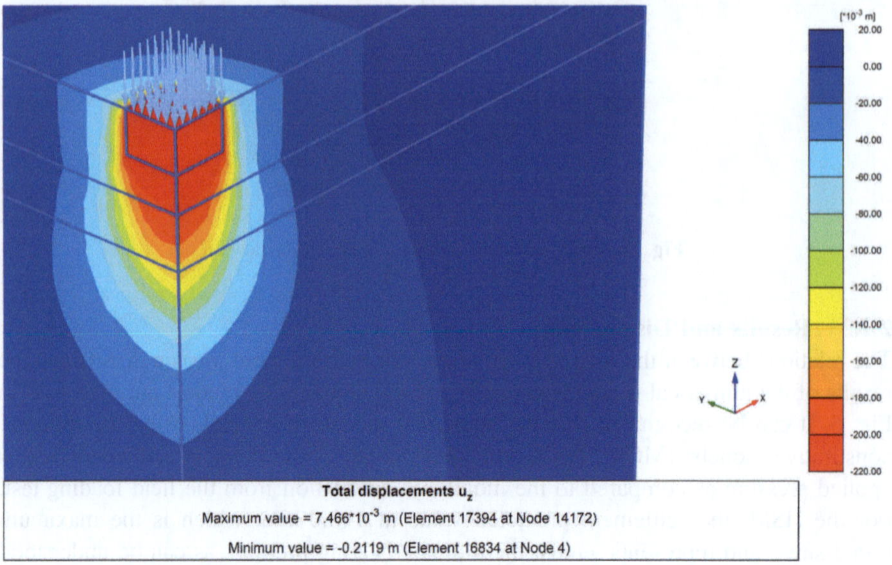

Fig. 7. Displacement field beneath the loaded pad at stress level of 140 kPa, using the HSM.

settlement values of the loaded footings. For this case study, it can be concluded that the three investigated constitutive models can reasonable simulate the actual Bothkennar clay behavior, with the preference to the HSM at high stress levels. Figure 7 presents the total deformations field at an applied stress value of 140 kPa that is almost the ultimate pressure sustained by the loaded pad utilizing the HSM.

Case Study No (2)

The second case study numerically analyzed in this research is fully detailed by Colin J. Serridge (2013) in his PhD thesis comprising a full-scale field testing of 8 models resting on Bothkennar clay. Eight models were constructed to represent the behavior of the rectangular and square footings in which seven models were resting on a treated soil with small groups of stone columns and one model was loaded without soil improvement. Table 2 summarizes the eight trial footings configurations, while Table 3 provides the loading stages of each model. The first and second loading stages were consolidated in 5 months each.

Table 2. Summary of the trial tests configurations used in case study no (2)

Trial footing no.	Dimensions (L × B) m	Foundation depth (m)	Ground treatment		
			Number of stone columns	Stone columns spacing (m)	Stone columns lengths (below footings) m
1	6.0 × 0.75	0.50	4	1.50	5.70
2	6.0 × 0.75	0.50	3	2.00	5.70
3	3.0 × 0.75	0.50	2	1.50	3.70
4	3.0 × 0.75	0.50	2	1.50	5.70
5	3.0 × 0.75	0.50	2	1.50	7.70
6	3.0 × 0.75	1.20	2	1.50	5.70
7	1.5 × 1.5	0.50	2	1.20	5.70
8	3.0 × 0.75	0.50	Untreated soil	–	–

Table 3. Summary of the loading increments of the trial footings for case study no (2)

Trial footing no.	Foundation depth (m)	1st loading increment kN/m^2	2nd loading increment kN/m^2
1	0.50	35.5	72.0
2	0.50	32.9	67.1
3	0.50	33.1	67.8
4	0.50	34.9	71.7
5	0.50	32.1	67.0
6	1.20	34.2	69.6
7	0.50	32.7	67.0
8	0.50	34.3	71.6

2.2.3 Finite Element Modelling

The full-scale models configurations, shown in Table 2, and with the loading stages, summarized in Table 3, have been numerically modeled using PLAXIS 3D 2016. The models represent rectangular as well as square footings configurations. In addition, the foundation depths vary from 0.5 m to 1.2 m in order to check whether the crust layer would affect the results or not. The effect of the number of stone columns and the columns spacing (S) have also been investigated through these full-scale field tests. One model (Footing no (8)) has been loaded without adding stone columns, to examine the load-settlement behavior with time for the untreated soil.

2.2.4 Effect of Constitutive Laws on the Load-Settlement-Time Relationship

Figure 8 and Table 4 present the load-settlement-time relations pertaining to Footing no (8) for both the field and numerical models. It is noticed that, the MCM can't mimic the behavior of the Bothkennar clay with time especially at the high stress levels and results in underestimated settlement values. On the other hand, the SSM provides a better estimation of the load-settlement relation especially as the duration of loading increases. The error in predicting the settlement, as shown in Table 4, ranges between approximately (−10%) and (−18%) compared to the error values ranging between approximately (−25%) & (−40%) for the MCM. The HSM predicted settlement values are, in general, more close to the field measured deformations with error values ranging between (−5%) & (+14%), as can be noticed from Table 4. Accordingly, it can be concluded that the HSM can represent the load-settlement behavior of the footings founded on the untreated Bothkennar clay in a better manner than both the MCM and the SSM. However, the HSM overestimates the settlement values by (+10% to +14%) compared to the field results of stress levels above 50% of the approximate ultimate level.

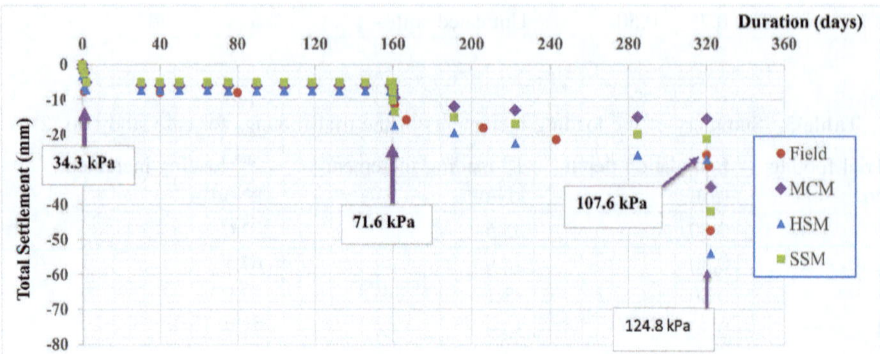

Fig. 8. Measured and predicted load-settlement relations with time for footing no (8) without ground improvement.

Table 4. Measured and predicted settlement at different stress levels and durations for case study no (2).

Applied stress (kPa)	Time (day)	Measured settlement (mm)	Numerical model results					
			MCM		SSM		HSM	
			Estimated settlement value (mm)	Error (%)	Estimated settlement alue (mm)	Error (%)	Estimated settlement Value (mm)	Error (%)
34.3	0	−7.80	−7.05	**−09.62**	−7.25	**−07.05**	−7.40	**−05.13**
	40	−7.90	−7.15	**−09.49**	−7.30	**−07.59**	−7.55	**−04.43**
	120	−8.00	−7.25	**−09.38**	−7.40	**−07.50**	−7.80	**−02.50**
	160	−8.10	−7.30	**−09.88**	−7.50	**−07.41**	−7.90	**−02.47**
71.6	160	−15.80	−9.50	**−39.87**	−13.50	**−14.56**	−17.50	**+10.76**
	220	−20.40	−13.20	**−35.29**	−17.15	**−15.93**	−22.50	**+10.29**
	280	−22.85	−15.25	**−33.26**	−19.90	**−12.91**	−25.95	**+13.57**
	320	−25.90	−15.60	**−39.77**	−21.30	**−17.76**	−27.30	**+05.41**
107.6	321	−29.10	−20.00	**−31.27**	−25.50	**−12.37**	−32.50	**+11.68**
124.8	322	−47.40	−35.25	**−25.63**	−42.30	**−10.76**	−54.15	**+14.24**

2.2.5 Effect of the Stone Columns on the Load-Settlement-Time Relations

The first footing with stone columns (Footing no (1)) was numerically modelled representing a small group of 4 stone columns with spacing of 1.5 m and length of 5.70 m with a foundation depth of 0.5 m. The stone columns are modelled using the Mohr Coulomb criterion, while the soil layers were simulated using the three types of soil constitutive models; MCM, HSM and SSM. Figure 9 presents the discretization of the model. Figure 10 shows the relation between the stresses and estimated average settlement with time along with the field measurements. From the results, it can be concluded that both the MCM and SSM constitutive models can't mimic the softy soil behavior properly and results in underestimated predicted values for the settlement

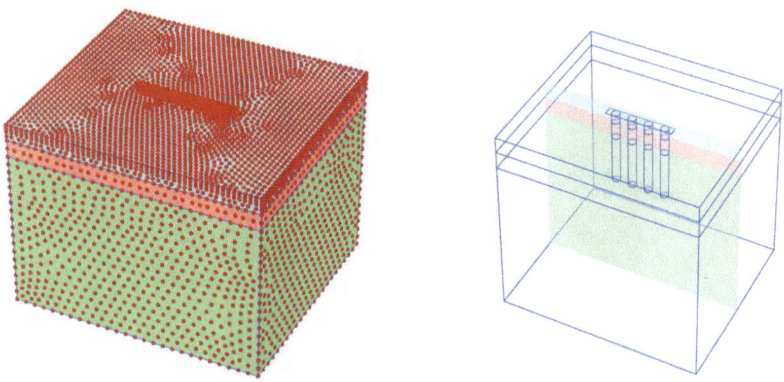

(a) Meshing and nodes. **(b) Cross-section to show the stone columns**

Fig. 9. Finite element models of footing no (1) for the second case study.

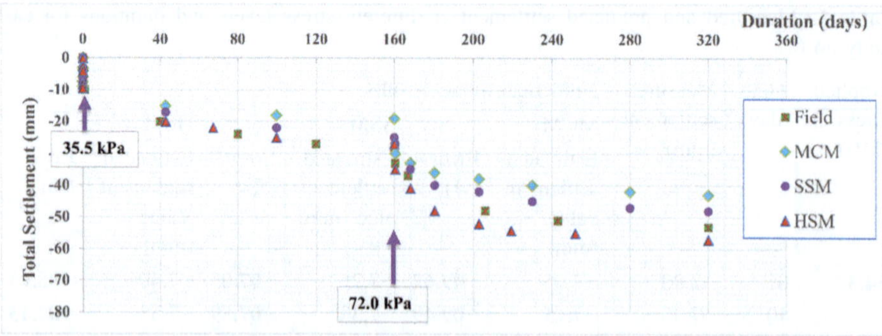

Fig. 10. Measured and predicted load-settlement relations with time for trial footing no (1) of the second case study (footing with four stone columns).

with time. However, all the three soil models could be used at a low stress levels, in which the stress is lower than 30% of the ultimate bearing pressure. The same as the footing without improvement (Footing 8), the HSM could represent the soil layers of the Bothkennar soil profile, however it results in higher predicted values of settlement in the range of (+12% to +17%) above the field measurements. Figure 11 shows the deformed shape of the stone columns under an applied stress of 72.0 kPa and using the HSM, in which it can be noticed that bulging partially occurs with some tilting of the edge columns.

In the light of the previous results, the trial model (Footing 5) was modelled, which represents a different group of only 2 stone columns with the same spacing of 1.5 m

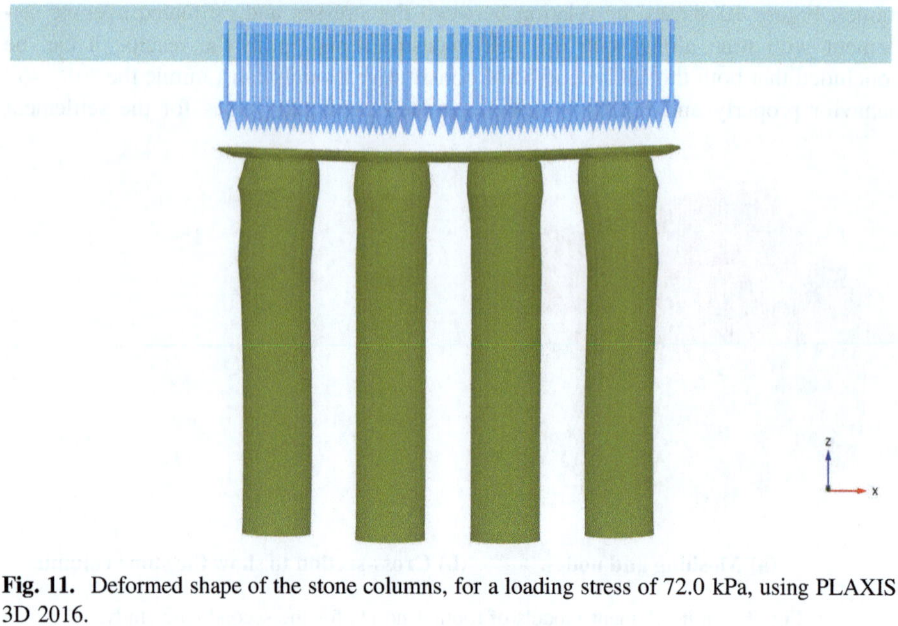

Fig. 11. Deformed shape of the stone columns, for a loading stress of 72.0 kPa, using PLAXIS 3D 2016.

and foundation depth of 0.5 m but with longer stone columns lengths of 7.7 m. Only the HSM has been adopted to represent the soil layers behavior. Figure 12 shows the relation between the stress and average settlement for both the numerical model and field measurements. It could be noticed that the HSM can represent the loading process, in general, but results in an overestimation of the settlement values by about (+13% to +18%) than the field measurements.

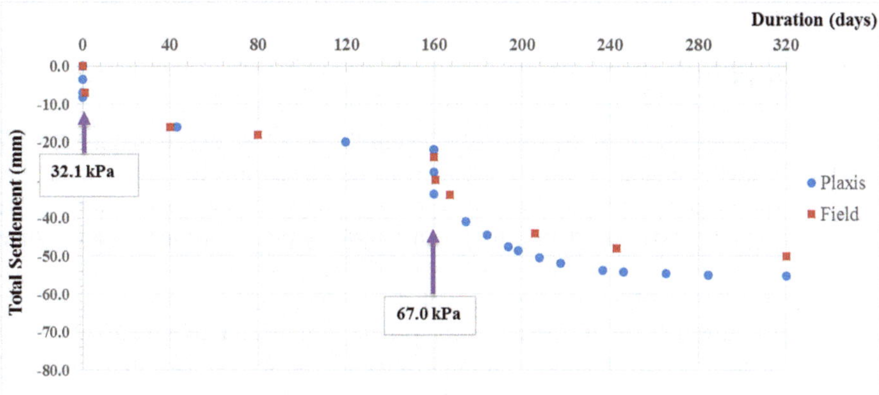

Fig. 12. Measured and predicted load-settlement relations with time for trial footing no (5) of the second case study (footing with two stone columns) utilizing the HSM.

Footing no. (6), as detailed in Table 2, has been also simulated using the same conditions, which represents a different foundation depth of 1.2 m to rest directly on the upper carse clay, and with using the stone column's spacing of 1.5 m and length of 5.7 m. The first construction stage was to excavate the overall crust layer, which is 1.2 m in height, in order to lay the footing directly on the upper carse clay. The second

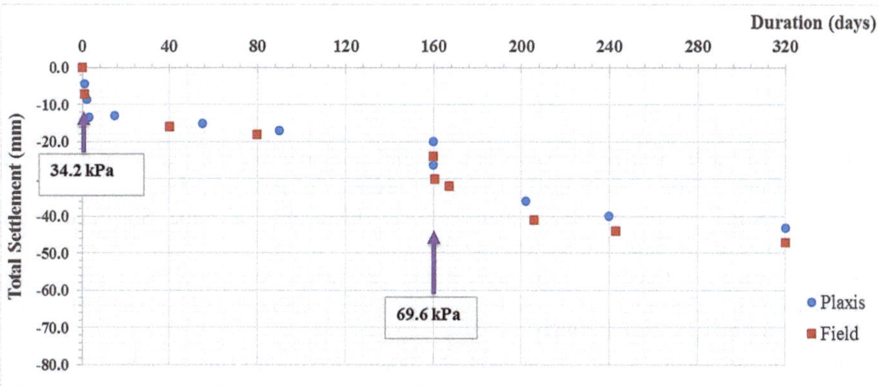

Fig. 13. Measured and predicted load-settlement relations with time for trial footing no (6) of the second case study (footing with two stone columns).

construction stage was to construct the stone columns and with the configuration given in Table 2. Due to the time duration gap between the two stages, significant heave has been occurred. Figure 13 shows the results of both the numerical model and the field measurements. In this case, and due to the heave resulted from the stages of construction, lower values of the settlement, ranging between (−5% to−9%), have been predicted using the HSM. However, this range of error is considered still small and would not greatly affect the results. Hence, the HSM could be used to reasonably simulate the behavior of the Bothkennar soil profile in predicting the settlement values.

3 Conclusions

This paper presents results of the study of effect of the adopted constitutive models for soft clays behavior simulation on the accuracy of numerical models in predicting the settlement values associated with loading of footings supported on stone columns. The finite element code Plaxis 3D 2016 is used in the performed analyses in which the three common constitutive models; MCM, SSM and HSM, have been used to investigate their suitability in modelling the behavior of the studied soft clay. Two case studies were analyzed in which the first case was for a short term loading of a footing founded on the Bothkennar clay without ground improvement. The second case study included modelling of several trial footings with and without ground improvement and with different configurations of the stone columns for the ground treated cases. The performed analyses showed that the three investigated constitutive models can capture the stress-settlement behavior of the Bothkennar soft clay with the preference of the HSM over other models especially for the long term loading conditions. Using the HSM resulted in predicting the settlement values associated with loading footings supported on stone columns in the Bothkennar soft clay with an approximate deviation in the accuracy ranging between (−9% to +20%) compared with the actual field measured values. Both the MCM and SSM can be used in simulating the behavior of the studied clay formation but with less accuracy ranging between approximately (−25% to −40%) and (−10% to −18%), simultaneously.

References

Black, J.A.: The settlement performance of a footing supported on soft clay reinforced with vibrated stone columns. Ph.D. thesis, Queen's University of Belfast (2006)

Brinkgreve, R.B.J., Vermeer, P.A.: Plaxis Finite Element Code for Soil and Rock Analysis-Version 7. Balkema, Rotterdam (1997)

Brinkgreve, R.B.J.: Selection of soil models and parameters for geotechnical engineering application. In: Yamamuro, J.A., Kaliakin, V.N., (eds.) Geotechnical Special Publication No. 128, ACSE, pp. 69–98 (2005)

Serridge, C.J.: An evaluation of partial depth dry bottom-feed vibro stone columns to support shallow footings in deep soft clay deposits. Ph.D. thesis, Anglia Ruskin University, UK (2013)

Elshazly, H.A., Hafez, D.H., Mossaad, M.E.: Reliability of conventional settlement evaluation for circular foundations on stone columns. Geotech. Geol. Eng. **26**(3), 323–334 (2009)

Gäb, M., Schweiger, H.F., Kamrat-Pietraszewska, D., Karstunen, M.: Numerical analysis of a floating stone column foundation using different constitutive models. In: Proceedings of the 2nd International Workshop on Geotechnics of Soft Soils, Glasgow, pp. 137–142 (2008)

Hight, D.W., Bond, A.J., Legge, J.D.: Characterisation of the Bothkennar clay: an overview. Geotechnique **42**(2), 303–347 (1992)

Institution of Civil Engineers: Bothkennar soft clay test site: characterization and lessons learned. Geotechnique **42**(2), 161–378 (1992)

Jardine, R.J., Lehane, B.M., Smith, P.R., Gildea, P.A.: Vertical loading experiments on rigid pad foundations at Bothkennar. Geotechnique **45**(4), 573–597 (1995)

McCabe, B.A., Nimmons, G.J., Egan, D.: A review of field performance of stone columns in soft soils. In: Proceedings of ICE Geotechnical Engineering (2009)

Nash, D.F.T., Sills, G.C., Davison, L.R.: Onedimensional consolidation testing of soft clay from Bothkennar. Geotechnique **42**(2), 241–256 (1992a)

Nash, D.F.T., Powell, J.J.M., Lloyd, I.M.: Initial investigations of the soft clay site at Bothkennar. Geotechnique **42**(2), 163–181 (1992b)

Paul, M.A., Peacock, J.D., Wood, B.F.: The engineering geology of the Carse clay of the national soft clay research site. Bothkennar. Géotechnique **42**(2), 183–198 (1992)

Schanz, T.: Zur modellierung des mechanishen verhaltens von reibungsmaterialen. Habilitation. Stuttgard University (1998)

Auto-Controlled Ménard Pressuremeter: A Novel Tool for Optimal Use of the Pressuremeter

Wissem Frikha[1,2](✉) and Serge Varaksin[2]

[1] Université Tunis El Manar, Ecole Nationale D'ingénieurs De Tunis LR14ES03, Ingénierie Géotechnique, Tunis, Tunisia
frikha_wissem@yahoo.fr
[2] APAGEO, ZA de Gomberville-rue Salvador Allende, 78114 Magny les Hameaux, France
s.varaksin@apageo.com

Abstract. The pressuremeter test is an in situ controlled load-deformation test that is performed on the wall of a borehole using a radially expanded cylindrical probe. From the test readings (volume variation based on controlled pressure), a plane deformation stress-strain curve can be obtained for the soil.

Since the initial prototype, the pressuremeter has consistently been improved in its design, and the latest version of the pressuremeter, which is called the "auto-controlled Pressuremeter" has been developed to address the issues of repeatability and accumulation of approximations in a test. This apparatus is fully automatic and autonomous, and manages all steps of the test as instructed by the operator. The auto-controlled Pressuremeter simplifies the work procedure for the operator, reinforces the reliability of the results, and reduces the time of set up.

The present paper describes the auto-controlled pressuremeter, and compares it with the manual pressuremeter. The first part of the paper compares the deficiencies of the manual and auto-controlled Pressuremeter. In the second part, test results of the two types of pressuremeter that have been performed in similar conditions are presented and commented.

1 Introduction

The pressuremeter test is an in situ controlled load-deformation test that is performed on the wall of a borehole using a radially expanded cylindrical probe. From the test readings (volume variation based on controlled pressure), a plane deformation stress-strain curve can be obtained for the soil.

The first pressuremeter prototype was developed by Louis Ménard in January 1955. Since then, the pressuremeter has consistently been improved in its design.

The first pressuremeter prototype, which is called Type A, consists of a hand pump for injecting constant increments of water and a large probe with a diameter of 140 mm. After this first model, over the years, several prototypes have been designed by selecting better materials for pressure application, minimizing approximations, and

W. Frikha et al. (eds.), *Soil Testing, Soil Stability and Ground Improvement*,
Sustainable Civil Infrastructures, DOI 10.1007/978-3-319-61902-6_20

improving the sensor and recording system; i.e. Types B, C, D, E, F, G, etc. (Cassan 2005).

Type G pressuremeter refers to devices with nested cell probes. Old notations GA, GC used in 1965–1966, correspond to pressure-volume controllers interchangeable with type G. For these devices, the guard cells are inflated with compressed air. The pressuremeter GB, has a totally different design, characterized by two guard cells inflated with water and by two volumeters.

From 1984, a pressuremeter identical to the GA was put into service, except that there is no differential pressure gauge. The latter is calculated by the operator from the pressures measured by gauges from the measuring cell and the guard cells. At present, it is the most used pressuremeter. Over three thousand pressuremeters have been built and used in all the word (over hundred countries).

Recently, there has been technological developments such as pressure gauges that connect to the circuit and allow working with automatic data acquisition (Geospad®). This system contains electrical recording components connected to pressure and volume sensors.

The last generation of pressuremeter follows the electronic and automatic technological developments. It is called the "auto-controlled pressuremeter" and has been developed to address the issues of repeatability and accumulation of inaccuracies in the test. According to ISO 22476-4 standard, this apparatus is fully automatic, autonomous, and manages all steps of the test as preselected by the operator. The auto-controlled pressuremeter simplifies the work procedure for the operator, enhances the reliability of the results, and reduces the time of preparation.

The first part of this paper describes the manual and the auto-controlled pressuremeters. The second part of the present paper presents the process of correction of pressure loss provided by the auto-controlled pressuremeter. Finally, a comparison between tests that have been performed (in situ and in artificial ground) using the two types of pressuremeter, in similar conditions, are presented and commented.

2 Manual Ménard Pressuremeter –Type G

The manual Ménard pressuremeter consists of a CU (control unit), a plastic tubing and a 3-cell probe, which allow performing in situ tests according to the ISO 22476-4 and ASTM D4719-07 standards. The borehole is drilled to minimize wall disturbance and to keep the diameter of the hole in accordance with the selected probe size.

The CU is an aluminium box (86 cm × 43 cm × 26 cm), with a protective cover, standing on a 65 cm high tripod. It is handle transported, its weight is 24.5 kg (tripod, 3.5 kg) (Fig. 1). The tripod and level allow the vertical setting of the pressuremeter in all sites.

The CU consists of devices allowing the regulation of the applied pressure and the reading of the volume change. It includes an 800 cm^3 sight tube volumeter for reading the volume changes of the measuring cell, regulators for both main and differential pressures, pressure gauges of 0 to 25 bars and 0 to 60 bars for both guard and measuring cells, and several valves and connectors (Fig. 1).

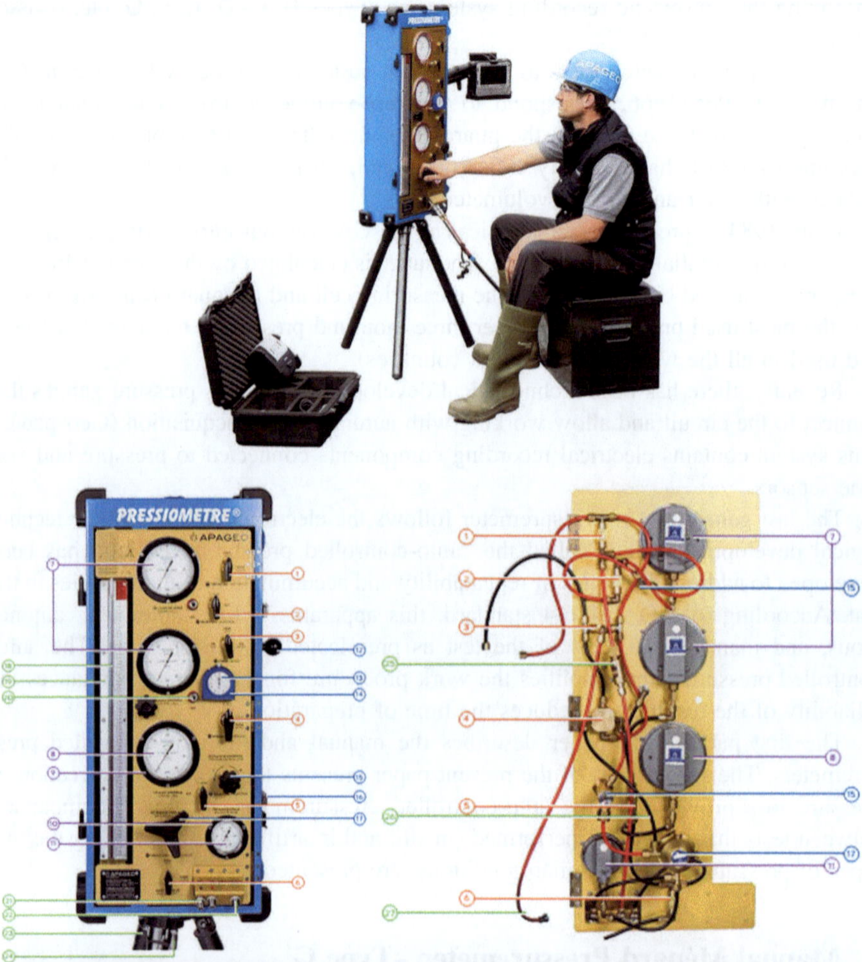

Num	DESCRIPTION	Num	DESCRIPTION
1	Gas circuit stop valve (complete)	15	Pressure regulator (complete)
2	25/100 bar gas selection valve (complete)	16	Fixation for lid type 3
3	Bleed valve for both gas and water (complete)	17	Main pressure regulator
4	Inversion valve 0-10m	18	Sight tube complete (new model)
5	25/100/60 bar water selection valve (complete)	19	Polycarbonate cover plate for sight tube
6	Water circuit stop valve (complete)	20	Quick female socket for extra gauge
7	0-25 bar gauge (vertical outlet)	21	Quick female socket for hose outlet (water and gas)
8	0-60 bar gauge (vertical outlet)	22	Quick female socket for Nitrogen bottle inlet
9	Ø 100 gauge ring	23	Tripod
10	Ø 60 gauge ring	24	Stainless steel tripod axis
11	0-250 bar gauge (vertical outlet)	25	Filter housing complete
12	Ball Fixation for GeoBOX®	26	Rilsan tubing 3x6 Pressuremeter inner black lead
13	Stop watch 1 minute	27	Rilsan tubing 3x6 Pressuremeter inner red lead
14	Protection case for stop watch		

Fig. 1. Ménard pressuremeter equipped with central unit computer (GeoBOX®) and central acquisition tool (Geospad®)

The probe is totally protected by a rubber cover (different types according to the soil stiffness), which is inflated by a gas in the 2 guard cells and by water in the measuring cell. The different cells' applied pressures are controlled by the differential regulator to ensure a cylindrical deformation along the measuring cell and to avoid any adverse boundary condition. The pressure source is provided through an external nitrogen gas cylinder.

The pressuremeter can be equipped with system tools that allows instant display of test results via a central unit computer (GeoBOX®) and central data acquisition tools (Geospad®). It permits the automatic record of the test data and the specific test conditions and the visualisation of the evolution of the data during the test. The pressure of the measuring cell, the differential pressure, the volumes variation, the number of increments and time are displayed during the performance of the test. The records are automatically carried out after 0, 15, 30 and 60 s with optimized accuracy: 0.10 cm^3 on the volume and 10 kPa on the pressures (in accordance to procedure B of ISO 22476-4).

The acquisition tool (Geospad®), integrated into Ménard pressuremeter, is a waterproof box including 2 pressure sensors of 0 to 100 bars and an ultrasonic sensor for volume.

To perform the test, a borehole is drilled to minimize wall disturbance and to keep the cavity diameter in accordance with the probe size. The probe is lowered into the borehole to the required test depth and the pressure is applied by equal increments. Pressure and volume is read from the control unit. As soon as the probe is lowered into the borehole to the required test depth, the operator can start the test by pressure increments with the control unit.

2.1 Auto-controled Ménard Pressuremeter (Geopac®)

The main differences between the auto-controlled pressuremeter and manual pressuremeter is the method of application of the pressure and the volume and their measurement.

The pressurization of the guard cell is performed by a modulation of nitrogen gas from an external cylinder using a control pressure regulator and sets of solenoid valves. Two pressure intervals are considered: 0 to 7 bars with an accuracy range of 0.01 bars and 7 to 100 bars with an accuracy range of 0.1 bars.

Water pressurizing in the measuring cell is performed using electro-mechanical tools consisting of a DC electric motor of 24 V and a reducer with a normal torque of 24 N/m, a high ratio gearbox with an epicycloid stage gear train, a ball screw torque, a transmitter of pushing force, a pressure-transmitting sealing piston, etc. The electronic control allows the variation of the power associated with the electric motor to command the speed by using an incremental encoder of small resolution.

The mechanical part of the auto-controlled pressuremeter permits the transformation of the electrical energy into work pressure applied on the water circuit.

The performance tests, the analysis of operations and the approval of the model (electrical and mechanical) were performed and certified by an independent national organisation CETIM (Centre Technique des Industries Mécaniques, France) in 2010.

The auto-controlled pressuremeter (GeoPAC®) consists of an adapted piston in a cylinder driven by a precision micro-motorization and is entirely piloted by a control unit (GeoBOX®) with a wireless connection (Fig. 2). The pressure sensors chosen for the device are of a high-precision class, allowing the target pressure to be reached and regulated quickly without drift. The regulation ensures the simultaneous recording of data at the imposed times. The precision on the measurement of the volumes, of an order of magnitude never previously reached in manual pressuremeter test, is 1/100 cm^3. The deformation at the interface between the probe and the borehole wall is therefore measurable with a resoluteness, depending on the type of probe used and its degree of expansion, relative to the loaded surface, in the order of 1 µm. Even before starting the soil tests, this precision is enhanced, due to the reliability and repeatability of the measurements, when the calibrations are carried out in series.

Fig. 2. Auto-controlled pressuremeter (GéoPAC®)

When the test is performed, the operator records the initial parameters and specific conditions in the control unit, such as borehole number, depth of the test, pressure of first stage, etc. As soon as the probe is lowered into the borehole to the required test depth, the control unit can begin the test. All the sequences of the whole process of standard test are also automated: pressure loss, volume loss, pressure increments and pressure steps settings. During the entire process, the control unit computer shows the monitoring of the actual test on its screen (progression, view of the results in real time, line graph etc.). The operator can stop or modify the test progress from the control unit at any time.

The auto-controlled pressuremeter can measure volume and pressure respectively with the precision of 0.05 cm^3 and 0.025 bar.

2.2 Principle of Pressure Loss Measurement

The pressure loss in a pressuremeter test is related to the tubing dimensions and the flow rate of the liquid circuit. For the classical Ménard pressuremeter, the regulation is usually performed after the test. Only auto-controlled pressuremeter is able de correct instantaneously the measurements by taking into account the regulation of the pressure loss. The adjustment of the pressure measurements depend on the pressure increment between two steps. In this phase, pressure should be increased to the next increment in an optimal time that does not exceed 20 s with adequate differential pressure between the liquid and gas circuits.

The differential pressure in the probe cannot be relied upon due to the overestimation of the liquid pressure when the pressure loss is not taken into account during the test. Indeed, at the nearest hydrostatic pressure, for a target differential pressure of 0.1 MPa and with a pressure loss of 0.2 MPa, the real differential pressure in the probe is then −0.1 MPa. The pressure regulation undergoes a strong disturbance around the target pressure. This regulation tends to decrease the flow rate when the measured pressure is near the set point, but the reduction in the flow rate causes a drop in the measured pressure (Arsonnet et al. 2013).

The control responds to this reduction by applying again the flow, which is translated by a phenomenon of oscillation going until the hydraulic circuit is excited (Arsonnet et al. 2013).

In the case of creep in certain levels (in the expansion of probe phase), the adjustment of flow in the liquid circuit is necessary to maintain the pressure constant. This flow creates a pressure loss, which adds to the probe pressure and results in the overestimation of the pressure by the sensor. The main consequence during the trial is that the differential pressure in the probe is not respected due to the overestimation of the liquid pressure and the pressurizing level is not made at a lower pressure that varies according to the instantaneous flow rate injected during this phase. This result is more evident in a calibration test where the creep is maximal and can cause intense oscillation in tests with long tubing.

To overcome these problems, the first approach for compensating the pressure drop was to identify the characteristics of the tubing, through laboratory tests, and to determine the coefficients of pressure loss per unit length (Fig. 3). The results show different tendencies lines as a function of tubing length. The auto-controlled pressuremeter also take into account the singular pressure loss in the entire hydraulic system, which can be varied significantly with the occasional or permanent sealing phenomenon: this is related to the in situ extended use of the hydraulic circuits tubing, their deformation during loading, and their repair by shortening and connector installation (Arsonnet et al. 2013).

Therefore, the pressuremeter test always requires a minimum volume of liquid to be injected at the beginning of test and before the probe comes into contact with the wall of the drilled hole. The auto-controlled pressuremeter consists of an automatic identification procedure that instantaneously computes the coefficients of the loss pressure in the hydraulic circuit. This procedure, which is called the "pressure loss calibration procedure" is carried out automatically without intervention of the operator at the beginning of the test, based on the control of hydraulic flow.

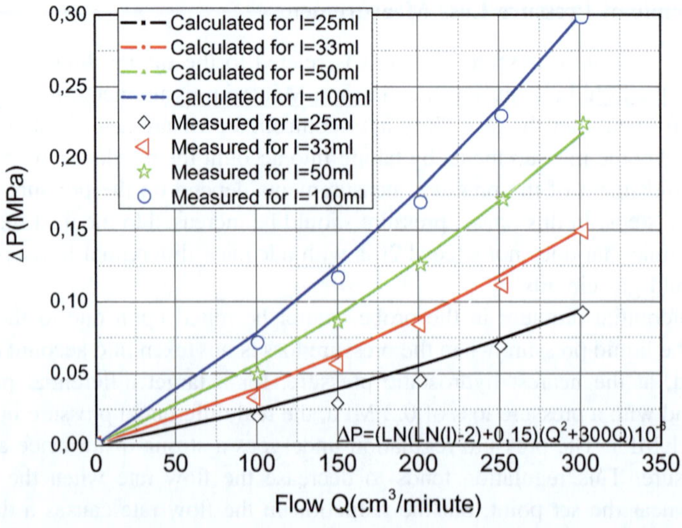

Fig. 3. Pressure loss as a function of flow rate for different tubing length (l = 25, 33, 50, 100 m) (from Arsonnet et al. 2013)

Several "pressure/flow" points are measured at different flow rates after stabilization of the hydraulic circuit (Fig. 3). These points are then used to characterize the pressure losses by calculating the coefficients of their representative function. This function is used to correct the pressure measurements of the liquid circuit as a function of the instantaneous flow rate.

This solution, which has been developed exclusively by APAGEO for GéoPAC® auto-controlled pressuremeter, has many advantages and is of great interest. In fact, it is integrated into the test and runs seamlessly at start-up without affecting the test progress itself. The pressure loss covers the entire liquid circuit. The accuracy of pressures level of soil that represent a certain creep is significantly improved in the case for tests between the beginning of the creep and the end of the failure, which increase in repeatability and accuracy. The oscillatory phenomena are completely attenuated and controlled and the automatic regulation becomes more efficient and accurate.

From the report of CETIM (2010), the flow rate used in the design of GéoPAC® is 270 cm^3/min. It is calculated from theses parameters: engine torque, engine reduction torque, piston travel speed and surface area, etc. Figure 4 shows the changes of pressure loss ΔP as a function of tubing length. The values of ΔP range from 0.07 to 0.26 bars.

The stiffness of auto-controlled pressuremeter is defined as the ratio of the applied pressure to the corresponding injected volume. The injected volume is important at low pressures (several tens of cm^3); whereas, for high pressure, the injected volume is very low (less than one cm^3). An algorithm has been developed to control the injected volume at any instant and to compensate the variation in pressure.

The gain (defined as the ratio of the voltage to the pressure) is adapted so that the process of the control increases with the decrease of the errors in order to command the

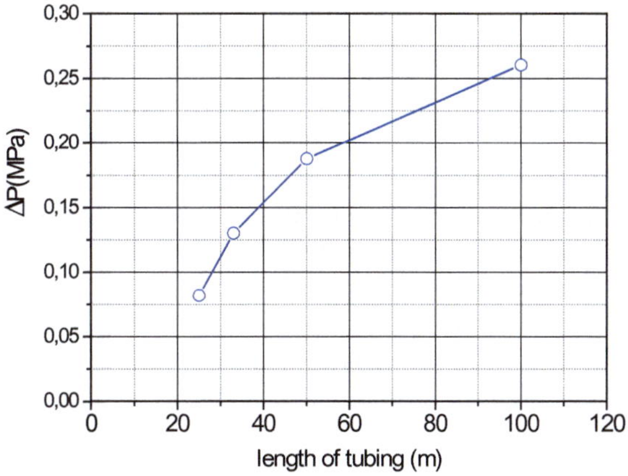

Fig. 4. Pressure loss ΔP as a function of tubing length

rotating the motor even for small deviations. An adaptive position control algorithm is also involved. A further synchronization is also implemented to maintain a constant pressure difference between the gas and the liquid pressures.

The inaccuracy obtained by the calibration is very important in manual pressuremeters, but is insignificant when using the GéoPAC®. In the auto-controlled pressuremeter, the difference between the pressure loss recorded for 10 m and 50 m of tubing is not important, is equal to 0.20 MPa for a volume exceeding 70 cm^3, and corresponds to the value obtained from the equation shown in Fig. 3 that relates flow to pressure loss (0.187 MPa). The pressure loss is double this value in the manual pressuremeter (Fig. 5).

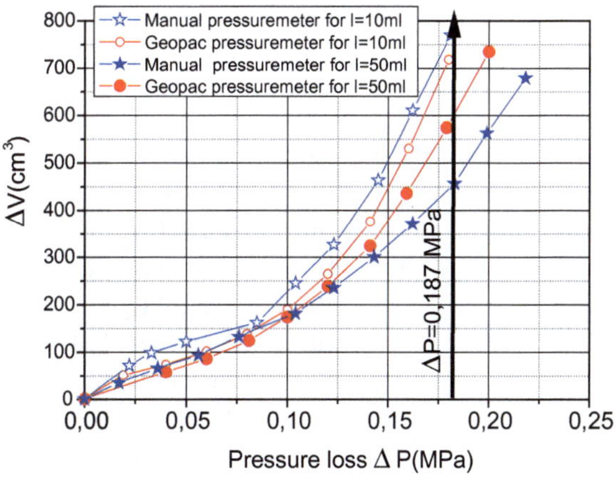

Fig. 5. Volume variation as a function of pressure loss

2.3 Test Performed on Artificial Ground

The first set of tests were performed by the auto-controlled and manual pressuremeters on artificial grounds to eliminate errors due to the heterogeneity of soil and the effect of boring and installation of the probe. The artificial grounds were high density moulded sleeves of polyurethane elastomer that were 63 mm × 155 mm × 500 mm (Fig. 6). Two types of artificial ground were tested (70 and 90 Shore A) using tubing lengths of 10 and 50 m. According to the indentation hardness of materials (like polymers, elastomers, and rubbers) by using the durometer (ASTM D2240 scales), the 70 Shore A is equivalent to 20 bars of resistance for a probe that has a diameter of 60 mm reach a volume of 600 cm^3. The 90 Shore A is equivalent to 30 bars of resistance. Table 1 summarizes the details of all pressuremeter tests that were performed on the artificial grounds.

Fig. 6. Photograph of an artificial ground that was used for testing

Table 1. Performed test on artificial ground

Tests	Ground resistance for 600 cm^3 of volume (bars)	Length of tubing (m)	Pressuremeter type	E_M (MPa)	P_f (MPa)	P_{LM} (MPa)
Press_AG_20b_10 m	20b	10	Manual type G	11.6	0.89	1.99
Press_AG_20b_50 m	20	50	Manual type G	13	0.65	1.86
Press_AG_30b_10 m	30	10	Manual type G	17.6	1.87	3.2
Press_AG_30b_50 m	30	50	Manual type G	18.9	1.74	3.5
Geopac_AG_20b_10 m	20	10	Auto-controlled	11.6	0.86	2.0
Geopac_AG_20b_50 m	20	50	Auto-controlled	12.1	0.85	2.06
Geopac_AG_30b_10 m	30	10	Auto-controlled	17	1.86	3.08
Geopac_AG_30b_50 m	30	50	Auto-controlled	21	1.75	3.4

Figures 7(a) and (b) show volume variations as a function of corrected limit pressure obtained from pressuremeter tests performed respectively on artificial grounds of 20 and 30 bars (70 and 90 shore A). Both tests show comparable results and the pseudo-elastic and plastic behaviours of the artificial grounds. All pressuremeter tests were performed using at least pressure increments. The volume variation measured with respect to time was recorded for each pressure increment at 15, 30 and 60 s after the application of pressure. The first levels (six to seven points) were derived with pressure increments of 0.05 MPa. For the final points, the pressure increment used was 0.1 MPa. The tests were completed when volume reached 600 cm^3.

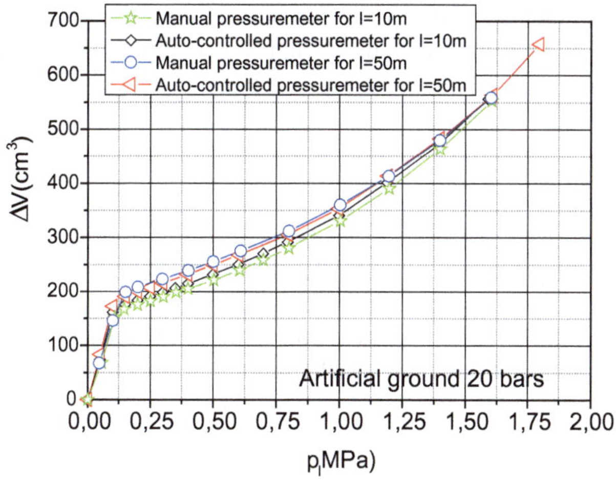

(a)

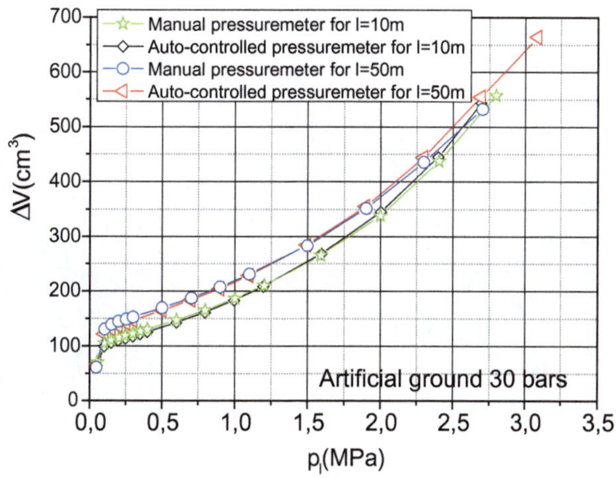

(b)

Fig. 7. Volume variation as a function of corrected limit pressure from pressuremeter tests performed in artificial grounds of (a) 20 bars and (b) 30 bars

The limit pressure (2 MPa) obtained from either the auto-controlled or manual pressuremeter for 10 m of tubing corresponds to the real value of the limit pressure of the 20-bar artificial ground. The limit pressure value that was derived varied for the manual and auto-controlled pressuremeters respectively from 1.86 to 2.06 bars when tubing length was 50 m. For the 30-bars artificial ground, the limit pressure derived from the automatic pressuremeter shows more precision than what was measured by the manual device. The inaccuracy is not very significant and does not exceed 10%.

The auto-controlled and manual pressuremeters show comparable pressuremeter moduli for 10 m of tubing lengths for the two types of the artificial grounds. The values are slightly different when the length of tubing is longer (50 m), but the difference does not exceed 11%.

3 In Situ Pressuremeter Test

The second set of tests consisted a series of in situ auto-controlled and manual pressuremeter tests in Site located at Salvador Allende Road, Magny Les Hameaux, France. For both devices, the probes were initially calibrated (Fig. 8). The pressure and volume readings are then adjusted to compensate for the head of water (pressure loss) in the measuring cell tubing and the inertia of the probe (membrane and probe cover). The calibrations are required to determine the strength of the probe covers and membrane assemblies as functions of the expansion of the probe (volume variation) and the parasitic increase in volume variation in the entire apparatus as a function of the applied pressure.

Fig. 8. In situ pressuremeter testing

The type of membrane and the number of cycles can affect the result. Authors' experience shows that the strength of the probe (membrane) decreases as a function of the number of tests that are carried. The recorded volume-pressure curves allowed the strength of the membranes to be determined for injection volumes of 700 cm^3 using 60 mm diameter probes. The pressure increment used was 25 kPa. The calibration curve obtained at the end of the test was the volume variation, reading at 60 s, as a function of the applied pressures.

The pre-expansion test of the probe is carried out in situ each time a new probe is used. The differential valve is adjusted to 0 m depth.

The expansion coefficient "*a*" is determined using a steel calibration tube that is usually 1.05 m in length and 65 mm in diameter. The test is carried out in 500 kPa pressure increments to 5 MPa.

The initial volume of the measuring probe cell is (AFNOR, NF P 94-110):

$$V_s = \left(\frac{\pi d_i}{4} l_s\right) - V_c \tag{1}$$

Where:

ls = the length of the membrane (210 mm),
di = inside diameter of the calibration tube,
Vc = the ordinate at the origin of the straight line, which gives the best fit with the second portion of the expansion curve (Fig. 9).

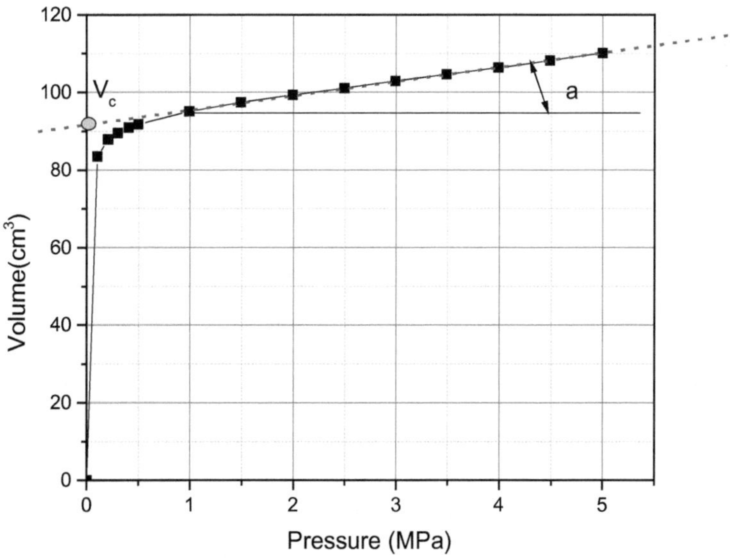

Fig. 9. Expansion curve of a pressuremeter probe in a calibration tube

The dilation coefficient is then:

$$a = \frac{\Delta V}{\Delta p} \quad (2)$$

The coefficient a must be less than 6 cm^3/MPa (0.6 cm^3/bar) for a device that is equipped with 50 m of tubing. The recorded values of the expansion coefficient for the auto-controlled and manual pressuremeters were respectively 3.668 cm^3/bar and 3.112 cm^3/bar. The recorded Vc for the auto-controlled and manual pressuremeters were respectively to 91.795 cm^3 and 83.857 cm^3.

Table 2 summarizes the details of the second series of auto-controlled and manual pressuremeter tests that were performed in situ. The tests are carried to the depth of 4 m every one meter in pre-bored holes. The holes are drilled using a continuous flight auger with a diameter of 63 mm. The upper layers of soil consisted of sandy to clayey silt underlain by plastic clay. All pressuremeter tests were performed in at least 9 pressure increments. The auto-controlled test was performed with more increments to achieve a higher precision.

The diagrams in Fig. 10 show examples of the software output that was used to determine the various parameters of the pressuremeter (Geovision®): modulus, creep pressure and limit pressure from two tests carried out at 1 m using the auto-controlled pressuremeter. Both modulus and pressure derived from the auto-controlled pressuremeter, for the sandy silt, were slightly higher than those obtained from the manual pressuremeter.

Table 2. Detail of performed in situ tests

Tests	Depth	Soil	No of points	Pressuremeter type	E_M (MPa)	P_f (MPa)	P_{LM} (MPa)	E_M/P_{LM}
Press_Situ_1 m	1 m	Sandy Silt	12	Manual type G	13	0.33	0.785	16.56
Press_Situ_2 m	2 m	Sand clay	9	Manual type G	3.9	0.12	0.291	13.56
Press_Situ_3 m	3 m	Clay	11	Manual type G	11.2	0.28	0.933	11.98
Press_Situ_4 m	4 m	Clay	15	Manual type G	34.2	1.24	1.886	18.15
Geopac_ Situ_1 m	1 m	Sandy silt	13	Auto-controlled	16.6	0.43	0.801	20.67
Geopac_ Situ_2 m	2 m	Sand clay	10	Auto-controlled	4.4	0.17	0.334	13.24
Geopac_ Situ_3 m	3 m	Clay	11	Auto-controlled	16.5	0.29	0.812	20.29
Geopac_ Situ_4 m	4 m	Clay	18	Auto-controlled	27.3	1.44	2.754	9.922

Figure 11 show the results of several different tests. It can be clearly observed that there are differences between the values that are obtained by the auto-controlled and manual pressuremeters at the end of the plastic phase or near the limit pressure.

Figure 12 compares the results of the auto-controlled and manual pressuremeters. It can be observed that all moduli values derived from auto-controlled pressuremeter are

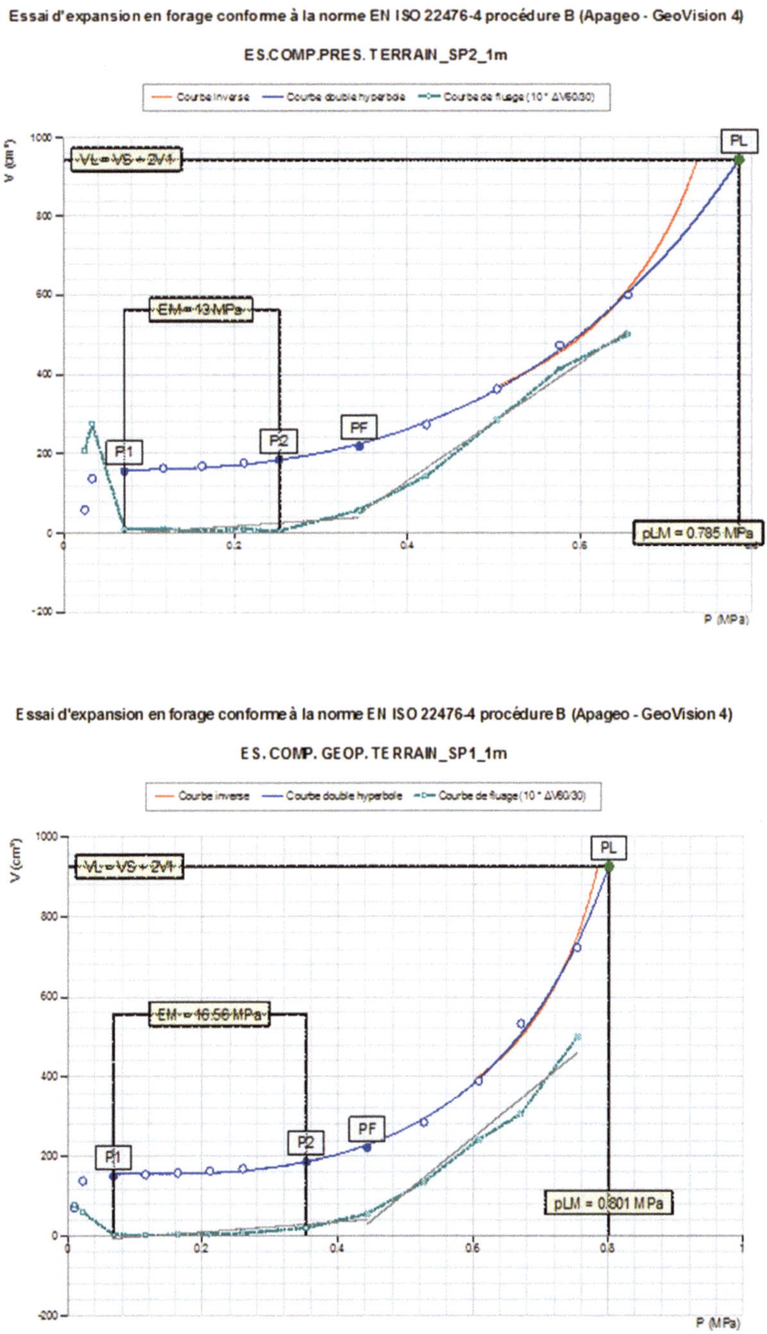

Fig. 10. Example of determination of soil proprieties using Geovision® software

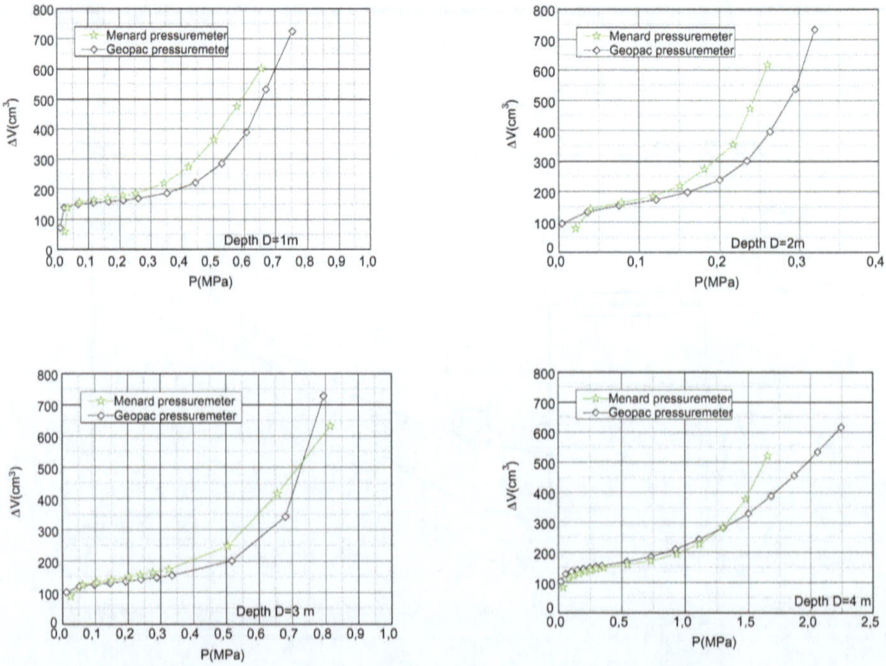

Fig. 11. Results of the pressuremeter tests (expansion curve)

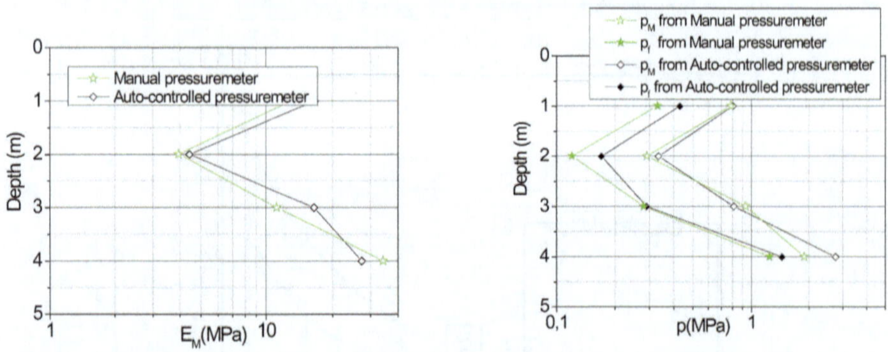

Fig. 12. Modulus, creep pressure, limit pressure versus depth (auto-controlled and manual pressuremeters)

slightly higher than those obtained from the manual pressuremeter in the first three meters. The creep pressures derived from the manual pressuremeter are consistently lower than those obtained from the auto-controlled one. The limit pressures are comparable in the first three meters; however, a small difference is noticeable at 4 m depth.

Figure 13 shows the software outputs used for the analyses of data (Geovision®) with pictures of extracted specimens and equipment used.

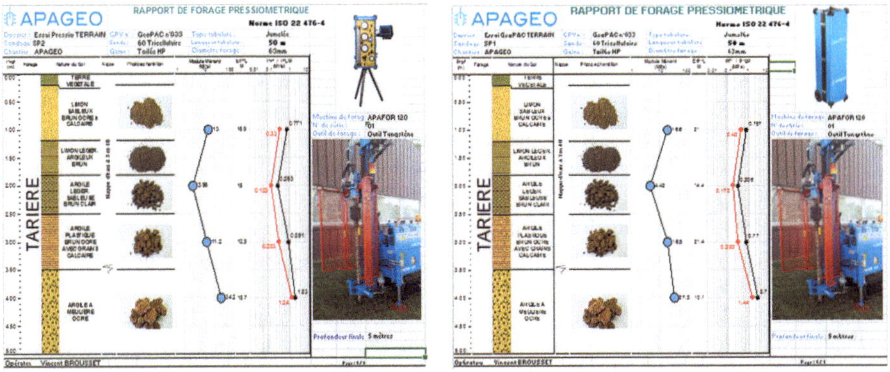

Fig. 13. Output of software used for analysis of data (Geovision®)

4 Discussion and Conclusion

The main purpose of the present paper is the description of a new apparatus that is able to automatically perform a pressuremeter test. Such an automatic apparatus enables the geotechnical engineer to obtain repeatable measurements, and reduces inaccuracies that might be induced by the operator, uncertainties, and loss of head. Both in situ tests and tests performed on artificial ground demonstrate the feasibly of the auto-controlled test and the possibility to obtain good results. The performance tests, the analysis of operations and the approval of the model (electrical and mechanical) were conducted by an independent organisation CETIM (Centre Technique des Industries Mécaniques, France) in 2010. The inaccuracy obtained by the calibration is substantial when the manual pressuremeter is used; however, this value is insignificant when using the novel generation pressuremeter.

The solution of self-correction of pressure loss, developed exclusively for GéoPAC® auto-controlled pressuremeter, has many advantages and is of great interest especially in deep soils when using long tubing. In fact it is integrated into the test and runs at start-up without affecting the test progress. The accuracy of pressures of soil that represent a certain creep is significantly improved. The accuracy of pressures level of soil that represent a certain creep is significantly improved in the case for tests between the beginning of the creep and the end of the failure, which increase in repeatability and accuracy. The oscillatory phenomena are completely attenuated and controlled and the automatic regulation becomes more efficient and accurate.

The auto-controlled pressuremeter also makes it possible to carry out true cyclic tests, which can be performed by varying the amplitude or the frequency of the test in either cases of pressure or volume control. The cyclic(s) loading(s) can be easily programmed and the record is also possible using the central unit.

Acknowledgments.. The authors gratefully appreciate the help provided by Damien Brechot, Vincent Brousset, Lionel Daré and Jean Pierre Arsonnet from APAGEO Company during the carried out experimental work and Hamidi Barak for the review of the present paper.

References

AFNOR, NF P 94-110: Sols, Reconnaissances et essais-Essai pressiométrique Ménard. French Standard, Edited by afnor, Paris-La défense, Publisher Location, July 1991 (1991)

AFNOR, NF P 94-110-1: Sols, Reconnaissances et essais-Essai pressiométrique Ménard, Partie 1: Essai sans cycle. French Standard, Edited by afnor, Paris-La défense, Publisher Location, January 2000 (2000)

AFNOR, XP P 94-110-2: Sols, Reconnaissances et essais-Essai pressiométrique Ménard, Partie 1: Essai avec cycle. French Standard, Edited by afnor, Paris-La défense, Publisher Location, December 1999 (1999)

ASTM Standards D2240-15: Standard Test Method for Rubber Property - Durometer Hardness. Published August 2015. Originally approved in 1964. Last previous edition approved in August 2015 as D2240 – 15. Developed by Subcommittee: D11.10 (2015)

ASTM Standards D 4719-07: Standard Test Methods for Prebored Pressuremeter Testing in Soils. Current edition approved Feb. 15, 2007. Published April 2007. Originally approved in 1987. Under the jurisdiction of ASTM Committee D18 on Soil and Rock and the direct responsibility of Subcommittee D18.02 on Sampling and Related Field Testing for Soil Evaluations (2007)

Arsonnet, G., Baud, J.-P., Gambin, M., Youssef, W.: Le GéoPAC®, un contrôleur pression volume automatisé pour les essais pressiométriques de qualité. The Geopac®, an Automated Control Unit for Quality Ménard PMTs. In: Proceedings of the 18th International Conference on Soil Mechanics and Geotechnical Engineering, Paris 2013. Parallel session ISP 6 (2013)

Cassan, M.: Les essais pressiométriques et leurs applications en France. Rappels Historiques et état des connaissances. The pressuremeter test and their applications in France. Historical Summary and Sate of the art. 50 ans de pressiomètres. Vol. 1 Gambin, Magnan et Mestat (ed.) Presses de l'ENPC/LCPC, Paris (2005)

Rapport of CETIM: Conception d'un contrôleur pression/volume automatique. CETIM, Centre Technique des Industries Mécaniques, France, 19 Février 2010 (2010)

ISO 22476-4: Geotechnical investigation and testing—Field testing—Part 4: Ménard pressuremeter test. Reconnaissance et essais géotechniques – Essais en place – Partie 4: Essai au pressiomètre Ménard. International Standard. First edition 2012-12-01. Published in Switzerland (2012)

Assessment of Relationship Between Static and Dynamic Load Using Regression Analysis and Artificial Neural Network Model

Ahmed H. Abulkareem(✉)

Civil Engineering Department, Engineering College,
University of Anbar, Ramadi, Iraq
ahm1973ed@gmail.com

Abstract. The Light Falling Weight Deflectometer (LFWD) load test has been developed to directly estimate the in-situ elastic modulus of near surface profiles as foundation, and subgrade layers is presented in this paper. For this purpose, field tests were conducted on selected sections from landfill project within Anbar province. In addition, forty test sections were constructed and tested at the Civil Engineering Department- University of Anbar. All sections were tested using the ZFG 3000 model - LFWD in companion with the Plate Load Test (PLT) that were used as reference measurements. Regression analyses were conducted to determine the best correlations between the elastic modulus obtained from LFWD and PLT tests., E_{vd}. ANN model is used to calculate dynamic deformation modulus, E_{vd} and comparing with the regression statistical model. The results indicate that ANN model have the capability of predicting dynamic deformation modulus, E_{vd} with a high degree of accuracy. Good correlations were obtained, which demonstrated that the LFWD can be reliably used to predict the modules obtained from plate load test and degree of compaction values, and hence can be used to evaluate the stiffness/strength parameters of shallow subgrade layers.

Keywords: Plate load test · LFWD · Dynamic modulus · ANN · Regression analysis

1 Introduction

The light falling weight Deflectometer (LFWD) is used in landfill project in Iraq. The German device (ZFG 3000) from Zorn is a compaction control device used in this research. It is able to measure the dynamic load bearing capacity of subgrades, subsoils, embankment layers and backfills.

Extensive application of these apparatus still has not been achieved since the dynamic modulus in not accepted in the quality assessment and quality control process of embankments and subgrade layers. Only marginal use of these devices can be noticed, mainly on areas of low importance (e.g. road shoulders) or trenches where performing a static plate load test could be complicated (Zoltán 2008).

W. Frikha et al. (eds.), *Soil Testing, Soil Stability and Ground Improvement*,
Sustainable Civil Infrastructures, DOI 10.1007/978-3-319-61902-6_21

For being able to use these dynamic devices on subgrade and embankment layers for landfill project, the research for converting the measured dynamic modulus into static modulus has been initiated.

The main objective was to determine the correlation between static and dynamic modules. Since direct conversion formulas are not frequently used in practice, introduction of an easy to use table with the required static and dynamic target values has been aimed. This is achieved by conducting field tests on constructed layers using the investigated device LFWD along with standard in-situ test device Plate Load Test (PLT). Otherwise new quality assessment based on dynamic modulus might be able to substitute the exclusive usage of the slow and complicated static plate load test in the near future. With the help of these results, new dynamic design methods can be worked out and applied.

2 Light Falling Weight Deflectometer (LFWD)

Light Falling Weight Deflectometer (LFWD) is a portable falling weight deflectometer that has been developed in Germany as an alternative in-situ testing device to the plate load test. Different types of LFWD exist in the market. All types exhibit many similarities in their mechanics of operation although there are many differences in design and mode of operation, which lead to variations in the measured results. Generally, the LFWD consist of a loading device that produces a defined load pulse, a loading plate, and one center geophone sensor (electric deflection data device) to measure the center surface deflection (Garcia and Thompson 2004).

The ZFG 3000 LFWD from Zorn Instruments was used in this study Fig. 1 is a compaction control device, according to ASTM E2835-11 and ASTM E2583-07 (2011). Dynamic modulus of deformation, E_{vd} is the most accurate and independent means for judging deformation (stiffness) and, thus, a material's level of compaction (www.ticservicegroup.com.au, 2013).

A 10 kg falling weight is dropped onto a 300 mm diameter plate from a height of 72 cm through guide rod; the vertical displacement of the plate (δ_c) is recorded by an

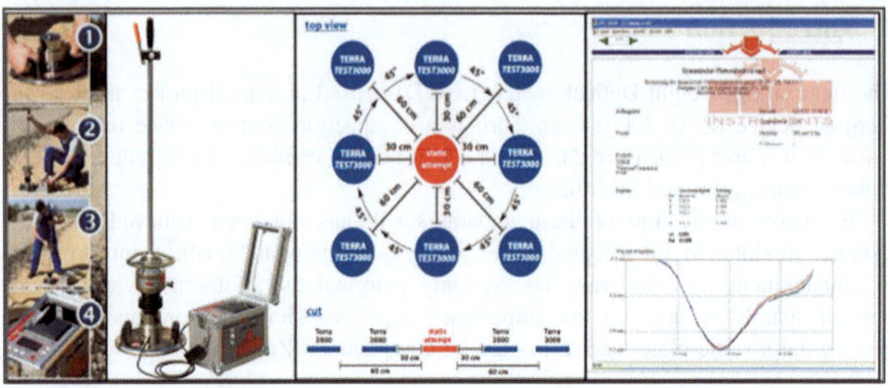

Fig. 1. Light Falling Weight Deflectometer (LFWD) (TIC Service Group 2013).

accelerometer built in a steel case on the top of the plate. The drop weight, drop height and plate diameter are constants. The plate coefficient (c) and the Poisson's ratio (μ) are also set constant, therefore the dynamic subgrade modulus, (E_{vd}) is calculated by a simplified Boussinesq equation (Zoltán 2008):

$$E_{vd} = 22.5/\delta_c \quad (1)$$

To estimate dynamic modulus of deformation, E_{vd} for each test in this study, eight position (nearly 45° apart) surrounding the location of the static plate load test were selected and dynamic plate load test were performed. Each test was repeated three times, in each time three preconsolidation tests (pre-compaction to remove any bedding errors, and are ignored) were conducted as shown in Fig. 1.

3 Static Plate Load Test (PLT)

The Static Plate Load Test (PLT) has been a useful site investigation tool for many years and has been used for proof testing of pavement structure layers in many countries. Currently it is used for both rigid and flexible pavements. The test was conducted by the procedure recommended by ASTM D1196-93as shown in Fig. 2. The test consists of a circular plate (450 mm diameter) that is in close contact with the layer to be tested and measuring deflections under load increments. The load increments were applied via a hydraulic jack with a suitable load capacity. The load was applied in increments up to a final value of externally applied stress of 700 kPa. The corresponding settlement was monitored and recorded, by using three suitable dial gauges (120° apart), for each increment until the settlement has ceased.

Fig. 2. Static Plate Load Test (PLT).

Plate loading tests can be used to estimate the modulus of subgrade reaction (k). Determination of the modulus of subgrade reaction is made in the field on the selected subgrade soil at its natural moisture content. This test is conducted by subjecting the subgrade to a known stress at a predetermined rate of speed using a loading system, and recording the resulting settlement. The modulus of subgrade reaction, k, can be calculated using the following relation (Yoder and Witczak 1975):

$$k = p/\delta \tag{2}$$

Where

P = unit load on plate (kN)
δ = Settlement of the plate (mm)

The value of Young's modulus was obtained from the well-known relationship that correlates the young's modulus to the modulus of subgrade reaction (Bowels 1997):

$$E_s = kB(1 - \mu^2) \tag{3}$$

Where B in this case represents the diameter of the testing plate and μ is the Poisson's ratio and its value can be 0.5 for the assumption of flexible base plate (Yoder and Witczak 1975).

4 Existing Relationships

The Institute for Transport Sciences in Hungarian (KTI) launched a research program in 1995 aiming to convert the dynamic modulus obtained by that device (E_{vd}) into the well-known static plate load test modulus (E_2) obtained by conventional measurements (research report, ÁKMI Kht 1995). After collecting 64 measurement results performed on different subgrade and subsoil materials, a general conversion formula was suggested as following:

$$E_{vd} = 0.52 \cdot E_2 + 9.1 \tag{4}$$

Several correlation results between E_2 and E_{vd} are available in the international literature. The most relevant results are summarized in Fig. 3.

Figure 3 shows that the value of the static plate load test modulus clearly exceeds at least two times that of the E_{vd} modulus. Some of the results show even higher ratios. Only two publications give a ratio less than two, but both of them are based on modulus values measured only at few points and within small intervals (Zoltán 2008).

Nazzal (2003) present a correlation study between the PLT and the LFWD on cement treated soils, lime treated soils, unstabilizsed fine-grained soils and granular soils. The relationship for modulus (surface modulus for LFWD) thus developed is as follows:

$$E_{PLT} = -20.9 + 0.69(E_{LFWD}) \quad (R^2 = 0.94, \text{for } 2.5\,\text{MPa} < E_{LFWD} < 865\,\text{MPa} \tag{5}$$

These regression models are illustrated in Fig. 4.

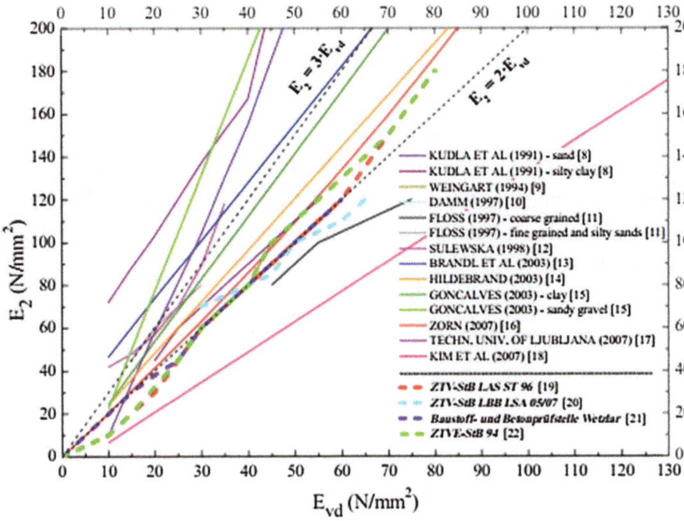

Fig. 3. Correlation results between E_2 and E_{vd} (Zoltán 2008).

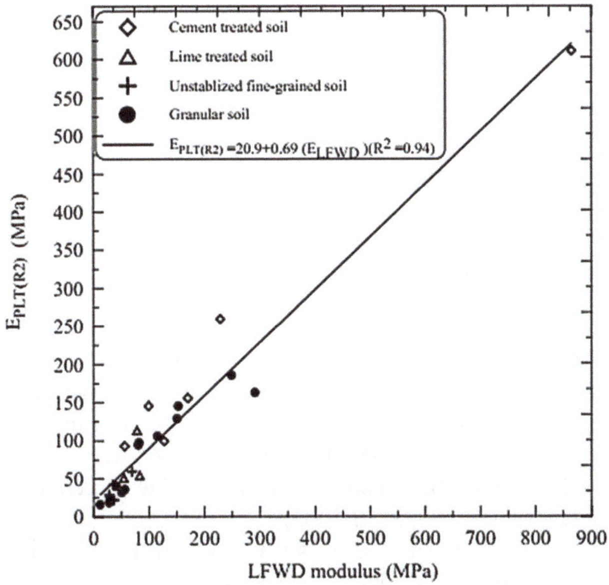

Fig. 4. Relationship between $E_{PLT(R2)}$ and E_{LFWD} (Nazzal 2003).

5 Analysis of Results

5.1 Linear Regression Model

In this study, collecting 40 measurement results for plate load test (PLT) performed on subgrade crushed limestone material in excess of (38 mm) in diameter by the field laboratory for civil engineering department at University of Anbar. After the division of project area as strips and establish the location of each point by mark, it has performed LFWD testing as shown in Fig. 5.

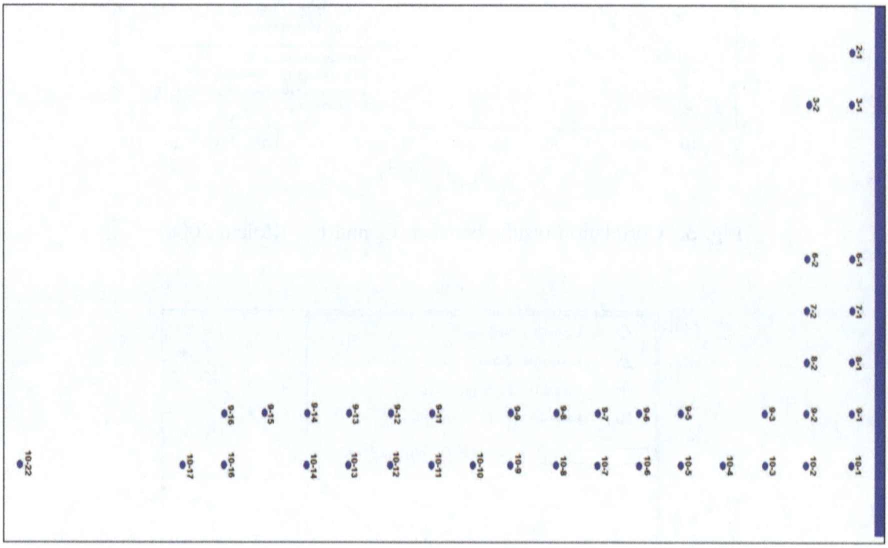

Fig. 5. Profile of the project area.

In order to characterize the variation of E_{vd} used as an independent value, descriptive statistics such as; minimum, maximum, mean, mode, median, variance, standard deviation, skewness and kurtosis etc. were calculated using the SPSS Version 20 (2012) package. Table 1 shows that the independent statistics value. Depending on the kind of soil and the degree of compaction, it can possible exceeding the limits of the lower and upper E_{vd} values as (Nazzal 2003).

In this study, simple regression analysis was performed in the analysis. The relations between the measured E_{vd}, moduli values with conventional static E_s, moduli values were analyzed employing linear, power, logarithmic and exponential functions. Statistically significant and strong correlations were found to be linear.

The objective of this regression analysis is to determine the parameters in the least square error models, which is used to predict the E_{vd} moduli from E_s moduli, with their corresponding coefficient of determination, R^2. In the linear regression models, the dependent variable is assumed to be a linear function of one or more independent

Table 1. Descriptive statistics for E_{vd} as an independent value.

N	40
Mean	49.87
Std. error of mean	4.7394
Median	46.28
Mode	27.61
Std. deviation	29.97452
Variance	898.472
Skewness	3.957
Std. error of skewness	0.374
Kurtosis	20.714
Std. error of kurtosis	0.733
Range	187.7
Minimum	21.3
Maximum	209
Sum	1994.91

variables plus an error introduced to account for all other factors, a typical form of a regression linear model is as follow:

$$Y_i = \beta_o + \beta_1 x_{i1} + \ldots\ldots \beta_k x_{ik} \tag{6}$$

Where Y_i is the dependent variable, and x_{i1}, …, x_{ik} are the independent or explanatory variables, and β_o is the disturbance or error term (SPSS Inc. 2009).

The coefficient of determination, R^2, represents the proportion of variation in the dependant variable that is accounted by the regression model and has values from 0 to 1. If it is equal to one, the entire observed points lie on the suggested least square line, which means a perfect correlation exists. Significance level is the result of the statistic test with null hypothesis $\beta_1 = 0$; it is expressed in percent. The greater the significance level the more supportive the model to alternative hypothesis ($\beta_1 \neq 0$), which indicates that a relation does exist between the dependent and independent variable. Finally, the standard error is the square root of the mean square errors (MSE).

After collecting forty measurement results performed on subgrade layer, a general conversion formula was suggested as follows:

$$E_{vd} = 7.384 + 0.527\,E_s \tag{7}$$

This formula can be used to convert the measured E_{vd} dynamic moduli values into conventional static E_s moduli values. The coefficient of correlation gives a value of $R^2 = 0.801$ and standard error = 13.56 for 40 sites, which seems to be acceptable in geotechnical testing. This regression model is illustrated in Fig. 6. It can be seen that there are some values of E_{vd} and E_s outlier range caused by the low correlation coefficient (R^2).

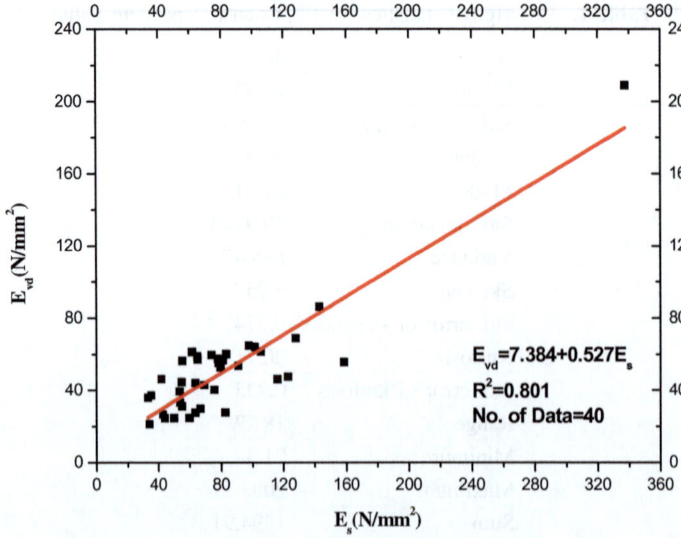

Fig. 6. Relationship between E_{vd} and E_s.

5.2 Multiple Regression Model

In this study, 13 test results were selected from 40 sites. These sites have been tested by PLT and LFWD and have conducted field density test by sand cone method according to ASTM D1556-00. The modified proctor compaction test according to ASTM D1557-12 also showed that the maximum dry unit weight and the optimum moisture content for crushed limestone material in excess of (38 mm) in diameter were 17.6 kN/m^3 and 5.6%, respectively. Summary of results for this section is presented in Table 2.

Table 2. Summary of test results for crushed limestone layer.

Strips	E_{vd} MPa	E_s MPa	MC (field) %	γ_{dry} (field) kN/m^3	$D_{pr,\ Modified}$ %
6–2	37.08	35.21	7.59	16.99	96.56
8–1	59.63	73.93	13.57	17.27	98.12
9–5	21.3	34.01	16.2	16.25	92.36
10–8	44.2	63.84	14.08	16.54	93.96
9–1	68.8	127.90	15.6	16.42	93.30
10–3	24.53	50.30	15.7	16.23	92.20
10–4	46.33	41.91	13.08	16.72	94.99
10–5	26.55	43.05	13.8	16.13	91.64
10–6	52.63	79.56	14.08	16.89	95.95
10–7	47.53	123.00	13.8	16.74	95.12
9–2	64	101.57	16.11	16.22	92.18
9–6	24.6	59.97	15.32	16.23	92.21
9–7	86.2	143.12	16.2	16.92	96.14

Multiple regression, a time-honored technique going back to Pearson's 1908 use of it, is employed to account for (predict) the variance in an interval dependent, based on linear combinations of interval, dichotomous, or dummy independent variables. The general purpose of multiple regression is to learn more about the relationship between several independent or predictor variables and a dependent or criterion variable. The typical form of the multiple regression model is as follow (SPSS Inc. 2009):

$$y = b_1 x_1 + b_2 x_2 + \ldots . + b_n x_n + c \tag{8}$$

Where b_1, b_2, ..., b_n are the regression coefficients, representing the amount the dependent variable *y* changes when the corresponding independent changes 1 unit. *c* is a constant, where the regression line intercepts the *y* axis, representing the amount the dependent *y* will be when all the independent variables are 0. The standardized versions of the *b* coefficients are the beta weights, and the ratio of the beta coefficients is the ratio of the relative predictive power of the independent variables. The major conceptual limitation of all regression techniques is that one can only ascertain relationships, but never be sure about underlying causal mechanism (SPSS Inc. 2009).

Table 3. Model summaries of multiple regressions for prediction of E_{vd}.

Independent variables	Coefficient	Std. error	t-Value	Sig. level
Constant	−345.76	193.484	−1.787	0.108
E_s, MPa	0.375	0.1	3.733	0.005
MC, %	0.754	1.855	0.406	0.694
D_{pr}, %	3.75	1.902	1.971	0.08

Multiple regression analysis was carried out to correlate the measured E_{vd} to three parameters, namely, static modulus (E_s), moisture content (MC) and degree of compaction (D_{pr}) (Table 3). Multiple regression model to predict E_{vd} is given:

$$E_{vd} = 0.375(E_s) + 0.754(MC\%) + 3.75\left(D_{pr}\%\right) - 345.76 \tag{9}$$

The coefficient of correlation between the measured and predicted values is a good indicator to check the prediction performance of the model. Figure 7 shows the relationships between measured and predicted values obtained from the Multiple regression model for E_{vd}, which good correlation coefficient with $R^2 = 0.804$.

5.3 Target Values for Dynamic Models

Direct conversion between dynamic, static models and degree of compaction is not frequently used in practice. Generally target values are given for different embankment and subgrade layers, more often depending on the required degree of compaction of the tested layer. E_{vd} modulus target values are fixed in Germany, and some other countries.

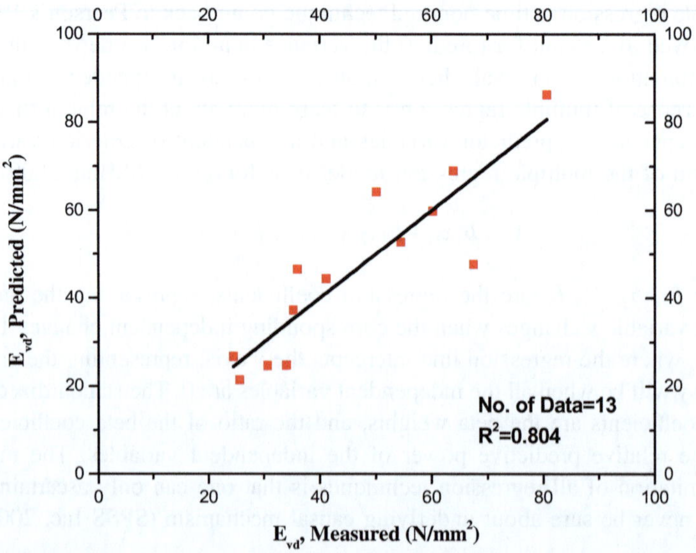

Fig. 7. Relationship of predicted and measured values of E_{vd} for Multiple regression model.

Based on the results above, a table of target values can be introduced. Different E_s and degree of compaction values are given for required E_{vd} values in Table 4. It can be seen that E_{vd} values reduce in access 10% compared with E_s due to kind of soil layers and degree of compaction. Interpolation between given values is acceptable (Zoltán 2008).

Table 4. Target values for crushed limestone subgrade layers.

E_s MPa	E_{vd} MPa	D_{pr} % Modified	E_s/E_{vd} ratio
	Crushed limestone subgrade layers		
180	100	≥ 100	2
150	85	≥ 99	2
120	70	≥ 98	2
100	60	≥ 98	2
80	50	≥ 97	2
60	35	≥ 95	2
45	30	≥ 95	2
20	15	≥ 94	1.5

5.4 Artificial Neural Network (ANN) Modeling of Overview

When the materials are natural, there is always an uncertainty sourced from the nature of the materials. This could be the main reason why soft computing methodologies such as artificial neural networks, Fuzzy systems, genetic algorithms have been developed in recent years and adopted but used for some other application. These techniques attract more and more attention in several research fields because they tolerate a wide range of uncertainty (Jin and Jiang 1999). Multilayer feed-forward neural network model is the most widely used network for its efficient generalization capabilities. Figure 7 presents typical multi-layer feed-forward neural networks. This type of neural network consists of an input layer, one or more hidden layer(s) and an output layer. Layers are fully connected by arrows, and comprise a number of processing units, the so-called nodes or neurons. The strength of connections between neurons is represented by numerical values called weights. Each neuron has an activation value that is a function of the sum of inputs received from other neurons through the weighted connections (Demuth et al. 2006). The optimum number of hidden layers and the number of neurons in each hidden layer is problem specific. Therefore, trial and error should be carried out to choose an adequate number of hidden layers and the number of neurons in each hidden layer.

Back propagation is the most successful and widely used in neural network applications. In this method, the input is propagated from the input layer through the hidden layers to the output layer. The network input is connected to every neuron in the first hidden layer while each network output is connected to each neuron in the last hidden layer. In this case, this would call full connection ANN. The network weights were originally set to random values and new values of the network parameters (weights) are computed during the network training phase. The neurons output are calculated using (Abdul-Razzak and Yousif, 2007):

$$O_i = F\left(\sum_j I_j \times W_{ij} + b_i\right) \tag{10}$$

Where

Oi = the output of the neuron i, Ij are the input of j neurons of the previous layer,
Wij = the neuron weights, bi is the bias for the modeling, and
F = the activation function.

The activation function is the portion of the neural network where all the computing is performed. The activation function maps the input domain (infinite) to an output domain (finite). The range to which most activation functions map their output is either in the interval [0, 1] or the interval [−1, 1]. The most common activation functions belong to five families as follows: (1) linear activation function; (2) step activation function; (3) ramp activation function; (4) sigmoid activation function; and (5) Gaussian activation function. The network error is then back propagated from the output layer to the input layer in which the connection weights are adjusted. This process is repeated until the error is minimized to a preference level. The error incurred during the learning can be expressed as Mean Squared Error and is calculate using (Abdul Razzak and Yousif 2007):

$$MSE = \frac{1}{nm}\sum_{i=1}^{n}\sum_{j=1}^{m}\left(t_{ij} - y_{ij}\right)^2 \tag{11}$$

Where t is the target value, y is the output value.

5.5 ANN Model for Prediction of E_{vd}

The use of ANN provides an alternative way to estimate dynamic modulus of deformation, E_{vd}. In this work a multi-layered feed-forward neural network with a back-propagation algorithm was adopted. (MATLAB 7.1 2005) software was used in neural network analyses having a three-layer feed-forward network (Alvarez and Babuska 1999). Forty cases of actual measured were extracted from experimental tests used in this study. The databases is randomly divided into three sets such as; training (70% of all data), test (15% of all data), and verification (15% of all data). The training set is used to adjust the connection weights, whereas the testing set is used to check the performance of the model at various stages of training and to determine when to stop training to avoid over-fitting. The validation set is used to estimate the performance of the trained network in the deployed environment (Shahin et al. 2008). A typical structure of artificial neural networks consists of many processing elements that are arranged in layers: an input layer, an output layer, and one or more layers in between, called intermediate or hidden layers (Fig. 8). Each processing element in a specific layer is interconnected to all the processing elements in the next layer via weighted connections. The scalar weights determine the strength of the connection between interconnected neurons. A zero weight refers to no connection between two neurons and a negative weight refers to a prohibitive relationship (Shahin et al. 2008). The model has one input parameter and one output parameter. The model has two hidden layers with nine nodes each (MATLAB software uses for determining the optimal number of hidden nodes rather than assuming a fixed number of hidden nodes in advance), and output layer with one node giving dynamic modulus of deformation, E_{vd}.

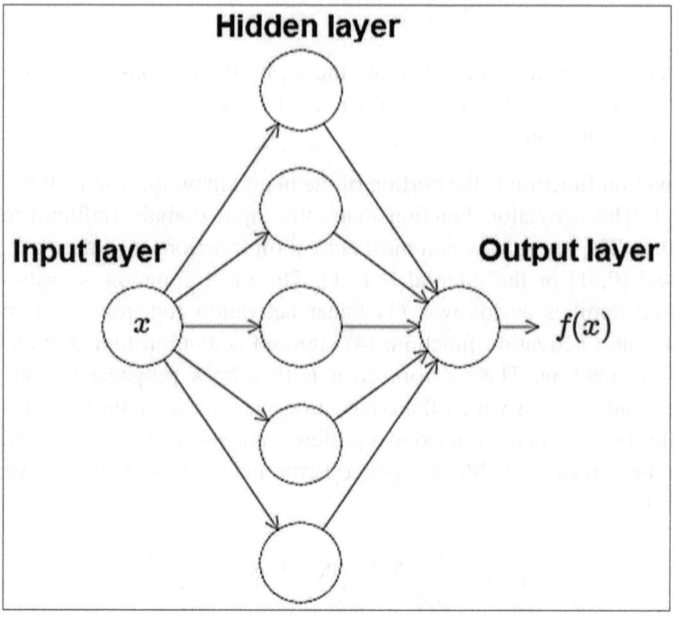

Fig. 8. Structure of Multi-layered feed-forward network (Michael 2015)

By increasing the number of hidden neurons they can typically get a better approximation and do still better by further increasing the number of hidden neurons. In the analyses network of learning parameter, momentum parameters and networks training function, which is an activation (transfer) function for all layers, have typical values of 0.01, 0.9, trainLm (training Levenberg-Marquardt function) and tansig (transfer function) respectively. As in many other network training methods, models and parameters were used in order to reach the minimum Mean Square Error (MSE) values.

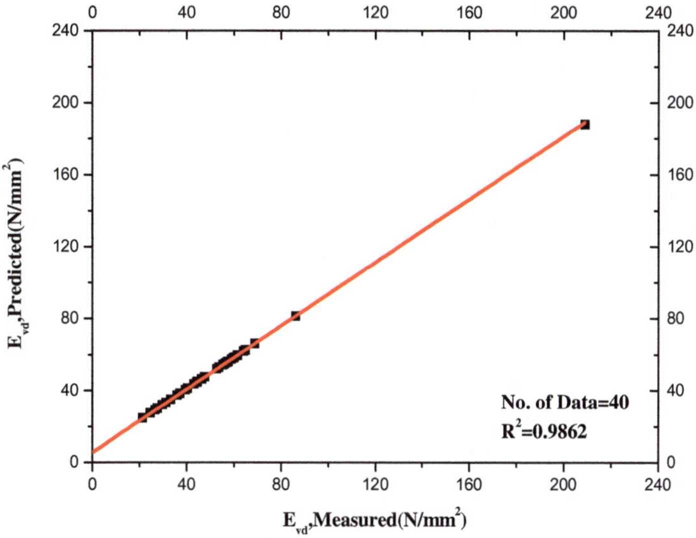

Fig. 9. Relationship of predicted and measured values of E_{vd} for ANN Model.

In fact, the coefficient of determination between the measured and predicted E_{vd} values is a good indicator to check the prediction performance of the model. Figure 9 shows the relationships between measured and predicted values obtained from the models for E_{vd}, with good coefficient of determinations. In this study, variance account for VAF (Eq. 12) and root mean square error (RMSE) (Eq. 13), indices were also calculated to control the performance of the prediction capacity of predictive models developed in the study as employed by (Alvarez and Babuska 1999):

$$VAF = \left[1 - \frac{var(y - y')}{var(y)}\right] \times 100 \tag{12}$$

$$RMSE = \sqrt{\frac{1}{N}\sum\nolimits_{i=1}^{N} (y - y')^2} \tag{13}$$

Where y and y' = the measured and predicted values, respectively.

If the VAF is 100 and RMSE is 0, then the model will be excellent. The obtained values of VAF and RMSE given in Table 5 indicated a high prediction performance.

Table 5. RMSE, VAF and R^2 values used to predict E_{vd}.

Model	Predictive model	RMSE	VAF%	R^2
LRM	$E_{vd} = 7.384 + 0.527\ E_s$	9.25	75.09	0.81
ANN		2.57	98.31	0.98

6 Conclusions

The LFWD proved to be a truly light weight FWD and is highly transportable. It is very easy to operate and changes from the 10 kg to the 20 kg drop weights or loading plates (200 mm and 300 mm) are quick and easy to do. All tests are saved automatically on a chip card, simply insert the card and the data can be transferred to the PC immediately. The basic information provided by the LFWD clearly proved to be very useful for construction quality control and assurance purposes.

The objective of this study is to evaluate the potential use of non-destructive testing device to measure the stiffness/strength parameters of highway materials and embankment soils during and after construction for landfill project. The field testing program included conducting tests using the investigated devices, in addition to standard tests, which included the static Plate Load Test (PLT), field density test by the sand cone method, and modified proctor compaction test). The field testing performed on subgrade crushed limestone material in excess of (38 mm) in diameter.

The results of the statistical analysis show that a good correlation do exist between the device under evaluation LFWD and the standard tests PLT, and degree compaction depending on standard tests. The relations obtained from statistical analysis, were linear regression to model and multiple regression for another. All regression models had an adjusted, correlation coefficient (R^2) greater than 0.8.

A multi-layered feed-forward neural network with a back-propagation algorithm was used to demonstrate the feasibility of ANNs to predict the dynamic modulus of deformation, E_{vd}. Forty cases of actual field measurements were used for model development and verification. The predicted E_{vd} obtained by utilising ANNs were compared with the measured E_{vd}. The results indicate that ANN model have the capability of predicting E_{vd} with a high degree of accuracy. From *VAF*, *RMSE* indicators and correlation coefficient (R^2) results, it can be seen that the ANN model is more accurately than regression analysis to predict E_{vd} as in Table 5.

The result of this study suggests that LFWD can be reliably used to predict the modules obtained from PLT, and degree of compaction values, and hence can be used to evaluate the stiffness/strength parameters of subgrade layers.

The new dynamic target values could open up the opportunity to perform the quality control and assess the bearing strengths of the tested layer, not only by static plate load test, which proved to be time-consuming and labour intensive, but by dynamic devices too.

The widespread use of mentioned dynamic devices referred to above, may facilitate for contractors, laboratories and engineers in the highway and railway construction industry to perform quick and continuous quality control of embankments, subgrade and subsoil layers and backfills.

Acknowledgment. The authors express his sincere thanks to the staff of the field laboratory for civil engineering department at University of Anbar for help in the conduct the field tests.

References

Zoltán, T.: Conversion between static and dynamic load bearing capacity moduli and introduction of dynamic target values. Periodica Polytech. Civil Eng. **52**(2), 97–102 (2008)

Garcia, G., Thompson, M.R.: Subgrade strength/stiffness evaluation. University of Illinois, Dept of Civil and Environmental Engineering, Technical Note, 1215NCEL, MC-250, Urbana, IL 61801 (2004)

TIC Service Group: Test Equipment for Geotechnics and Pavements (2013). www.ticservicegroup.com.au

Yoder, E.J., Witczak, M.: Principles of Pavement Design. Wiley, New York (1975)

Bowels, J.E.: Foundation Analysis and Design, 5th edn. The McGraw-Hill Companies, New York (1997)

Institute for Transport Sciences: Measurement of load bearing capacity with lightweight deflectometers (Wemex/ ZFG). Research report, ÁKMI Kht (1995)

Nazzal, M.D.: Field evaluation of in-situ test technology for QC/QA during construction of pavement layers and embankments. M.Sc. thesis in civil engineering, Louisiana State University and Agricultural and Mechanical College (2003)

SPSS Inc.: PASW Statistics for Windows, Version 18.0. Chicago: SPSS Inc. (2009)

Jin, Y., Jiang, J.: Techniques in neural-network based fuzzy system identification and their application to control of complex systems. In: Leondes, C.T. (ed.) Fuzzy Theory Systems, pp. 112–128. Techniques and Applications, Academic Press, New York (1999)

Demuth, H.B., Beale, M., Hagan, M.: Neural network toolbox for use with matlab User's guide. MathWorks, Incorporated (2006)

Abdul-Razzak, A.A., Yousif, S.T.: Artificial neural network model for predicting nonlinear response of uniformly loaded fixed plates. Eng. Technol. **25**(3), 334–348 (2007)

MATLAB 7.1 Software for technical computing and Model-Based Design. The MathWorks Inc. (2005)

Alvarez, G.M., Babuska, R.: Fuzzy model for the prediction of unconfined compressive strength of rock samples. Int. J. Rock Mech. Min. Sci. **36**, 339–349 (1999)

Shahin, M.A., Jaksa, M.B., Maier, H.R.: State of the art of artificial neural networks in geotechnical engineering. Electron. J. Geotech. Eng. **8**, 1–26 (2008)

Michael, N.: Neural Networks and deep learning. online book (2015) http://neuralnetworksanddeeplearning.com/chap4.html

Scanning Electron Microscope Analysis of Fly Ash, Quarry Dust Stabilized Soil

P. Indiramma[(✉)] and CH. Sudharani

Department of Civil Engineering, S.V. University College of Engineering, Tirupati, India
indu_svuce2000@yahoo.com, sudhajawahar@gmail.com

Abstract. Soil is a complex material with different minerals present in variant proportions, since it is formed due to weathering of rocks. The proportion of mineral governs the behaviour of soils. The soils with Montimorillonite mineral are more expansive in nature. The structures build with or on these soils are prone to problems like cracking, foundation failures due to their Swelling and Shrinking behaviour. Stabilization of expansive soil using admixtures is effective way of improving the intended soil properties and strength in specific. This paper focuses on the performance of Fly Ash and Quarry Dust as admixtures to the selected expansive soil. Laboratory investigations for determining Atterberg Limits, Compaction Characteristics, Deformation behaviour and Unconfined Compressive Strength and mineralogical analysis are performed by Scanning Electron Microscope (SEM) technique. The general behaviour of soil alone and after admixing with different percentages of Fly Ash and Quarry Dust are analyzed from laboratory test results and cross checked from Scanning Electron Microscope studies on the selected expansive soil.

Keywords: Expansive soil · Fly Ash · Quarry Dust · Scanning Electron Microscope (SEM) · Unconfined Compressive Strength

1 Introduction

Around 20% of total land area in the world are covered by expansive soils. Expansive soils have low Bearing Capacity, high Swelling and Shrinkage characteristics and high water susceptibility. Presence of Montimorillonite mineral is responsible for this behaviour of expansive soils. Structural cracks, premature failure of the pavement occurs when constructed on or with expansive soils. Stabilization of expansive soils using different additives such as Fly Ash, Quarry Dust, Lime, Cement and Bitumen are economical and effective. Fly Ash is an industrial by-product which is generated from Thermal Power Plants. Fly Ash is classified into class F and class C Fly Ash. Fly Ash is generally grey in colour, abrasive, mostly alkaline and refractory in nature. Safe disposal of Fly Ash is a major problem and creating severe environmental problems. Quarry Dust is a mineral solid waste, obtained from crushing of stone aggregates. Researchers used industrial waste admixtures namely, Quarry Dust, Fly Ash and Steel Slag etc. as soil stabilizers so that there is a value addition to the industrial wastes and environmental pollution is also minimized. Many research works are reported in

W. Frikha et al. (eds.), *Soil Testing, Soil Stability and Ground Improvement*,
Sustainable Civil Infrastructures, DOI 10.1007/978-3-319-61902-6_22

literature using Fly Ash Phanikumar and Sharma (2004, 2007); Phanikumar (2009) and Quarry dust Sridharan and Soosan (2005) in their study areas to investigate the variation of properties of soils to analyze them as beneficial Geotechnical materials.

An attempt has been made by in this investigation to use industrial waste admixtures namely, Quarry Dust and Fly Ash as soil stabilizers to the selected expansive soil. The effect of Fly Ash and Quarry Dust on expansive soils, laboratory tests are conducted for determining Plasticity characteristics, Differential Free Swell Index, Compaction Characteristics, and Unconfined Compressive Strength. Soil behaviour mainly depends on minerals present in soil, so a micro level investigation is carried out using Scanning Electron Microscope (SEM) technique in order to arrive the mineral composition and to understand the mechanism involved in soil stabilization.

2 Materials and Methods Used

2.1 Materials

1. An expansive soil collected from Tirupati Airport site, near Renigunta.
2. Fly ash from Vijayawada Thermal Power Station (VTPS), Vijayawada and
3. Quarry Dust from Tirupati Quarry.

2.2 Admixture Proportions

In the present study two series of tests are conducted. In first series the expansive soil is mixed with Fly Ash varying from 5% to 25% in increments of 5% by weight of soil taken. In second series expansive soil is mixed with Quarry Dust varying from 5% to 25% in increments of 5% by weight of soil taken.

2.3 Methods Adopted

The following tests are conducted on selected soil alone and Soil-Fly Ash mixtures and Soil-Quarry Dust mixtures:

Liquid Limit Test (IS: 2720 (PART V)-1975), Plastic Limit Test (IS: 2720 PART V-1975), Differential Free Swell Index (DFSI) Test (I.S: 2720 (Part-XL)-1977), Compaction Test IS: 2720 (PART-7)-1980), Unconfined Compression Test (IS: 2720 (Part-10)-1991) and Scanning Electron Microscope studies (SEM) using FEI Quanta 200 unit (Tables 1, 2 and 3).

3 Analysis of Test Results

The variation of Liquid Limit, Plastic Limit, Plasticity Index, Differential Free Swell Index, Optimum Moisture Content, Maximum Dry Unit Weight and Strength of soil admixtures are given in Figs. 1, 2, 3, 4, 5, 6 and 7 in the form of plots and the quantified values of respective properties are given in Tables 4, 5, 6, 7, 8 and 9.

Table 1. Properties of tested soil

Sl.No.	Property	Value
1.	Grain Size Distribution	
	Gravel (%)	2.76
	Sand (%)	30.0
	Silt and Clay (%)	67.9
2.	Atterberg Limits	
	Liquid Limit (%)	98.0
	Plastic Limit (%)	34.0
	Plasticity Index (%)	64.0
3.	Differential Free Swell Index (%)	377.0
4.	Specific Gravity	2.61
5.	IS Classification	'CH'
6.	Compaction Characteristics	
	Maximum Dry Unit Weight (kN/m^3)	16.62
	Optimum Moisture Content (%)	14.0
7.	California Bearing Ratio Value	
	@ 2.5 mm Penetration	1.83
	@ 5.0 mm Penetration	1.71
8.	Unconfined Compressive Strength (kN/m^2)	118.00

Table 2. Properties of Fly Ash

Sl.No.	Property	Value
1.	Grain Size Distribution	
	Gravel (%)	0.0
	Sand (%)	0.0
	Silt (%)	93.0
	Clay (%)	7.0
2.	Specific Gravity	2.15
3.	Compaction Characteristics	
	Maximum Dry Unit Weight (kN/m^3)	12.5
	Optimum Moisture Content (%)	21.2
4.	Color	Grey
5.	Free Swell Index (%)	0.0

Table 3. Properties of Quarry Dust

Sl. No.	Property	Value
1	Grain size distribution	
	Gravel (%)	3.0
	Sand (%)	94.0
	Silt and Clay (%)	3.0
2	Specific Gravity	2.71
3	Compaction Characteristics	
	Maximum Dry Unit Weight (kN/m^3)	19.1
	Optimum Moisture Content (%)	12.6
4	Shear parameters	
	Angle of Shearing Resistance (deg)	4.1
	Cohesion (C) (kN/m^2)	0.00

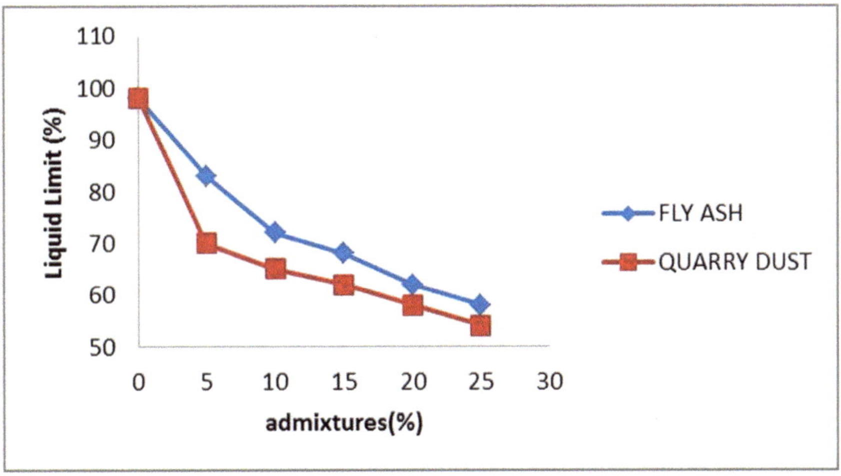

Fig. 1. Variation of Liquid Limit with percentages of admixtures

3.1 Plasticity Characteristics

The variation of Liquid Limit and Plastic Limit with different of percentages of admixtures as represented in Figs. 1 and 2. It can be observed that as the percentage of admixtures i.e. Fly ash alone/Quarry dust increases, Liquid Limit and Plastic Limit decreases. At 25% of admixtures, the maximum percentage of decrease in Liquid Limit and Plastic Limit can be observed in both the test series. The same trend observed in variation of Plasticity Index with the addition of soil-admixtures in increasing percentages. The same trend can be observed in variation of Plasticity Index with the addition of soil-admixtures in increasing percentages.

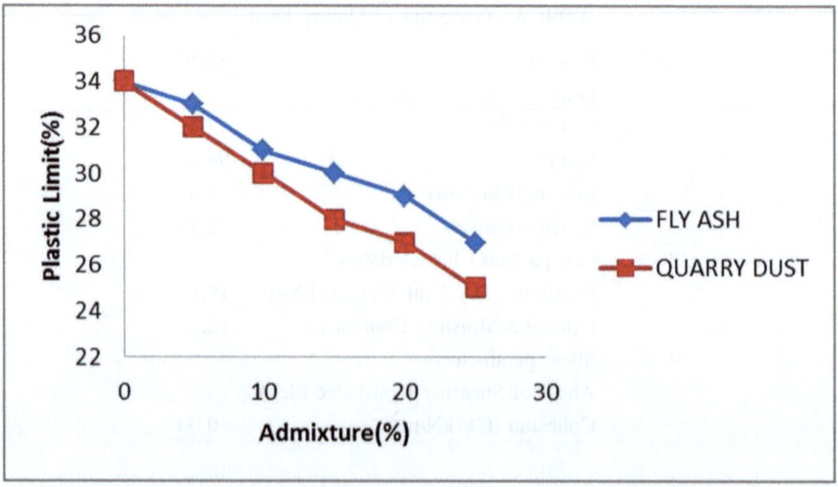

Fig. 2. Variation of Plastic Limit with percentages of admixtures

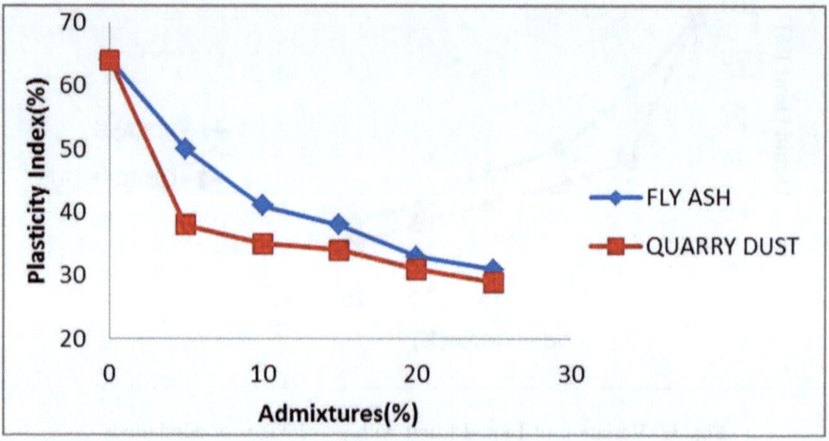

Fig. 3. Variation of Plasticity Index with percentages of admixtures

Reduction in Diffused Double Layer thickness and Clay percentage are responsible for decrease in Liquid Limit with increase the percentage of admixtures. The reduction in Plastic Limit of soil admixtures is due to filling of voids of flocculated soil thereby reducing water holding capacity.

3.2 Differential Free Swell Index

The swelling tendency of soils can be directly assessed by conducting Differential Free Swell Index test. The swelling tendency of selected expansive soil is decreasing with

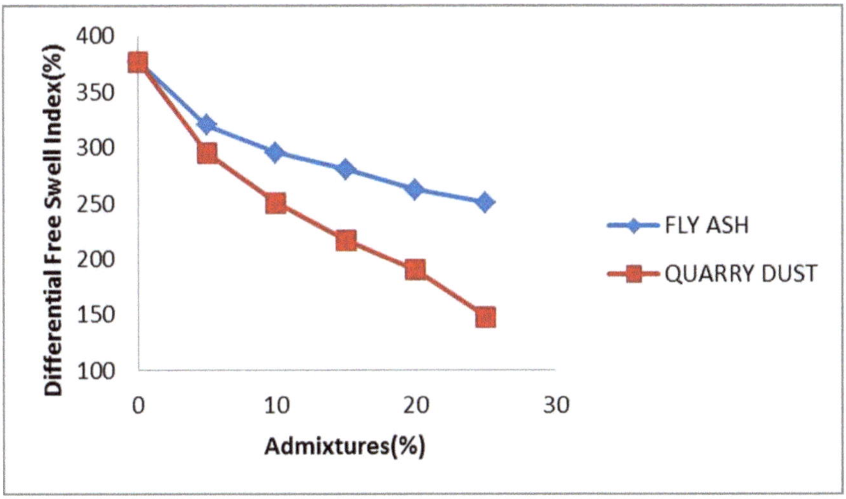

Fig. 4. Variation of Differential Free Swell Index with percentages of admixtures

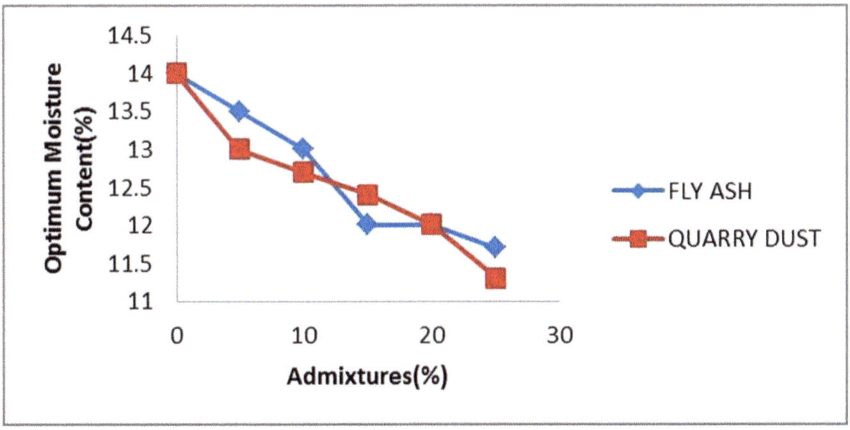

Fig. 5. Variation of Optimum Moisture Content with percentages of admixtures

addition of admixtures Fly ash or Quarry Dust as shown in Fig. 4. The Differential Free Swell Index is observed to decrease with increase in percentage of admixtures due to prevailing of Divalent and Trivalent cations (Ca^{2+}, Al^{3+}, Fe^{3+} etc.,) increasing flocculation of clay particles therby reducing surface area and water affinity of the soil sample.

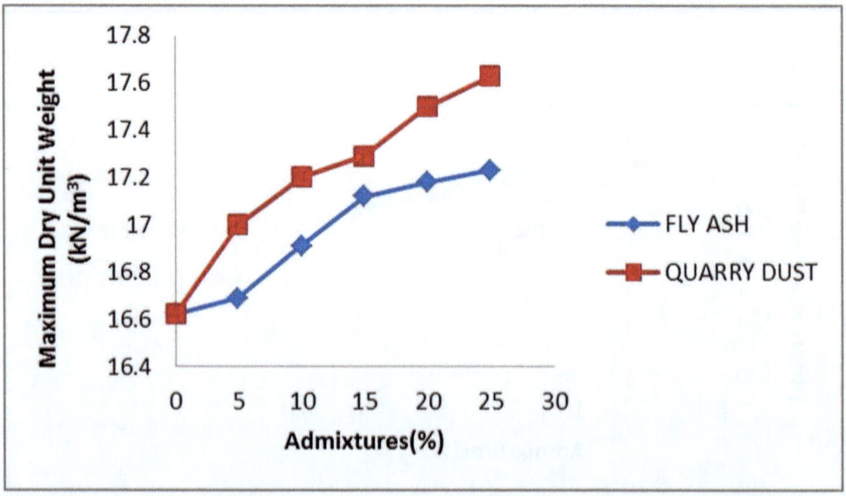

Fig. 6. Variation of Maximum Dry Unit Weight with percentages of admixtures

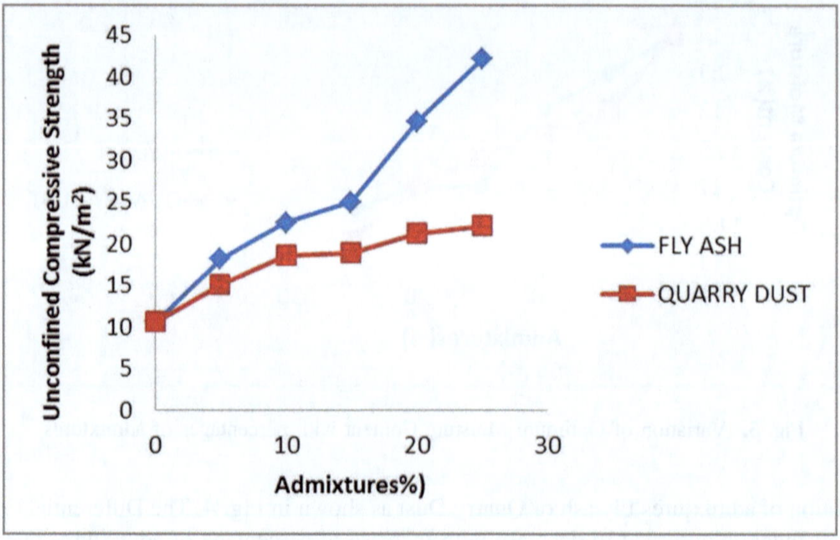

Fig. 7. Variation of Unconfined Compressive Strength with percentages of admixtures

Table 4. Liquid Limit at different percentages of Fly Ash and Quarry Dust

Fly Ash (%)	Liquid Limit (%)	Quarry Dust (%)	Liquid Limit (%)
0	98	0	98
5	83	5	70
10	72	10	65
15	68	15	62
20	62	20	58
25	58	25	54

Table 5. Plastic Limit at different percentages of Fly Ash and Quarry Dust

Fly Ash (%)	Plastic Limit (%)	Quarry Dust (%)	Plastic Limit (%)
0	34	0	34
5	33	5	32
10	31	10	30
15	30	15	28
20	29	20	27
25	27	25	25

Table 6. Plasticity Index at different percentages of Fly Ash and Quarry Dust

Fly Ash (%)	Plasticity Index (%)	Quarry Dust (%)	Plasticity Index (%)
0	64	0	64
5	50	5	38
10	41	10	35
15	38	15	34
20	33	20	31
25	31	25	29

Table 7. Differential Free Swell Index at different percentages of Fly Ash and Quarry Dust

Fly Ash (%)	Differential Free Swell Index (%)	Quarry Dust (%)	Differential Free Swell Index (%)
0	377	0	377
5	320	5	295
10	295	10	250
15	280	15	217
20	262	20	190
25	250	25	148

Table 8. Optimum Moisture Content at different percentages of Fly Ash and Quarry Dust

Fly Ash (%)	Optimum Moisture Content (%)	Quarry Dust (%)	Optimum Moisture Content (%)
0	14	0	14
5	13.5	5	13
10	13	10	12.7
15	12	15	12.4
20	12	20	12
25	11.7	25	11.3

Table 9. Maximum Dry Unit Weight at different percentages of Fly Ash and Quarry Dust

Fly Ash (%)	Maximum Dry Unit Weight (kN/m^3)	Quarry Dust (%)	Maximum Dry Unit Weight (kN/m^3)
0	16.62	0	16.62
5	16.69	5	17.00
10	16.91	10	17.20
15	17.12	15	17.29
20	17.18	20	17.50
25	17.23	25	17.63

3.3 Compaction Characterstics

Figures 5 to 6 the Optimum Moisture Content is found to decrease and MDD to increase with addition of Fly ash or Quarry Dust to the expansive soil. The increasing attractive pressure of soil particles minimizes dispersion thereby reducing and Optimum Moisture Content increasing Dry Unit Weight

4 Strength Characterstics

Figure 7, depicts the variation of unconfined compressive strength with increasing percentages of Fly Ash or Quarry Dust to the soil. Maximum Unconfined Compressive strength is observed at 25% of admixture addition to the soil due to reduction of cohesion. Soil-Fly Ash strength is more compared to Soil-Quarry Dust strength.

5 Scanning Electron Microscope (SEM) Studies

Scanning Electron Microscope studies are conducted using Quatum 200 unit on soil alone, soil-Fly Ash and Soil-Quarry Dust admixtures for understanding the mechanism of behavior of soils with different additives at different percentages. Figures 8, 9 and 10 shows Microscope analysis for soil alone, 10% Fly Ash + 90% Soil and 10% Quarry Dust + 90% soil respectively. It is observed that structure of the soil has dispersed fabric in natural state as given in Fig. 8, the texture of an soil alone consists of number of flaky-like particles. Soils with flaky shapes are likely to have low strength.

Flocculated structure is assumed after addition of admixtures. The extent of variation depends upon the type of admixtures. Figure 9 represents the texture of soil when Fly Ash is used as a stabilizer. The surfaces seems to be smooth and voids in soils occupied by Fly Ash. The texture of soil with addition of Quarry Dust is shown in Fig. 10 where rough surfaces with sharp edges and presence of Quarry Dust particles on surfaces can be observed. Compared to Fly Ash, Quarry Dust has more voids, so soil-Fly Ash with Fly Ash mixes gives more strength compared to soil-Quarry Dust mixtures. These observations strengthen the improvement in Strength, decrease in Swelling nature and Plasticity Characteristics of Soil-Fly Ash and soil-Quarry Dust mixtures found from laboratory tests (Table 10).

Fig. 8. Microanalysis report of soil

6 Conclusions

A series of laboratory tests were conducted to study the effect of Fly Ash and Quarry Dust on Plasticity Characteristics, Differential Free Swell Index, Compaction Characteristics (MDD, OMC) Compressive Strength on selected expansive soil. Fly Ash and Quarry Dust were added in different percentages from 0–25% in increments of 5%. Based on the test results the Plasticity Characteristics indicated by Plasticity Index, Optimum Moisture Content and Differential Free Swell Index are observed to decrease whereas Maximum Dry Unit Weight and Strength of soil admixtures increased with increase in percentage of Fly Ash and Quarry Dust.

Table 10. Unconfined Compressive Strength at different percentages of Fly Ash and Quarry Dust

Fly Ash (%)	Unconfined Compressive Strength (kN/m^2)	Quarry Dust (%)	Unconfined Compressive Strength (kN/m^2)
0	10.6	0	10.6
5	18.1	5	15.0
10	22.5	10	18.6
15	25.0	15	18.9
20	34.6	20	21.2
25	42.2	25	22.1

Fig. 9. Microanalysis report of (Soil + 10% Fly Ash)

Fig. 10. Microanalysis report of (Soil + 10% Quarry Dust)

The SEM observations for soil alone depicts flaky structure and addition of Fly Ash and Quarry Dust the same soil represents Flocculated structure with voids filled by admixtures resulting in decrement of Plasticity Characteristics, Optimum Moisture Content and Differential Free Swell Index and increment in Maximum Dry Unit Weight thereby Strength of Soil-Admixtures compared to soil alone. Fly Ash or Quarry Dust can be used as effective stabilizing admixtures for expansive soil considered in this investigation.

References

Sridharan, A., Soosan, T.G.: Utilization of quarry dust to improve the geotechnical properties of soils in highway construction. Can. Geotech. J. **28**(4), 391–400 (2005)

Phanikumar, B.R., Sharma, R.S.: Effect of fly ash on Engg properties of expansive soil. J. Geotech. Geoenviron. Eng. **130**(7), 764–767 (2004)

Phanikumar, B.R., Sharma, R.S.: Volume change behavior of fly ash-stabilized clays. J. Mater. Civ. Eng. (ASCE) **19**(1) (2007). doi:10.1061/(ASCE)0899-1561(2007)19:1(67)

Phanikumar, B.R.: Effect of lime and fly ash on swell, consolidation and shear strength characteristics of expansive clays: a comparative study. Geomech. Geoeng. Int. J. **4**(2), 175–181 (2009).

Analysis of Cylindrical Cavity Expansion in Modified Cam Clay with *Ko* Consolidation

Vincenzo Silvestri[1(✉)] and Claudette Tabib[2]

[1] École Polytechnique, Montreal, Canada
vincenzo.silvestri@polymtl.ca
[2] Montreal, Canada
claudette.tabib@yahoo.ca

Abstract. This paper presents explicit expressions for the principal effective stresses generated around a cylindrical cavity expanded in plane strain and undrained conditions in Modified Cam Clay. The assumption made in the present analysis is that Poisson's ratio ν' remains constant throughout the shearing process. Theoretical expressions are applied to the simulation of a cylindrical cavity expansion test in K_o normally consolidated remoulded Boston Blue Clay modelled as Modified Cam Clay. The results, which are compared to those obtained by assuming that the shear modulus G' remains constant, show that the two approaches are quite similar.

1 Introduction

Chen and Abousleiman (2012) and Silvestri and Abou-Samra (2012) recently obtained semi-analytical solutions for the plane strain undrained expansion of cylindrical cavities in Modified Cam Clay. Chen and Abousleiman (2012) employed small strains, but Silvestri and Abou-Samra (2012) used finite natural strains in both the elastic and plastic phases of the expansion. The latter authors also considered Almansi strains for obtaining the limiting radial expansion pressure. In addition, Silvestri and Abou-Samra (2012) adopted the restrictive assumption that the hardening parameter p'_c, which controls the size of the yield loci, remained constant during shearing. Such simplifying assumption permitted the determination of explicit expressions for the principal effective stresses generated in the soil around the cavity, but it led to approximate responses.

Silvestri and Abou-Samra (2012) also assumed the shear modulus G' to remain constant throughout the expansion process, following the approach of Randolph et al. (1979). From a theoretical point of view, it is preferable to assume a constant shear modulus G', as Zytynski et al. (1978) showed that the use of a constant Poisson's ratio ν' would lead to a non-conservative model in the sense that it may not conserve energy during closed-stress cycles (Yu 2006). However, this effect was not important in the case treated by Silvestri and Abou-Samra (2012) since there were no unloading-reloading cycles. As for Chen and Abousleiman (2012), these authors assumed Poisson's ratio ν' to remain constant and obtained semi analytical expressions for the principal effective stresses.

W. Frikha et al. (eds.), *Soil Testing, Soil Stability and Ground Improvement*,
Sustainable Civil Infrastructures, DOI 10.1007/978-3-319-61902-6_23

Cao et al. (2001) also obtained an approximate closed-form solution of the undrained cavity expansion in Modified Cam Clay by combining a large strain theory in the plastic zone and a small strain theory in the elastic zone. In the present paper, as the soil is modelled as a non-linear elastic plastic material, elastic and plastic zones are not treated separately. Thus, the analysis cannot be compared directly with Cao et al.'s approach.

In the present paper, both the shear modulus G' and the hardening parameter p'_c vary during undrained shearing, but that Poisson's ratio ν' remains constant. Indeed, Gens and Potts (1988) pointed out that a constant shear modulus did not agree well with experimental measurements and might imply negative values of Poisson's ratio at low stresses, which is physical unreasonable (Yu 2006).

As Poisson's ratio ν' remains constant, such approach allows obtaining explicit expressions for the principal effective stresses generated in the soil around the expanding cylindrical cavity. However, both the total radial pressure and the natural shear strain in the horizontal plane must still be determined numerically. The results obtained in the present paper are applied first to a well-known benchmark case and are thereafter compared with those found by assuming that the shear modulus G' remains constant during the expansion, but that the hardening parameter varies during shearing.

The results also show that the vertical effective stress, which represents the major principal stress at the beginning of the expansion, becomes the intermediate principal stress during the latter stages of the expansion. Similarly, while one of the horizontal effective stresses (i.e., the radial stress) becomes the major principal stress, the other horizontal effective stress (i.e., the tangential stress) becomes the minor principal stress during the expansion. Thus, failure occurs on vertical planes. Such response is different from the one analysed by Monnet (2007). Indeed, this author found that failure in a cylindrical cavity expansion could also occur on inclined planes with the vertical and tangential stresses being the major and minor principal stresses, respectively. Such situation typically arises in expansion tests carried out at shallow depth.

2 Approach

2.1 Modified Cam Clay Model

The general effective stress invariants used in the Modified Cam Clay model are the mean effective stress p' and the deviator stress q, which are defined as:

$$p' = \frac{\sigma'_r + \sigma'_\theta + \sigma'_z}{3} \tag{1a}$$

and

$$q = \left[\frac{(\sigma'_r - \sigma'_\theta)^2 + (\sigma'_\theta - \sigma'_z)^2 + (\sigma'_z - \sigma'_r)^2}{2}\right]^{1/2} \tag{1b}$$

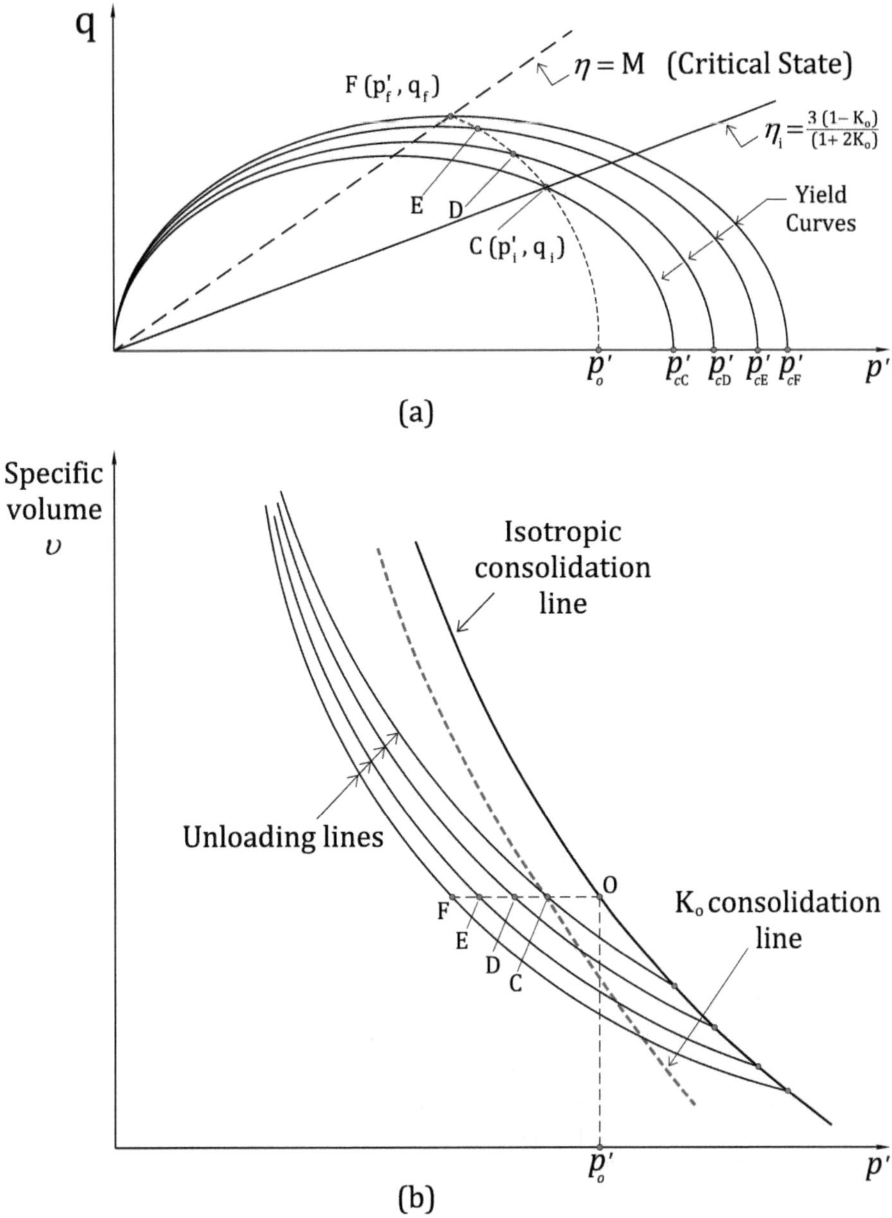

Fig. 1. Undrained compression test on normally consolidated clay: (a) $p' : q$ plane; (b) $\upsilon : p'$ plane (adapted from Wood (2007))

where $\sigma'_r, \sigma'_\theta, \sigma'_z$ = principal radial, tangential, and axial effective stresses. Figure 1a presents various yield curves of the Modified Cam Clay model. Each curve is described by the following expression:

$$q = p'M\left[(p'_c/p') - 1\right]^{1/2} \tag{2}$$

where $M = q/p'$ at critical state $= 6sin\phi'/(3 - sin\phi')$, with $\phi' =$ friction angle and $p'_c =$ hardening parameter. The latter parameter controls the size of the yield locus. An effective stress path (ESP) followed by a K_o normally consolidated specimen in an undrained compression test is also reported in Fig. 1a. The ESP is described by the expression

$$q = p'\left[\left(M^2 + \eta_i^2\right)\left(p'_i/p'\right)^{1/\Lambda} - M^2\right]^{1/2} \tag{3a}$$

where

$\eta_i = q_i/p'_i = 3(1 - K_o)/(1 + 2K_o)$,
$q_i = \sigma'_{zi} - \sigma'_{ri} = \sigma'_{zi} - \sigma'_{\theta i} = (1 - K_o)\sigma'_{zi}$,
$p'_i = (\sigma'_{ri} + \sigma'_{\theta i} + \sigma'_{zi})/3 = (1 + 2K_o)\sigma'_{zi}/3$,
$\sigma'_{ri} = \sigma'_{\theta i}, \sigma'_{zi} =$ initial horizontal and vertical effective stresses, respectively,
$K_o = \sigma'_{ri}/\sigma'_{zi} = \sigma'_{\theta i}/\sigma'_{zi} =$ in situ coefficient of lateral earth pressure at rest,
$\Lambda = (\lambda - \kappa)/\lambda$,
$\lambda =$ slope of $\upsilon : ln\, p'$ line in loading (Fig. 1b),
$\kappa =$ slope of $\upsilon : ln\, p'$ line in unloading (Fig. 1b),
$\upsilon = 1 + e =$ specific volume, and
$e =$ void ratio.

The ESP may also be described by the following equation:

$$q = p'M\left[\left(p'_o/p'\right)^{1/\Lambda} - 1\right]^{1/2} \tag{3b}$$

where $p'_o =$ value of p' for $q = 0$, as shown in Fig. 1a. The ESP of the K_o normally consolidated specimen begins at point C (p'_i, q_i) and ends at point F $\left(p'_f, q_f\right)$ on the critical state line, whose coordinates are $p'_f = 2^{-\Lambda}p'_o$ and $q_f = Mp'_f = 2^{-\Lambda}Mp'_o$ (Wood 2007). The undrained shear strength S_u is equal to $q_f/\sqrt{3}$. As the expansion occurs under undrained conditions, the specific volume υ remains constant along the ESP, as shown in Fig. 1b, and each point on the stress path lies on a new yield locus. The test thus moves across progressive higher yield loci, which are associated with expansion of the yield locus and with decrease of the mean effective stress p' for the normally consolidated specimen.

2.2 Modified Cam Clay Stress-Strain Relationships

Silvestri and Abou-Samra (2012) showed that the incremental elastic-plastic relationships of the Modified Cam Clay model in undrained shearing are given by:

$$d\sigma'_r = 2G' d\varepsilon_r + dp'\left[1 + \frac{6G'\kappa(\sigma'_r - p')}{\upsilon M^2 p'(2p' - p'_c)}\right] \tag{4a}$$

$$d\sigma'_\theta = 2G' d\varepsilon_\theta + dp'\left[1 + \frac{6G'\kappa(\sigma'_\theta - p')}{\upsilon M^2 p'(2p' - p'_c)}\right] \tag{4b}$$

$$d\sigma'_z = 2G' d\varepsilon_z + dp'\left[1 + \frac{6G'\kappa(\sigma'_z - p')}{\upsilon M^2 p'(2p' - p'_c)}\right] \tag{4c}$$

where $d\varepsilon_r, d\varepsilon_\theta, d\varepsilon_z$ = incremental radial, tangential, and axial natural strains. However, as $d\varepsilon_z = 0$ in plane strain and $d\varepsilon_r + d\varepsilon_\theta = 0$ in undrained shearing, then Eqs. 4a–4c become:

$$d\sigma'_r = -2G' d\varepsilon_\theta + dp'\left[1 + \frac{6G'\kappa(\sigma'_r - p')}{\upsilon M^2 p'(2p' - p'_c)}\right] \tag{5a}$$

$$d\sigma'_\theta = 2G' d\varepsilon_\theta + dp'\left[1 + \frac{6G'\kappa(\sigma'_\theta - p')}{\upsilon M^2 p'(2p' - p'_c)}\right] \tag{5b}$$

$$d\sigma'_z = dp'\left[1 + \frac{6G'\kappa(\sigma'_z - p')}{\upsilon M^2 p'(2p' - p'_c)}\right] \tag{5c}$$

The shear modulus G' and the hardening parameter p'_c are given by (See, for example, Wood 2007):

$$G' = \frac{3}{2}\frac{(1 - 2v')}{(1 + v')}\mathrm{K} = \frac{3}{2}\frac{(1 - 2v')}{(1 + v')}\frac{\upsilon p'}{\kappa} \tag{6}$$

and

$$p'_c = p'\left(p'_o/p'\right)^{1/\Lambda} \tag{7}$$

where v' = Poisson's ratio and $K = \upsilon p'/\kappa$ = bulk modulus. According to Eq. 6, the shear modulus G' increases as the mean effective stress p' increases. Substitution of G' and p'_c in Eqs. 5a–5c yields:

$$d\sigma'_r = -3\frac{(1-2v')}{(1+v')}\frac{\upsilon p'}{\kappa}d\varepsilon_\theta + dp'\left[1+\frac{9}{M^2}\frac{(1-2v')}{(1+v')}\frac{(\sigma'_r - p')}{\left(2-(p'_o/p')^{1/\Lambda}\right)p'}\right] \tag{8a}$$

$$d\sigma'_\theta = 3\frac{(1-2v')}{(1+v')}\frac{\upsilon p'}{\kappa}d\varepsilon_\theta + dp'\left[1+\frac{9}{M^2}\frac{(1-2v')}{(1+v')}\frac{(\sigma'_\theta - p')}{\left(2-(p'_o/p')^{1/\Lambda}\right)p'}\right] \tag{8b}$$

$$d\sigma'_z = dp'\left[1+\frac{9}{M^2}\frac{(1-2v')}{(1+v')}\frac{(\sigma'_z - p')}{\left(2-(p'_o/p')^{1/\Lambda}\right)p'}\right] \tag{8c}$$

Equation 8c may be rearranged to give:

$$d\sigma'_z - dp' = d(\sigma'_z - p') = \frac{9}{M^2}\frac{(1-2v')}{(1+v')}\frac{(\sigma'_z - p')}{\left(2-(p'_o/p')^{1/\Lambda}\right)}\frac{dp'}{p'} \tag{9a}$$

or

$$\frac{d(\sigma'_z - p')}{(\sigma'_z - p')} = \frac{9}{M^2}\frac{(1-2v')}{(1+v')}\frac{dp'}{p'\left(2-(p'_o/p')^{1/\Lambda}\right)} \tag{9b}$$

and, finally,

$$dln(\sigma'_z - p') = \mathrm{P}\frac{dp'}{p'\left(2-(p'_o/p')^{1/\Lambda}\right)} \tag{9c}$$

where $\mathrm{P} = \frac{9}{M^2}\frac{(1-2v')}{(1+v')}$.

Integration of Eq. 9c gives (See appendix):

$$\sigma'_z = p' + (\sigma'_{zi} - p'_i)\left[\frac{2-(p'_o/p')^{1/\Lambda}}{2-(p'_o/p'_i)^{1/\Lambda}}\left(\frac{p'_i}{p'}\right)^{1/\Lambda}\right]^{\mathrm{P}\Lambda/2} \tag{10}$$

The remaining principal effective stresses σ'_z and σ'_θ are found by introducing Eq. 10 into the expressions of the mean effective stress p' and the deviator stress q, as given by Eqs. 1a and 1b, resulting into:

$$\sigma'_r = p' - \frac{(\sigma'_{zi} - p'_i)}{2}\left[\frac{2-(p'_o/p')^{1/\Lambda}}{2-(p'_o/p'_i)^{1/\Lambda}}\left(\frac{p'_i}{p'}\right)^{1/\Lambda}\right]^{\mathrm{P}\Lambda/2} + \tau \tag{11a}$$

and

$$\sigma'_\theta = p' - \frac{(\sigma'_{zi} - p'_i)}{2}\left[\frac{2-(p'_o/p')^{1/\Lambda}}{2-(p'_o/p'_i)^{1/\Lambda}}\left(\frac{p'_i}{p'}\right)^{1/\Lambda}\right]^{P\Lambda/2} - \tau \tag{11b}$$

where the shear stress τ in the horizontal plane $(r:\theta)$ is equal to $(\sigma'_r - \sigma'_\theta)/2$ and is given by:

$$\tau = \frac{1}{2}\left\{\frac{4}{3}q^2 - 3(\sigma'_{zi} - p'_i)^2\left[\frac{2-(p'_o/p')^{1/\Lambda}}{2-(p'_o/p'_i)^{1/\Lambda}}\left(\frac{p'_i}{p'}\right)^{1/\Lambda}\right]^{P\Lambda}\right\}^{1/2} \tag{12}$$

The natural shear strain γ is found by integration of the incremental shear strain $d\gamma = |d\varepsilon_r - d\varepsilon_\theta| = 2d\varepsilon_\theta$, which is determined from Eqs. 5a and 5b, that is, from

$$d\gamma = \frac{d\tau}{G'} - \frac{6\kappa\tau dp'}{\upsilon M^2 p'(2p' - p'_c)} \tag{13a}$$

or

$$d\gamma = \frac{2}{3}\frac{(1+\nu')}{(1-2\nu')}\frac{\kappa d\tau}{\upsilon p'} - \frac{6\kappa\tau dp'}{\upsilon M^2 p'(2p' - p'_c)} \tag{13b}$$

where G' is given by Eq. 6. Integration of Eq. 13b gives:

$$\gamma = \frac{2}{3}\frac{\kappa(1+\nu')}{\upsilon(1-2\nu')}\int_0^\tau \frac{d\tau}{p'} - \frac{6\kappa}{\upsilon M^2}\int_{p'_i}^{p'} \frac{\tau dp'}{p'(2p' - p'_c)} \tag{14}$$

where the hardening parameter $p'_c = p'(p'_o/p'_i)^{1/\Lambda}$ from Eq. 7. It is apparent that the shear strain γ in Eq. 14 must be evaluated numerically due to the complex nature of the expression of the shear stress τ from Eq. 12.

Computations of total radial stress and pore pressure were carried out following the same approach of Silvestri and Abou-Samra (2012). For completeness, their expressions are briefly repeated herein.

The total radial stress σ_r acting in the clay around the expanding cylindrical cavity is given by (Yu 2000; Silvestri and Abou-Samra 2012):

$$\sigma_r = \int_0^\gamma \frac{\tau d\gamma}{e^\gamma - 1} + \sigma_{ri} \tag{15a}$$

which becomes

$$\sigma_{ra} = \int_0^{\gamma_a} \frac{\tau d\gamma}{e^{\gamma} - 1} + \sigma_{ri} \tag{15b}$$

at the wall of the cavity, where σ_{ri} is the initial total radial or horizontal stress, and the natural shear strains γ and γ_a are given by:

$$\gamma = ln[\left(\frac{r'}{r}\right)^2] \tag{16a}$$

and

$$\gamma_a = ln[\left(\frac{a'}{a}\right)^2] \tag{16b}$$

at the wall of the cavity. The pore pressure is determined by subtracting the radial effective stress σ'_r given by Eq. 11a from the radial total stress σ_r given by Eq. 15a. In Eqs. 16a and 16b, (r, r') and (a, a') represent respectively, generic radial distances and cavity radii, before and after the distortion has occurred. In order to obtain the limiting radial expansion pressure at the wall of the cavity, it is convenient to use the Almansi tangential strain which is defined as (Baguelin et al. 1978):

$$\alpha = \frac{1}{2}\left(\frac{r'^2 - r^2}{r'^2}\right) \tag{17a}$$

which becomes equal to

$$\alpha_a = \frac{1}{2}\left(\frac{a'^2 - a^2}{a'^2}\right) \tag{17b}$$

at the wall of the cavity. Substitution of Eqs. 17a and 17b into Eqs. 16a and 16b gives:

$$\gamma = -ln(1 - 2\alpha) \tag{18a}$$

and

$$\gamma_a = -ln(1 - 2\alpha_a) \tag{18b}$$

which, when introduced into Eqs. 15a and 15b yield:

$$\sigma_r = \int_0^{\alpha} \frac{\tau d\alpha}{\alpha} + \sigma_{ri} \tag{19a}$$

and

$$\sigma_{ra} = \int_0^{\alpha_a} \frac{\tau d\alpha}{\alpha} + \sigma_{ri} \tag{19b}$$

at the wall of the cavity, as also obtained by Baguelin et al. (1978). As mentioned above, the Almansi tangential strain was introduced for facilitating the determination of the limiting radial expansion pressure. Indeed, consideration of the upper limit of the integral in Eq. 15b indicates that γ_a must be equal to infinity for the radial pressure to reach the limiting expansion condition. Such a calculation may involve considerable computational problems. However, because Eq. 18b shows that $\gamma_a = \infty$ is reached for $\alpha_a = 0.5$, then it becomes relatively easy to carry on the integration process in Eq. 19b up to $\alpha_a = 0.5$ without experiencing any computational difficulties. This shows the superiority of the Almansi tangential strain over the natural strain in extending the integration process to infinity. There is also an additional advantage. In effect, in several software programs involving either finite elements or finite differences, it is often assumed that critical state is reached when the cavity radius has doubled in size, that is, when $a' = 2a$ in order to avoid numerical difficulties. As this condition corresponds to $\alpha_a = 0.375$ from Eq. 17b or to $\gamma_a = 1.386$ from Eq. 18b, it is apparent that the limiting state of $\alpha_a = 0.5$ is still far away and that by setting $\alpha_a = 0.375$ can only result in approximate limiting values.

3 Application

Before discussing in detail the application of the various theoretical expressions derived previously to the benchmark case presented below, it should be mentioned that the Modified Cam Clay model is known to give reasonable results only for isotropically normally consolidated clays (See, for example, Wood 2007). If either the initial stress state or the clay fabric, or both, are anisotropic, or if the soil is overconsolidated, better models should be resorted to, for example, such as the anisotropic Modified Cam Clay model (Dafalias 1987; Dafalias et al. 2002, 2006) or the Banerjee model (Banerjee and Yousif 1986; Banerjee et al. 1988). These two models have the advantage over more sophisticated and complex models that they can account for both inherent and induced anisotropy with relatively few model parameters. It is the authors' intention to apply one of these models to the problem at hand in the near future.

However, as the principal aim of the present study was to obtain the exact solution of the principal effective stresses generated around a cylindrical cavity in Modified Cam Clay under plane strain and undrained conditions, computations were carried out assuming that the Modified Cam Clay model could also be applied to a K_o normally consolidated clay.

The theoretical relationships derived previously will be applied to the simulation of a plane strain undrained expansion of a cylindrical cavity in K_o normally consolidated remoulded Boston Blue Clay modelled as Modified Cam Clay. The properties of the clay are the following (Randolph et al. 1979): OCR = 1, $K_o = 0.55$, $\upsilon = 2.16$,

$\lambda = 0.15, \kappa = 0.03, \Lambda = 0.8$, and $M = 1.2$ for $\phi' = 30°$. The OCR is based on the vertical effective stress.

The initial stress conditions are represented for illustration purposes by: $\sigma'_{zi} = 300$ [kPa], $\sigma'_{ri} = \sigma'_{\theta i} = 165$ [kPa], $u_i = 0$, $p'_i = 210$ [kPa], $q_i = 135$ [kPa], $p'_o = 256$ [kPa], $p'_c = 270$ [kPa], and Poisson's ratio $\nu' = 0.2855$. The choice of $\nu' = 0.2855$ was made so that the value of the initial shear modulus G'_i, calculated from Eq. 6, was equal to that obtained from the data of Randolph et al. (1979), that is, $G'_i = 7570$ [kPa]. These authors assumed $G' = G'_i =$ constant in the entire expansion process. The stress parameters at critical state are $p'_f = 147.6$ [kPa], $q_f = 177.1$ [kPa].

Figure 2 compares the shear stress-shear strain curve obtained in this study with that derived by assuming G' to remain constant. Examination of two curves shows that they are similar.

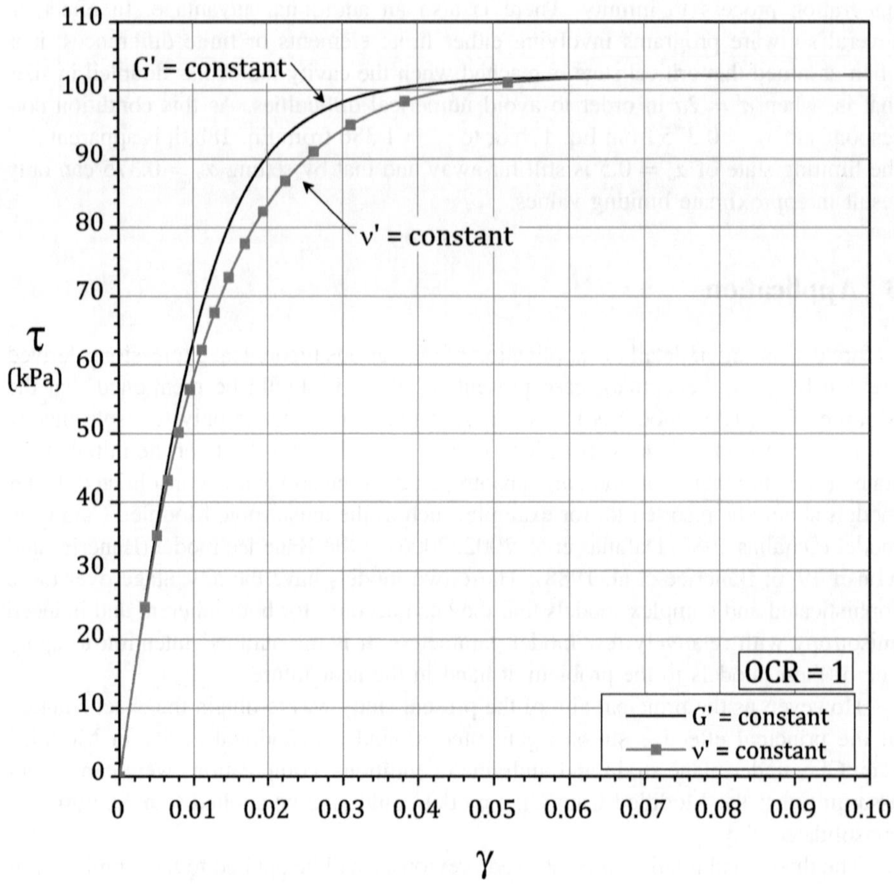

Fig. 2. Shear stress-shear strain curves

Figure 3 compares the total radial stress σ_{ra} and the excess of pore pressure u generated at the wall of the cavity as a function of the Almansi tangential strain α_a. Again, comparison between the two sets of curves indicates that the results are practically equivalent.

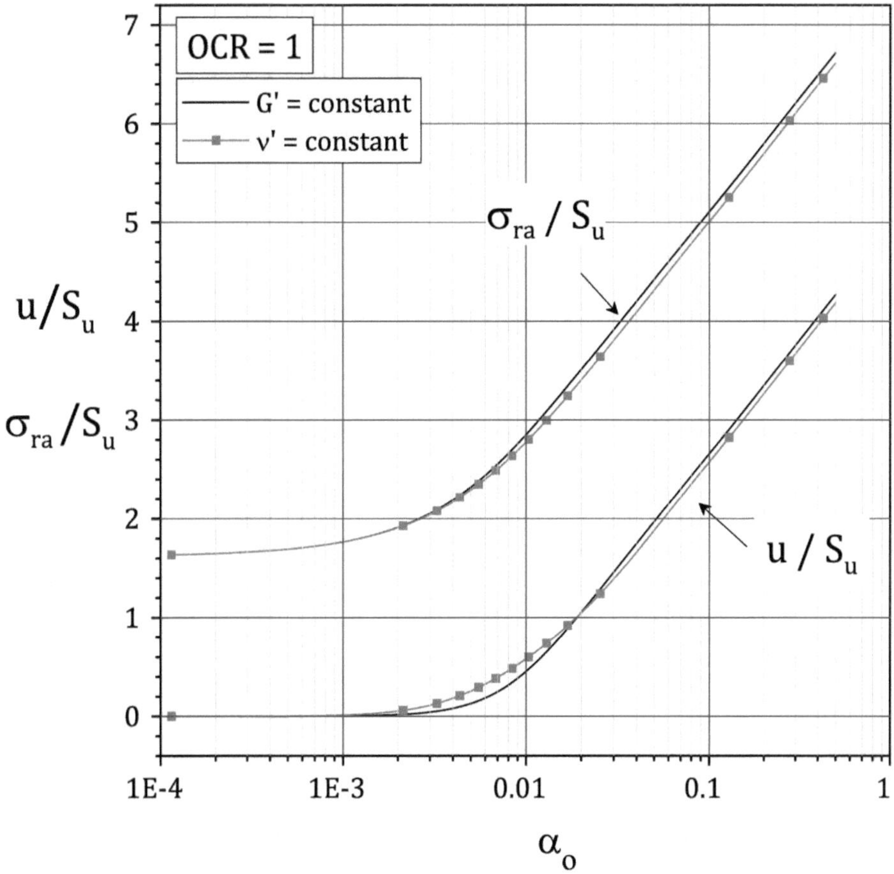

Fig. 3. Comparison of total radial stress and excess pore water pressure generated at wall of cavity as function of Almansi tangential strain α_a

Finally, Fig. 4 shows the distributions of the principal effective stresses and of the excess pore pressures around the cylindrical cavity at critical state. Once again, the two series of curves are quite similar.

Although the two sets of results presented in Figs. 2, 3 and 4 are practically equivalent, the approach followed in the present paper, by assuming Poisson's ratio ν' to remain constant during shearing, is superior to that in which the initial shearing modulus G' remains constant, because the present solution allows finding exact explicit expressions for the principal effective stresses.

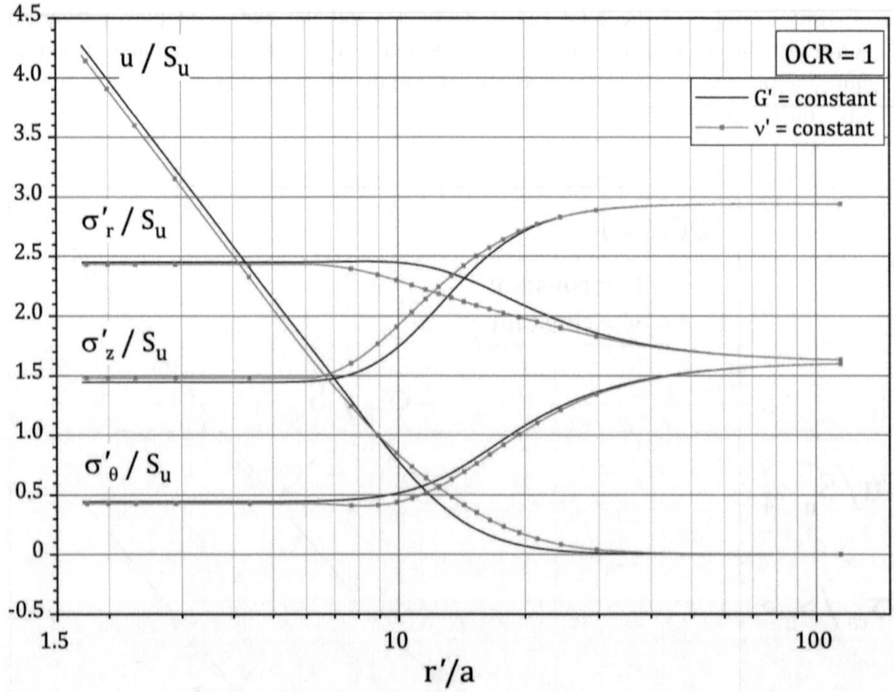

Fig. 4. Distributions of principal effective stresses and excess pore water pressures around cylindrical cavity at critical state

4 Conclusions

On the basis of the results reported in this paper, the following main conclusions are drawn:

1. The assumption of constant Poisson's ratio allows obtaining explicit expressions for the principal effective stresses generated around an expanding cylindrical cavity in Modified Cam Clay.
2. The theoretical relationships are applied to the simulation of an expanding cylindrical cavity in K_o normally consolidated remoulded Boston Blue Clay. The results, which are compared with those obtained by assuming that the shear modulus G' remains constant during the expansion, show that the two approaches are quite similar.

Acknowledgements. The author expresses his gratitude to the Natural Sciences and Engineering Research Council of Canada for the financial support received in this study.

Appendix: Derivation of Eq. 10

The integral in Eq. 9c reads

$$ln\left(\frac{\sigma'_z - p'}{\sigma'_{zi} - p'_i}\right) = \mathrm{P}\int_{p'_i}^{p'} \frac{dp'}{p'\left(2 - \left(p'_o/p'\right)^{1/\Lambda}\right)} \tag{A1}$$

where $\mathrm{P} = \frac{9}{M^2}\frac{(1-2v')}{(1+v')}$.

By letting $y = \left(p'_o/p'\right)^{1/\Lambda}$, then $p' = p'_o y^{-\Lambda}$ and $dp' = p'_o(-\Lambda)y^{-\Lambda-1}dy$, and the integral in Eq. A1 becomes

$$\mathrm{I} = -\mathrm{P}\Lambda\int_{\left(p'_o/p'_i\right)^{1/\Lambda}}^{\left(p'_o/p'\right)^{1/\Lambda}} \frac{dy}{y(2-y)} = \mathrm{P}\Lambda\int_{\left(p'_o/p'_i\right)^{1/\Lambda}}^{\left(p'_o/p'\right)^{1/\Lambda}} \frac{dy}{(y^2-2y)} \tag{A2}$$

Integration of Eq. A2 yields:

$$\mathrm{I} = -\frac{\mathrm{P}\Lambda}{2} ln\left(\frac{y}{y-2}\right)\Bigg|_{y_i}^{y} \tag{A3}$$

or

$$\mathrm{I} = -\frac{\mathrm{P}\Lambda}{2} ln\left[\left(\frac{y}{y_i}\right)\left(\frac{y_i-2}{y-2}\right)\right] = \frac{\mathrm{P}\Lambda}{2} ln\left[\left(\frac{y_i}{y}\right)\left(\frac{y-2}{y_i-2}\right)\right] \tag{A4}$$

where $y = \left(p'_o/p'\right)^{1/\Lambda}$ and $y_i = \left(p'_o/p'_i\right)^{1/\Lambda}$. As a result, Eq. A4 becomes:

$$\mathrm{I} = \frac{\mathrm{P}\Lambda}{2} ln\left\{\left[\frac{\left(p'_o/p'\right)^{1/\Lambda} - 2}{\left(p'_o/p'_i\right)^{1/\Lambda} - 2}\right]\left(\frac{p'_i}{p}\right)^{1/\Lambda}\right\} \tag{A5}$$

Combining Eq. A5 with Eq. A1 yields:

$$\sigma'_z - p' = \left(\sigma'_{zi} - p'_i\right)\left\{\left[\frac{\left(p'_o/p'\right)^{1/\Lambda} - 2}{\left(p'_o/p'_i\right)^{1/\Lambda} - 2}\right]\left(\frac{p'_i}{p'}\right)^{1/\Lambda}\right\}^{\frac{\mathrm{P}\Lambda}{2}} \tag{A6}$$

which, when rearranged, gives:

$$\sigma'_z = p' + \left(\sigma'_{zi} - p'_i\right)\left\{\left[\frac{2 - \left(p'_o/p'\right)^{1/\Lambda}}{2 - \left(p'_o/p'_i\right)^{1/\Lambda}}\right]\left(\frac{p'_i}{p'}\right)^{1/\Lambda}\right\}^{\frac{\mathrm{P}\Lambda}{2}} \tag{A7}$$

Eq. A7 is Eq. 10 in the main text of the paper.

References

Baguelin, F., Jézéquel, J.F., Shields, D.H.: The Pressuremeter and Foundation Engineering. Trans Tech Publications, Clausthal (1978)

Banerjee, P.K., Yousif, N.B.: A plasticity model for the mechanical behaviour of anisotropically consolidated clay. Int. J. Numer. Anal. Meth. Geomech. **10**(5), 521–541 (1986)

Banerjee, P.K., Kumbhojkar, A.S., Yousif, N.B.: Finite element analysis of the stability of vertical cut using an anisotropical soil model. Canad. Geotechn. J. **25**(1), 119–127 (1988)

Cao, L., Teh, C.I., Chang, M.-F.: Undrained cavity expansion in modified Cam clay I: theoretical analysis. Géotechnique **51**(4), 323–334 (2001)

Chen, S.L., Abousleiman, Y.N.: Exact undrained elasto-plastic solution for cylindrical cavity expansion in modified Cam clay. Géotechnique **62**(5), 447–456 (2012)

Dafalias, Y.F.: An anisotropic critical state clay plasticity model. In: Proceedings of the Second International Conference on Constitutive Laws for Engineering Materials. Theory and Applications, Tucson, Arizona, vol. 1, pp. 513–521. Elsevier, 5–8 January 1987

Dafalias, Y.F., Manzari, M.T., Akaishi, M.: A simple anisotropic clay plasticity model. Mech. Res. Commun. **29**(4), 241–245 (2002)

Dafalias, Y.F., Manzari, M.T., Papadimitriou, A.G.: SANICLAY: simple anisotropic clay plasticity model. Int. J. Numer. Anal. Meth. Geomech. **30**(12), 1231–1257 (2006)

Gens, A., Potts, D.: Critical state models in computational geomechanics. Eng. Comput. **5**, 178–197 (1988)

Monnet, J.: Numerical validation of an elasto-plastic formulation of the conventional limit pressure measured with the pressuremeter test in cohesive soil. ASCE J. Geotech. Geoenviron. Eng. **133**(9), 1119–1127 (2007)

Randolph, M.F., Carter, J.P., Wroth, C.P.: Driven piles in clay-the effect of installation and subsequent consolidation. Géotechnique **24**(4), 361–393 (1979)

Silvestri, V., Abou-Samra, G.: Analytical solution for undrained plane strain expansion of a cavity in modified Cam clay. Geomech. Eng. **4**(1), 19–37 (2012)

Yu, H.S.: Plasticity and Geotechnics. Springer, New York (2006)

Yu, H.S.: Cavity Expansion Methods in Geomechanics. Kluwer Academic Publishers, Dordrecht (2000)

Wood, D.M.: Soil Behaviour and Critical State Soil Mechanics. Cambridge University Press, Cambridge (2007)

Zytynski, M., Randolph, M.F., Nova, R., Wroth, C.P.: On modelling the unloading-reloading behaviour of soils. Int. J. Numer. Anal. Meth. Geomech. **2**(1), 87–93 (1978)

Modeling and Interpretation Wave Fields in Hierarchical Heterogeneous Medium

Olga A. Hachay[1(✉)], Andrey Y. Khachay[2], and Oleg Y. Khachay[2]

[1] Institute of Geophysics UB RAS, Ekaterinburg, Russia
[2] Ural Federal University, Ekaterinburg, Russia

Abstract. Geological medium is an open system which is influenced by outer and inner factors that can lead it to a unstable state. That non stability is as a rule occurred locally and these zones are named as dynamically active elements, which are indicators of potential catastrophic sources. These objects differ from the embedded geological medium by their structural forms, which often are of hierarchical type. The process of their activisation can be searched, using wave fields monitoring. For that purpose it is needed to develop new algorithms of modeling wave field's propagation through the local objects with hierarchical structure. Also it is needed to develop new theory of interpretation the wave field distribution for defining the contours of these local hierarchical objects. It had been constructed algorithms for 2D modeling of sound diffraction and linear polarized transversal seismic wave on an intrusion of hierarchic structure, located in the layer number J of N-layered elastic medium. We used the method of integral and integral-differential equations for a space frequency presentation of wave field's distribution. It is developed an algorithm for constructing the equation of theoretical inverse problem for 2-D electromagnetic field of E and H polarization and linear polarized longitudinal elastic wave by excitation of the N-layered conductive or elastic medium with hierarchic conductive or elastic inclusion located in the ν-th layer. From the theory it is obviously that for such complicated medium each wave field contains its own information about the inner structure of the hierarchical inclusion. Therefore it is needed to interpret the monitoring data for each wave field apart, and not to mix the data base. These results will be the base for constructing new systems of monitoring observations of dynamical geological systems. Especially it is needed to prevent rock shocks in deep mines by their exploitation.

Keywords: Hierarchic medium · Electromagnetic field · Seismic field · Algorithms of modeling · Equation of theoretical inverse problem

1 Introduction

The last decades are characterized by active development of Earth's sciences. We shall use the materials, published in the book (Dmitrievsky 2009) by Russian academician Dmitrievsky A.N., who suggested the conception about the development of oil-gas geology in Russian Federation. The modern research methods and technologies give the opportunity to obtain new data about the Earth's structure and processes, which

W. Frikha et al. (eds.), *Soil Testing, Soil Stability and Ground Improvement*,
Sustainable Civil Infrastructures, DOI 10.1007/978-3-319-61902-6_24

occur in its interior. The conception development about the nonlinear geodynamics practically coincides with research of nonlinear processes in different parts of physics. In geology soliton and auto wave conceptions are developed, principles of synergetic and self organization become be used, and in geodynamics the macro quantum behavior of large mass matter, which are in critical state, in geophysics the auto wave nature of geophysical fields is researched. A new of direction quantum geodynamics appeared. In contrary from traditional approaches in geodynamics, which are based on classical models of continuum, quantum geodynamics allows analyzing the Earth's energetic structure, which evolutes in time and penetrates in all natural phenomena and possesses macro quantum time features. The development of possible methods and approaches must be based on the considering energy of dynamical processes.

For understanding and analyzing the state of the geological medium it is needed to compare the key ideas of geophysics, which give the main ideas to research different dynamical events into the Earth.

"Geophysics of XX century" is the understanding of such features:

- geophysical fields are indicators of the processes, which occur in lithosphere; geophysical parameters, which are registered distantly can have functional or correlation relation with the matter-structural characteristics of geological medium (on macro- and micro levels);
- analysis of space-time and energetic distribution of geophysical field can give information about space-time distribution of geological medium properties;
- registration and analysis of geophysical field in the monitoring regime can give information about the geodynamical processes in near borehole space, in Earth's crust and lithosphere in more deep Earth's layers.

Practical problems of geophysics of XX century had been vigorous stimulus of evolution of theoretical and experimental physics of thin layered, porous and crack media. As a result it had been derived new classes of mathematical models fluid saturated heterogeneous media, it had been researched anisotropic effects of geological media, and it had been revealed different physical and physical-chemical effects which occur on the boundaries "solid skeleton-fluid". Geophysics for the first time set a question about possibility of construction physical-geological and mathematical models of geological objects and processes. Applied geophysics of XX century had realized the possibility of research one and the same geological objects on micro level (nuclear geophysics), meso level (electrical, heat, magnetic, acoustic fields) and macro level (fields of elastic waves and low frequency electromagnetic fields).

Key ideas in XXI century in geophysics: Geophysics of the XXI century is the understanding of such features: the Earth is a self evolutionary, self conditioned geo cybernetic system for which the geophysical fields fulfill the role of a driving mechanism. Hierarchic and quantum features of geophysical space can be used in new geophysical principles together with nonlinear effects and effects of re-emission of geophysical fields that allow us to create a new geophysics. At the same time we have to consider new aspects in theory and mathematical modeling of geophysical fields and new systems of data interpretation. That will lead to development new equation classes, which describe the distribution of elastic and electromagnetic fields in heterogeneous and hierarchic media, with account of various effects of nonlinearity of geological media and irreversibility of geophysical processes. New theories of inverse problems solution with account of the hierarchic structure of inclusions imbedded into the layered medium will be developed.

Now we shall investigate and study some results which are into the key direction of geophysics of XXI century, noted by Dmitrievsky (Dmitrievsky 2009).

In present day for more adequate understanding of the dynamics of processes, which occur in the geological medium on deep levels by the action of natural and man caused factors Academician V.E. Panin using the results, which had been derived together with his colleagues (Panin 2005), had introduced a new paradigm on the junction between physics and mechanics of deformed solid body, which is the base of physical mesomechanics.

1. Identification of mechanisms of plastic flow on different structural deformation levels, which lead to a fundamental change of the initial inner structure of the solid body and forming into it dissipative substructures as mesoscopic plastic deformation support.
2. Fixing a relation between the outer action, changing of the initial inner structure, forming a hierarchy of mesoscopic self matched structural levels of deformation and occurring as a result of it mechanical fields.
3. Synergetic approach in the methodology of describing the deformed solid body as a non equilibrium many leveled medium, which in the points of bifurcation losses its shift stability on different structural levels and becomes to be destroyed in the conditions of global losses of its shift stability on macro scale level.

For experimental research of deformation mechanisms of specimens on meso level it had been developed new methods with use of speckle interferometer and optical television devices of technical vision, measuring of fractal dimension of deformed solid body. It turns out, that on the meso level 3-D structural elements (meso volumes) move as a whole. In that case it is sufficient to consider a representative volume, which consists on some tens meso volumes, for writing the equations of mesomechanics, taking into account the inner structure of the deforming solid body.

For realizing the second point of the new paradigm Panin with his colleagues had written a system of equations, which describes the mechanical field in the deforming solid body on one level (Panin 2005). It turned out, that it is similarly the Maxwell system of equations for alternating electromagnetic fields. Similarly electromagnetic field, where alternating electric and magnetic fields are mutually linked, the common mechanical field occurs in the deforming solid body, which contains organic mutual linked the translational and rotating modes (Panin 2005).

That result was very significant for choice of geophysical methods, using for monitoring of structure and state of rock massif, which is under strong man based action, which is a component of the geological medium in seismic-tectonically areas.

2 Mathematical Modeling and Comparison of Seismic and Electromagnetic Response for a Block-Layered 2D Model with Homogeneous Inclusions

In that paper are derived integral equations and integral differential equations of 2D direct problem for the seismic field in the dynamical variant. We had been provided the joint analysis of the integral equations for 2D direct problems for electromagnetic and seismic fields. The received results can be used for definition of the complex criterions of achievement the research of high complicated medium both with seismic and electromagnetic methods. For the problem of sound diffraction on the 2D elastic heterogeneity, located in the j-th layer of the n-layered medium, using the approach from V.I. Dmitriev and V.D. Kupradze (Hachay et al. 2013) we can derive the integral differential equation for the distribution of the potential for the vector of elastic displacements inside the heterogeneity.

$$\begin{aligned}
&\frac{(k_{1ji}^2-k_{1j}^2)}{2\pi}\iint\limits_{Sc}\varphi(M)G_{Sj}(M,M^0)d\tau_M+\frac{\sigma_{ja}}{\sigma_{ji}}\varphi^0(M^0)\\
&-\frac{(\sigma_{ja}-\sigma_{ji})}{\sigma_{ji}2\pi}\oint\limits_{C}G_{Sj}\frac{\partial\varphi}{\partial n}dc=\varphi(M^0)\ by\ M^0\in S_C\\
&\frac{\sigma_{ji}(k_{1ji}^2-k_{1j}^2)}{\sigma(M^0)2\pi}\iint\limits_{Sc}\varphi(M)G_{Sj}(M,M^0)d\tau_M+\varphi^0(M^0)\\
&-\frac{(\sigma_{ja}-\sigma_{ji})}{\sigma(M^0)2\pi}\oint\limits_{C}G_{Sj}\frac{\partial\varphi}{\partial n}dc=\varphi(M^0)\ by\ M^0\notin S_C
\end{aligned}\tag{1}$$

Using the second integral-differential presentation we can define the potential of the elastic displacements in the arbitrary layer, and then we can calculate the distribution of the vector of elastic displacements in the arbitrary layer. Let us compare the derived expressions with the solution of the diffraction problem for electromagnetic field in the frame of the same geometrical model. That case corresponds to the problem of exciting by a plane wave H -polarization, the solution of which is done in the paper (Hachay et al. 2008a, b). Let us transform it to the form similarly to (1) and let us compare the derived equations for the solution of the inner 2-D seismic and electromagnetic problem. $\tilde{k}^2(M^0)=i\omega\tilde{\mu}_0\tilde{\sigma}(M^0)$ $\tilde{\mu}_0=4\pi10^{-7}\frac{H}{m}$, $\tilde{\sigma}(M^0)$ - conductivity in the point M^0, i - the imaginary unit, $H_x(M^0)$ - the summarized component of magnetic field, $H_x^0(M^0)$ - the component of magnetic field in the layered medium without heterogeneity, $\tilde{k}_{ji}^2(M^0)=i\omega\tilde{\mu}_0\tilde{\sigma}_{ji}$, $\tilde{k}_i^2(M^0)=i\omega\tilde{\mu}_0\tilde{\sigma}_i$, $\tilde{\sigma}_{ji}$ - conductivity into the heterogeneity, located into the j-the layer, $\tilde{\sigma}_i$ - conductivity of the i-th layer of the n-layered medium, $G_m(M,M^0)$ - the Green function of the 2-D problem for the case of H-polarization (Hachay 2007).

$$\begin{aligned}
&\frac{(k_{1ji}^2-k_{1j}^2)}{2\pi}\iint\limits_{Sc}\varphi(M)G_{Sj}(M,M^0)d\tau_M+\frac{\sigma_{ja}}{\sigma_{ji}}\varphi^0(M^0)\\
&-\frac{(\sigma_{ja}-\sigma_{ji})}{\sigma_{ji}2\pi}\oint\limits_C G_{Sj}\frac{\partial\varphi}{\partial n}dc=\varphi(M^0)\,by\,M^0\in S_C\\
&\frac{(\tilde{k}_{ji}^2-\tilde{k}_j^2)}{2\pi}\iint\limits_{Sc}H_x(M)G_m(M,M^0)d\tau_M+\frac{\tilde{k}_{ji}^2}{\tilde{k}_j^2}H_x^0(M^0)\\
&-\frac{(\tilde{k}_{ji}^2-\tilde{k}_j^2)}{\tilde{k}_j^2 2\pi}\oint\limits_C H_x\frac{\partial G_m}{\partial n}dc=H_x(M^0)\,by\,M^0\in S_C
\end{aligned}\tag{2}$$

The difference in the boundary conditions for the seismic and electromagnetic problems lead to different types of equations: in the seismic case- to the integral-differential equation, in the electromagnetic case to the load integral Fredholm equation of the second type. If for the solutions of the direct electromagnetic and seismic in dynamical variant problems we can establish the similarity in the explicit expressions for the components of electromagnetic and seismic fields by definite types of excitation then with complicating of the medium structure as can we see from the obtained result by the case of the seismic field linked with longitudinal waves the similarity vanishes. That means that the seismic information is additional to the electromagnetic information about the structure and state of the medium.

For the problem of diffraction of a linearly polarized elastic transverse wave on the 2-D heterogeneity located in the j-th layer of the n-layered medium, using the approach described in the paper (Hachay et al. 2008a, b) for the electromagnetic wave 2-D problem (case H-polarization), (the geometric model is similar to a that described higher in the previous problem) we obtain the expressions as follows for the components of the displacement vector:

$$\begin{aligned}
&\frac{(k_{2ji}^2-k_{2j}^2)}{2\pi}\iint\limits_{Sc}u_x(M)G_{Ss,j}(M,M^0)d\tau_M+\frac{\mu_{ja}}{\mu_{ji}}u_x^0(M^0)\\
&-\frac{(\mu_{ja}-\mu_{ji})}{\mu_{ji}2\pi}\oint\limits_C u_x(M)\frac{\partial G_{Ss,j}}{\partial n}dc=u_x(M^0)\,by\,M^0\in S_C\\
&\frac{\mu_{ji}(k_{2ji}^2-k_{2j}^2)}{\mu(M^0)2\pi}\iint\limits_{Sc}u_x(M)G_{Ss,j}(M,M^0)d\tau_M+u_x^0(M^0)\\
&-\frac{(\mu_{ja}-\mu_{ji})}{\mu(M^0)2\pi}\oint\limits_C u_x(M)\frac{\partial G_{Ss,j}}{\partial n}dc=u_x(M^0)\,by\,M^0\in S_C
\end{aligned}\tag{3}$$

The expressions (3) content the algorithm of seismic field simulation for distribution of transversal waves in the n-layered medium, which contain a 2D heterogeneity. The first expression is a Fredholm load integral equation of the second type the solution of which gives the distribution of the components of the elastic displacements vector inside the heterogeneity. The second of them is an integral expression for calculation of the elastic displacements vector in the arbitrary layer of the n-layered medium. Comparing the expressions (3) with correspondingly for the electromagnetic field (H-polarization) (2) we see that there is a similarity of the integral structure of these expressions. The difference is only for the coefficients of corresponding terms in the expressions (2) and (3). That we can account by choosing the system of observation with one or another field. We must also account the difference of the medium response frequency dependence from seismic or electromagnetic excitation. But keeping within the similarity of the coefficients the seismic field, excited by transversal waves, and the electromagnetic field will contain the similar information about the structure of the heterogeneous medium and state, linked with it. Those results are confirmed by the natural experiments described in the paper (Hachay 2007) Thus, it is showed that for more complicated, than horizontal-layered structures of the geological medium the similarity between the electromagnetic and seismic problems for longitudinal waves get broken. Therefore, these observations with two fields allow getting reciprocally additional information about the structure and especially about the state of the medium. These fields will differently reflect the peculiarities of the heterogeneous structures and response on the changing their state. If we can arrange seismic observations only with the transversal waves together with the magnetic component of electromagnetic one (H-polarization) in the 2-D medium, it will be establish the similarity, which can be used by construction of mutual systems of observation for magneto-telluric soundings and deep seismic soundings on exchanged waves.

3 Mathematical Modeling of Seismic Response for a Hierarchic Model

From the point of view of the paradigm of physical mesomechanics, which includes the synergetic approach to the change of rock massif state of different matter content, that problem can be solved with use of monitoring methods, which are settled on the research of hierarchic structured media (Panin 1995). For description of these effects it is needed to consider the wave process in the hierarchic blocked medium. Let us consider an algorithm of sound diffraction on 2-D elastic heterogeneity with hierarchic structure, located in the j-th layer of n-layered medium (Hachay et al. 2008a, b).

$$\begin{aligned}
&\frac{(k_{1jil}^2-k_{1j}^2)}{2\pi}\iint\limits_{Scl}\varphi_l(M)G_{Sp,j}(M,M^0)d\tau_{Ml}+\frac{\sigma_{ja}}{\sigma_{jil}}\varphi_{l-1}^0(M^0)\\
&-\frac{(\sigma_{ja}-\sigma_{jil})}{\sigma_{jil}2\pi}\oint\limits_{Cl}G_{Sp,j}\frac{\partial\varphi_l}{\partial n}dc_l=\varphi_l(M^0)\;by\;M^0\in S_{Cl}\\
&\frac{\sigma_{jil}(k_{1jil}^2-k_{1j}^2)}{\sigma(M^0)2\pi}\iint\limits_{Scl}\varphi_l(M)G_{Sp,j}(M,M^0)d\tau_{Ml}+\varphi_{l-1}^0(M^0)\\
&-\frac{(\sigma_{ja}-\sigma_{jil})}{\sigma(M^0)2\pi}\oint\limits_{Cl}G_{Sp,j}\frac{\partial\varphi_l}{\partial n}dc_l=\varphi_l(M^0)\;by\;M^0\notin S_{Cl}
\end{aligned}\tag{4}$$

$G_{Sp,j}(M,M^0)$ source function for the longitudinal wave, which coincides with the function from the paper (Hachay and Khachay 2013), $k_{1jil}^2=\omega^2(\sigma_{jil}/\lambda_{jil})$ – wave number for the longitudinal wave. In the equations, the index *ji* indicates the features within the heterogeneity membership, *ja* – out of the heterogeneity, $l = 1 \ldots L-1$ – number of the hierarchic level, $\vec{u}_l = grad\varphi_l$, φ_l^0 – potential of the normal seismic field in the layered medium, when the heterogeneity of the previous rank is absent. If $l = 2 \ldots L$ $\varphi_l^0=\varphi_{l-1}$, if $l = 1$, $\varphi_l^0=\varphi^0$, this coincides with the expression from the paper (Hachay et al. 2015).

$$\begin{aligned}
&\frac{(k_{2jil}^2-k_{2j}^2)}{2\pi}\iint\limits_{Scl} u_{xl}(M)G_{Ss,j}(M,M^0)d\tau_{Ml}+\frac{\mu_{ja}}{\mu_{jil}}u_{x(l-1)}^0(M^0)\\
&-\frac{(\mu_{ja}-\mu_{jil})}{\mu_{ji}2\pi}\oint\limits_{Cl} u_{xl}(M)\frac{\partial G_{Ss,j}}{\partial n}dc_l=u_{xl}(M^0)\ by\ M^0\in S_{Cl}\\
&\frac{\mu_{jil}(k_{2jil}^2-k_{2j}^2)}{\mu(M^0)2\pi}\iint\limits_{Scl} u_{xl}(M)G_{Ss,j}(M,M^0)d\tau_{Ml}+u_{x(l-1)}^0(M^0)\\
&-\frac{(\mu_{ja}-\mu_{jil})}{\mu(M^0)2\pi}\oint\limits_{Cl} u_{xl}(M)\frac{\partial G_{Ss,j}}{\partial n}dc_l=u_{xl}(M^0)\ by\ M^0\in S_{Cl}
\end{aligned}\tag{5}$$

$k_{2jil}^2=\omega^2(\sigma_{jil}/\mu_{jil})$ - wave number for the transversal elastic wave, $G_{Ss,j}(M,M^0)$ source function for a transversal elastic wave (Hachay and Khachay 2013). If l = 2… L $u_x^0=u_{x(l-1)}$, if l = 1, $u_{xl}^0=u_x^0$, this coincides with the expression from the paper (Hachay et al. 2015). If by transition on the next hierarchic level the axis of two-dimensionality does not change and only the geometry of the section of embedded structures change, then we can write the iteration process of modeling of the seismic field (case generation only longitudinal wave). The iteration process covers to modeling of the response of transition from the previous hierarchic level on the next level. Inside each hierarchic level the integral-differential equation and the integral-differential representation are calculated as it is written in the papers (Hachay et al. 2008a, b).

4 Inverse Problem for Electromagnetic Field Propagation Through a Layered Conductive Medium with Hierarchic Inclusions

In the paper (Hachay 1994) it had been considered a conception of a staged interpretation of the alternating electromagnetic field. On the first stage the parameters of the normal section or the parameters of the one dimensional non magnetic medium in which are embedded anomalous conductive or magnetic inclusions are defined. On the second stage it is developed a procedure of anomalous alternating field fitting by a system of singular sources, which are embedded into the horizontal layered medium with geoelectrical parameters, which had been defined on the first stage. On the third stage it is solved the theoretical inverse problem, that is for the given geoelectrical parameters for the embedded medium for the set of anomalous parameters, defined on the second stage, we define the contours of the inclusions. We had derived explicit integral-differential equations of the theoretical inverse problem for 2-D and 3-D alternating and 3-D stationary electromagnetic fields in a frame of the models: conductive or magnetic body in the J-th layer of the conductive layered half space. Here, using the approach, which was written in the papers (Hachay 1989, 1990), we had derived the equation of the theoretical inverse problem for the 2-D alternating electromagnetic field (scalar case) for the model: a conductive hierarchic inclusion of the k rank, located in the J-th layer of the conductive N-layered half space. $U^{+(k-1)}(M_k)$ - anomaly component $E_x^{+(k-1)}$ or $H_x^{+(k-1)}$ for the inclusion of k-th rank by $M_k\in\partial D_k$, U

(M) - sum field E_x or H_x, $U^{i(k-1)}(M)$ - normal field of the layered section E_x or H_x for k = 1, for k > 1 - the field E_x or H_x, must be calculated, using the direct problem solution with use the algorithm (Hachay et al., 2013). $G^{ak}(M_k, M_o)$ - Green function for the inner area of the heterogeneity of the k-th rank, $G(M_k, M_o)$ - Green function of the N-layered medium (Hachay 1989, 1990), b_v, b_i, b_{ak} - complex coefficients for the J-th layer and for the inner area of the heterogeneity of the k-th rank, which are introduced in (3).

$$\begin{aligned}2\pi U^{+(k-1)}(M_0) = \int_{\partial Dk} ((U_v^{+(k-1)}(M_k) \\ + U_v^{i(k-1)}(M_k))(\tfrac{\partial G^{ak}(M_k,M_0)}{\partial n} - (\tfrac{b_v}{b_i})\tfrac{\partial G(M_k,M_0)}{\partial n}) \\ - b_v(\tfrac{\partial U_v^{+(k-1)}(M_k)}{\partial n} + \tfrac{\partial U_v^{i(k-1)}(M_k)}{\partial n}) \\ \times ((\tfrac{1}{b_{ak}})G^{ak}(M_k, M_0) - (\tfrac{1}{b_i})G(M_k, M_0)))dl_k\end{aligned} \tag{6}$$

The hierarchic heterogeneities are approximated by embedded non axial conductive σ_{ak} cylinders, located along the axes OX. As a result of solution of the Eq. (1) for the function $r(\varphi)$, this describes the contour of the sought heterogeneity of the k-th rank.

5 Conclusions

We had developed an iterative algorithm for modeling electromagnetic and seismic fields in a frame of two models for layered-block 2D models with homogeneous and hierarchic inclusions. The last algorithms differ from the fractal model approach by a freer selecting of heterogeneities position of each rank. If the boundaries of the inclusion of the k rank are fractals (Mandelbrot 1982) the surface and contour integrals in the integral equations must be changed to repeated fractional integrals of Riman - Liuvill (Simko et al. 1987). We had derived the equation of the theoretical inverse problem for the 2D alternating electromagnetic field (scalar case) for the model: a conductive hierarchic inclusion of the k rank, located in the J-th layer of the conductive N-layered half space.

References

Dmitrievsky, A.N.: Selected works, V.2, Nauka, Moscow (2009)

Hachay, O.A.: Geophysical monitoring of the state of rock massif with use of paradigm of physical mesomechanics. Phys. Earth **4**, 58–64 (2007)

Hachay, O.A., et al.: Modeling of seismic and electromagnetic field in the hierarchic heterogeneous media. In: Proceedings of International Conference. IGF UB RAS, Ekaterinburg (2008a)

Hachay, O.A., et al.: Complex electromagnetic and seismic method of research of the crust and (2008b)

Hachay, O.A.: Mathematical modeling and interpretation of alternating electromagnetic field in heterogeneous crust and upper mantle of the earth. Professor Dissertation, IGF UB RAS, Sverdlovsk (1994)

Hachay, O.A.: About interpretation 2-D alternative and 3-D stationary anomalies of electromagnetic field. Phys. Earth **10**, 50–58 (1989)
Hachay, O.A.: About the solution of inverse problem of 3-D alternating electromagnetic fields. Phys. Earth **2**, 55–59 (1990)
Mandelbrot, B.B.: The Fractal Geometry of Nature. Freeman, San Francisco (1982)
Panin, V.E.: Physical mesomechanics and computer construction of materials. Novosibirsk, Nauka, SB RAS (1995)
Simko, S.G., Kilbas, A.A., Marichev, O.I.: Integrals and derivatives of partial rank and some their applications. Minsk, Technique (1987)

Strength and Stiffness Studies of Cement Stabilized Granular Lateritic Soil

Dipti Ranjan Biswal(✉), Umesh Chandra Sahoo, and Suresh Ranjan Dash

School of Infrastructure, IIT, Bhubaneswar 752050, Odisha, India
{drbll,ucsahoo,srdash}@iitbbs.ac.in

Abstract. A huge network of rural roads is being developed in India under the most ambitious Prime Minister's rural connectivity programme, *PMGSY (Pradhan Mantri Gramin Sadak Yojna)*. Under this programme, thousands of kilometers of rural roads are being constructed in the country, which require good quality pavement materials like crushed stone. The scarcity of natural aggregates has compelled to use marginal materials or locally available soils in structural layers of these pavements, which would reduced the cost of the project. Granular lateritic soils are widely available in many parts of India and presently, this is also used as sub-base material in different rural road projects, where it satisfies the code specifications. However, granular lateritic soils of some locations do not satisfy the strength and plasticity requirement of sub-base layer. But, they can probably be made suitable through stabilization. Though stabilization of soil by cement or lime is a well known process of improving the strength and stability of soil, the strength and stiffness parameters of stabilized lateritic soils in terms of modulus of rupture, resilient modulus, flexural modulus have very limited reference in literature. Therefore, in this study an attempt has been made to characterize the cement stabilized lateritic soils for use in sub-base and base layers in rural road pavements. A comprehensive laboratory testing programme has been conducted on cement stabilized granular lateritic soil samples collected from five different places of eastern India to study various strength parameters such as compressive strength, modulus of rupture and stiffness properties in terms of flexural modulus of cement stabilized granular lateritic soil. In this paper, strength and stiffness developments of cement stabilized granular lateritic soil in 7 days and 28 days have been studied and its suitability as a structural layer in rural roads has been investigated. Suitable modulus values of cement stabilized granular lateritic soils have been proposed which can be used as an input parameter for the input in mechanistic design of roads. Also relationships have been proposed to determine modulus of rupture and flexural modulus of cement stabilized granular lateritic soil from its compressive strengths.

1 Introduction

A huge network of rural roads is being developed in India under different schemes launched by the government including the Prime Minister's rural connectivity programme (*PMGSY*). The requirement of enormous amount of good quality pavement

W. Frikha et al. (eds.), *Soil Testing, Soil Stability and Ground Improvement*, Sustainable Civil Infrastructures, DOI 10.1007/978-3-319-61902-6_25

materials like crushed stones (whose source is depleting day by day), is becoming a concern for the administrators and concerned authorities of the country. In this scenario, investigation of the use of marginal aggregates and locally available granular materials in structural layers of the pavement is timely justified. Granular lateritic soil (GLS) is available in many parts of India and presently being used as a sub-base material in various road projects, where this meets the code specifications. However, GLS from various sources do not satisfy the strength and plasticity requirement for sub-base and base layers (Biswal et al. 2016). Hence, these soils need to be modified or stabilized to make them suitable for sub-base and base layers of pavement. Traditionally, cement and lime or both are used to stabilize the soils for road construction purpose (Gidigasu 1976). Indian Roads Congress (2010) provides guidelines are available for cement and lime stabilization (IRC 2010) of soils. Though stabilization of soil by cement or lime is a very well known process of improving the strength of soil, very few studies are available on strength and stiffness properties of cement stabilized granular lateritic soils (CLS). For the mechanistic design of pavements with stabilized layers, the input parameters usually required are modulus and flexural strength. But, studies related to characterization of CLS in terms of modulus, flexural strength etc. under repeated loading condition is scarce. Therefore, in the present study, strength properties of CLS such as unconfined compressive strength (UCS), modulus of rupture (MOR) i.e., flexural strength and stiffness in terms of flexural modulus have been investigated.

1.1 Studies on Stabilized Lateritic Soil

The objective of stabilization is to improve plasticity characteristics, alteration of grain size distribution, increase in mechanical strength or stiffness and durability under adverse conditions which may be done by mixing cement or lime and cement, fly ash or with any chemical stabilisers. A brief summary on stabilization of lateritic soils by different researchers is presented in Table 1. Properties of cement or lime stabilized lateritic soils have been studied by many researchers. However, most of the papers have explained the improvement of the geotechnical properties e.g., OMC, MDD, liquid limit, plastic limit, UCS, CBR of stabilized lateritic soil. It needs to be emphasized that, studies regarding flexural modulus, flexural strength including compressive strength of stabilized lateritic soil is very scarce The study carried out by Joel and Agbede (2010) on stabilization of lateritic soil with cement and sand, revealed that 45% sand and 6% cement is the optimum content for the mix design. Similarly, Portelinha et al. (2012), Jaritngam et al. (2012), Ravi et al. (2008), Joel and Agbede (2010) investigated the potential of stabilized lateritic soils of Brazil, Thailand, India and Nigeria respectively. It may be observed that cement and lime proves to be good modifiers or stabilizers for lateritic soils in order use as a base or sub-base course material.

Strength of Cement Stabilized Soils

As already mentioned, stabilization of soil by cement and lime is a well known process, and use of stabilized soil in pavements has also been in practice for decades. However, the proper characterisation of cement stabilized soil as sub-base and base course

Table 1. Some studies on stabilised lateritic soil

Location of laterite sample	Reference	Optimum content (%)	7 day UCS
Nigeria	Joel and Agbade (2010)	45% sand and 6% cement	2702 kPa, 270% improvement
Brazil	Portelinha et al. (2012)	3% cement or 3% lime	600 kPa for lime (118% improvement) 1000 kPa for cement (233% improvement)
Thailand	Jaritngam et al. (2012)	3% cement	2220 kPa
India	Ravi et al. (2008)	4% cement with pond ash	1000 kPa, 440% improvement

material is required as the mechanistic design of pavements requires input parameters like elastic modulus and Poisson's ratio etc. Unconfined compressive strength is one of the accepted methods for determining the strength of bound materials. UCS is used for determining the suitability of the mix to perform satisfactorily as a bound sub-base and base layer (Paige 1998; Vorobeiff 2004; Austroads 2008; Yeo et al. 2011). Addition of small percentage of stabiliser may improve the strength and plasticity properties of soil, but it may lack the required tensile strength and behave as an unbound granular (Vorobeiff 2004). A criteria developed by Austroads (2013) to differentiate modified soil from stabilized soil is given in Table 2.

Table 2. Typical properties of modified and bound materials (Austroads 2013)

Classification	Testing criteria	Test conditions
Modified	0.7 MPa < UCS < 1.5 MPa	28 day moist curing and 105 mm diameter and 115.5 mm height mould and 4 h soaking prior to testing
Bound	UCS > 1.5 MPa	

Many research reports and standard specifications have put the requirement of minimum strength in terms of UCS of stabilized soil for use in sub-base and base layer of pavement. As per Indian Road Congress (IRC:37-2012) guidelines, the minimum requirement of 7 day UCS value for cement stabilized soils are 4.5 MPa and 0.75–1.5 MPa for base and sub-base respectively. However, in case of low volume roads the minimum UCS value for stabilized or bound base is 3 MPa which has been mentioned in Table 3. Further, wet-dry durability tests need to be conducted on stabilized materials to assess the suitability of these materials for use as sub-base or base layer. The assessment is done in terms of mass loss after 12 wet-dry cycles. As per PCA (Portland Cement Association) (1992), the maximum permissible mass loss after 12 wet-dry durability cycles are 14% and 7% for granular materials and cohesive clays respectively Similarly, the performance of stabilized materials is studied by conducting shrinkage tests and load induced fatigue tests. However, durability tests, shrinkage tests and fatigue tests of cement stabilized lateritic soil are not within the scope of this paper. Table 3 summarises the strength requirement for stabilized layers for different types of pavement. Test conditions and sample size for determination of *UCS* vary as shown Table 3.

Table 3. Criteria in terms of UCS for suitability of stabilised subbase and base

References	UCS (MPa)		Curing period	Sample size
	Medium to high volume roads	Low volume roads		
Portland Cement Association (1992)	2.068 (for subbase/base)			
Gass (1993)	2.068 (for subbase/base)			
MEPDG (2004)	1.72 (for subbase) and 5.17 (for base)	1.72 (for subbase) and 5.17 (for base)	7 days for cement and 28 days for lime-flyash or cement-flyash	100 mm (d) × 115 mm (h)
Austroads (2008)	2	1–2	28 days curing and 4 h soaking prior to testing	105 mm (d) × 115.5 mm (h)
IRC 37 (2012)	0.75–1.5 (for subbase) and 4.5–7 (for base)	1.7 (for subbase) and 3 (for base)	7 days for cement and 28 days for lime-flyash	50 mm (d) × 100 mm (h)
Syed and Scullion (2001)	1.38 (for subbase/base)			

The strength criteria for lightly trafficked or low volume roads is low as compared to high volume roads, as there is less concern on fatigue cracking causing damaging effects on the life of thin bituminous surfacing. It can be observed that, the minimum 28 day *UCS* for stabilized materials for use in rural road is 1–2 MPa as per Austroads (2008). The minimum *UCS* value of 1.7 MPa and 3 MPa is considered suitable for sub-base and base course as per IRC:SP-72 (2015).

As the bound material is capable of developing tensile stress, which allows it to resist and transfer the traffic load to layers below, the *MOR* needs to be evaluated. Though bending tensile stress developed in bound layers due to traffic load is below the tensile strength of bound materials, the *MOR* is an important tensile strength parameter which is required for determination of cumulative fatigue damage of stabilized layer. The default value of *MOR* of soil cement and cement treated aggregate are taken as 0.69 MPa corresponding to minimum 28 days UCS value of 5.17 MPa (ARA 2004).

Stiffness of Cement Stabilized Soils
Elastic modulus of stabilized materials is an important parameter required in elastic layered analysis for determination of stresses and strains at critical locations of a pavement. Although elastic modulus of cemented materials can be estimated by monotonic compression test, third point flexure test is better for determination of modulus due to the similarity in development of stress-strain gradient generated in a pavement (Wen et al. 2014; Austroads 2008; Arulrajah et al. 2015; Yeo et al. 2011). Laboratory determination of stiffness or modulus values of cement stabilized materials depends on the type of test. The modulus values obtained under monotonic compression testing differs from that obtained from cyclic compression testing (Fall et al. 2008) which also differs with flexural modulus determined from beam testing. The modulus of cemented materials which influences development of tensile strain is the flexural modulus and therefore, this may be used as the design modulus. For the elastic analysis of pavement structure having soil-cement base layer, elastic modulus and Poisson's ratio are two important input material parameters. Though, the elastic modulus of cement stabilized base material depends on so many factors, the design elastic modulus have been presumed to be 5000 MPa, 3447 MPa and 3500 MPa by Austroads (2012), ARA (2004), IRC (2012) respectively. As the laboratory determination of elastic modulus is a tedious process, generally UCS of stabilized materials is used to determine the modulus value of cemented materials. Equations 1, 2 and 3 presents the models used to determine the design modulus values of soil cement by IRC (2012), ARA (2004), Austroads (2012) respectively.

$$E_f = 1000\, UCS \tag{1}$$

$$E_f = 1200\, UCS \tag{2}$$

$$E_f = 1000\ to\ 1250\, UCS \tag{3}$$

Where,

E = Flexural Modulus in MPa
UCS = Unconfined Compressive Strength in MPa

As, a range of stabilized materials are being used as bound layer in pavements, modulus of elasticity and strength cannot be uniquely related, (Williams 1972) which encourages to study the relationship between UCS and modulus values of stabilized granular lateritic soils.

2 Materials and Methods

2.1 Materials

The materials used in the present investigation are GLS collected from five different locations of Eastern India as shown in Fig. 1 which are marked as A, B, C, D and E. The physical properties of five different GLS are summarised in Table 4 and the chemical compositions of the soils are summarised in Table 5. The lateritic soils

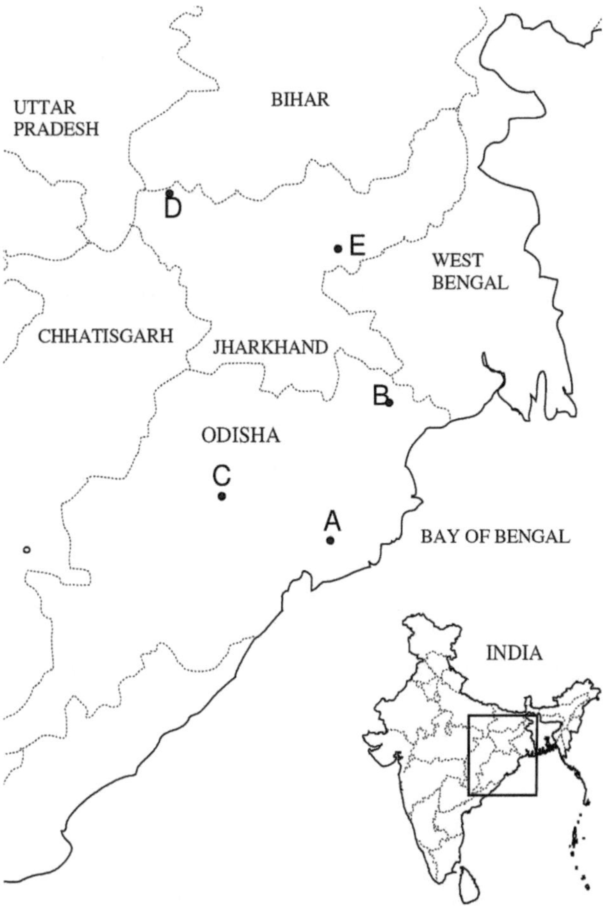

Fig. 1. Location map of collected granular lateritic soils in eastern India.

Table 4. Summary of the engineering properties of soil

Sample no	Region	Gravel	Sand	Silt & Clay	w_{omc} (%)	MDD (kN/m^3)	Soaked CBR (%)	UCS (kPa)	LL	PI	Soil type (ISCS)
A	Khurdha, Odisha	35	60	4	13	21			48	15	SW
B	Baripada, Odisha	95	4	0	9	22	20	760	34	10	GW
C	Sonepur, Odisha	42	37	17	9	22	27	950	68	38	SW
D	Hussainabad, Jharkhand	33	67	0	10	21	23	410	47	19	SW-SM
E	Bokaro, Jharkhand	24	75	0	8	21	24	360	52	29	SW

Table 5. Chemical Compositions of granular lateritic soil and cement used in this study.

Constituents	Lateritic Soil (% by mass)					Cement
	A	B	C	D	E	
Fe_2O_3	21.6	32.49	8.47	7.61	31.82	1.67
Al_2O_3	18.59	14.26	12.51	14.44	17.73	4.41
SiO_2	50.32	44.05	64.06	55.02	48.97	16.33
MgO	0.95	1.02	0.76	0.95	1.14	5.49
K_2O	2.1	1.34	1.69	7.69	2.45	1.09
CaO		1.03	0.3	0.82	0.66	67.99
TiO_2	2.49	2.97	1.36	1.34	3.62	
CuO						0.166
SO_2						0.53

selected under this study have a wide range of plasticity characteristics and gradation, in order to get a practical range of strength and stiffness properties of GLS. The lateritic soils have been air dried prior to testing or sample preparation. The gradation and plasticity characteristics of the studied soils suggests that both cement and lime can be used for stabilization of this soil as per IRC: SP89 (2010), However, cement has been chosen for the study of the stabilization of lateritic soils in the present study. The cement used in this study is bagged ordinary portland cement (OPC).

2.2 Methods

Unconfined Compressive Strength

The strength properties have been determined in two stages. In the first stage, UCS tests have been conducted on all soil samples at various cement content such as 2, 3, 4, 5, 6 and 8%. The air dried soil samples have been mixed with calculated amount of cement and water and samples of size 38 mm in diameter and 76 mm in height have been prepared and cured for 7 and 28 days. All the soil samples have been prepared at optimum moisture content and modified dry density at respective dosed state for the strength and stiffness study.

Flexure Test

The flexure test is conducted for determination of modulus of rupture. In this study, the test has been conducted as per ASTM D 1635. Beam samples of size 75 mm × 75 mm × 290 mm have been prepared for all five GLS mixed with two different cement content i.e., 2% and 5% to prepare stabilized materials for sub-base and base course respectively. The beam samples have been cured for a period of 28 days prior to the flexure test. Three specimens have been tested for each soil samples and the average value of the result has been used for analysis. The flexure test arrangement is shown in Fig. 2.

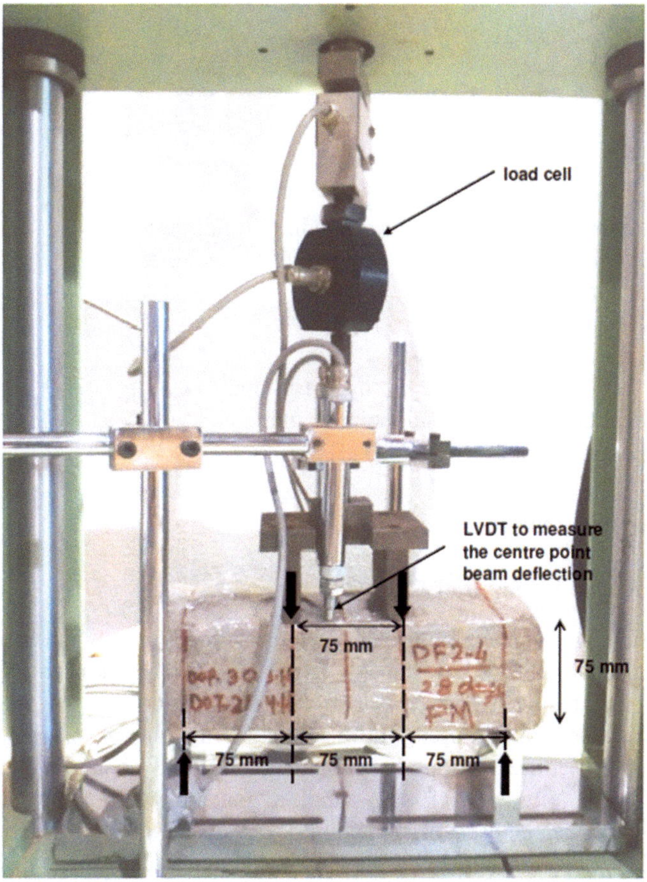

Fig. 2. Flexural Modulus test of beam sample

Flexural Modulus Test

In the absence of a standard method for determination of flexural modulus of stabilized beam samples, the protocol developed by Yeo et al. (2011) has been adopted for determination of flexural modulus. The flexure test arrangement which is shown in Fig. 2 has also been used for flexural modulus test. Two linear variable differential transformers (LVDTs) have been used for measuring deflection of the beam centre. The beam specimen have been subjected to haversine cyclic load of 100 cycles having a peak load which is equal to 20%, 30% and 40% of failure load of beam samples determined from flexure test. Modulus value corresponding to 40% of failure load have been taken for determination of flexural modulus (Wen et al. 2014; Yeo et al. 2008). The first 50 loading cycles are considered as conditioning stage and the average value of modulus calculated in last 50 cycles is reported and considered as modulus value of CLS. Following Equation has been used for determination of flexural modulus.

Flexural Modulus,

$$E_f = \frac{\sigma_t}{\varepsilon_t} = \frac{\frac{Pl}{bd^2}}{\frac{108\delta_d d}{23l^2}} = \frac{23Pl^3}{108bd^3\delta_d} \tag{4}$$

Where,

σ_t = Maximum bending stress corresponding to failure load P
ε_t = Maximum bending strain corresponding to failure load P
P = Cyclic load in kN
b = Width of beam in mm
d = Thickness of beam in mm
l = Span from centre to centre of support in mm
δ_d = Vertical deformation at midpoint in mm

3 Results and Discussions

3.1 Compressive Strength of CLS

The use of stabilized materials or soils for their suitability as a sub-base or base is generally decided by their UCS values (Yeo et al. 2011). Minimum requirement of UCS values for stabilized soils to satisfy the strength criteria are summarized in Table 3. UCS values are also used for differentiating mixtures of soil cement, lime, fly ash or other soil-chemical mixtures as modified soil, lightly bound or heavily bound soil. According to NAASRA (1986), stabilized material with 7 day UCS value of 0.8 MPa or higher can be used as a structural layers as per the strength criteria. As per IRC 37-2012, stabilized materials having UCS value more than 1.5 MPa can be used as a sub-base material and those having UCS value more than 4.5 MPa can be used as base material as per strength criteria. However, the durability, shrinkage and fatigue properties of stabilized granular lateritic soils are not within the scope of this paper. The 7 day and 28 day UCS values of CLS of samples A, B, C, D are shown in Figs. 3 and 4 respectively. As the UCS value of CLS with 3% cement is more than 1.5 MPa, it can be used as sub-base layer in low and high volume roads. UCS values of LSC of samples A, C, D and E at 5% cement content satisfies the requirement of base layer *i.e.,* UCS > 4.5 MPa. However, granular lateritic soil sample B needs minimum 6% cement to satisfy the IRC requirement for base layer. Further, as per IRC SP: 72-2015, a minimum laboratory UCS of 3 MPa is required for base whereas, minimum laboratory UCS of 1.7 MPa is required for sub-base for rural roads. Hence, CLS with 5% cement can be used for base layer in rural roads and CLS with 2% cement content can be used as sub-base layer in rural roads.

The relationship between MOR and UCS of 28 days cured cement stabilized granular lateritic soil samples is shown in Fig. 5 and presented by Eq. 5. A comparison of MOR predicted from UCS obtained by Eq. 4 (present study), Eq. 5 (Wen et al. 2014) and Eq. 6 (ARA 2004) have been illustrated in Fig. 6. The MOR of CLS having 2% cement and 5% cement varies in the range of 0.27–0.74 MPa and 0.74–1.69 MPa

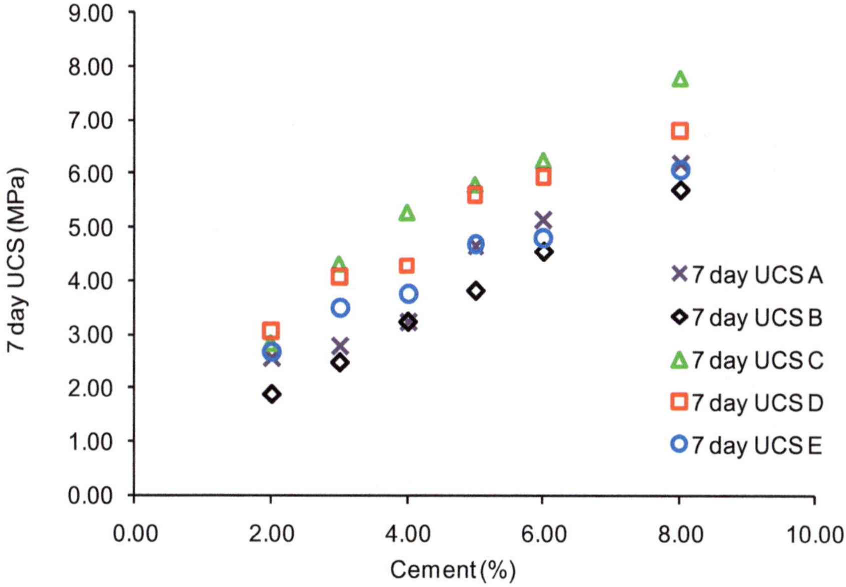

Fig. 3. 7 day UCS values of CLS (specimen size 38 mm diameter and 76 mm height)

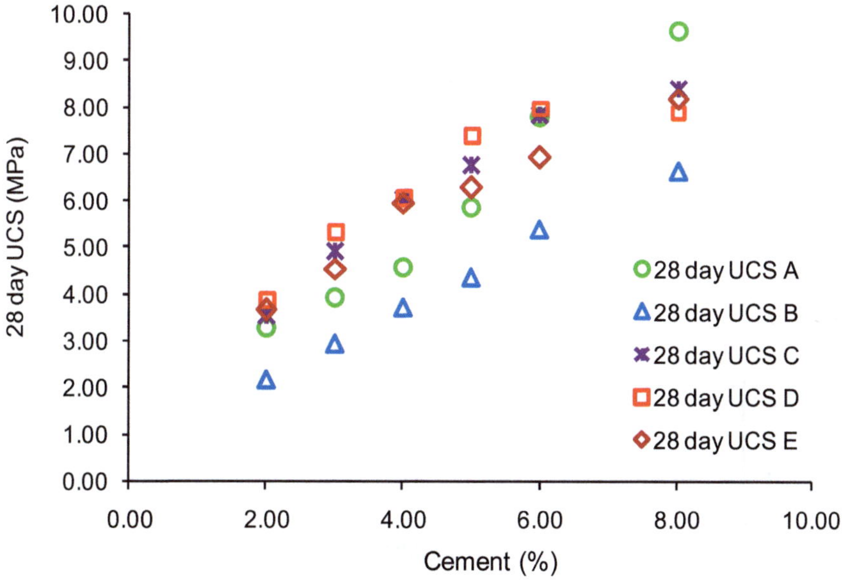

Fig. 4. 28 days UCS values of CLS (size 38 mm diameter and 76 mm height)

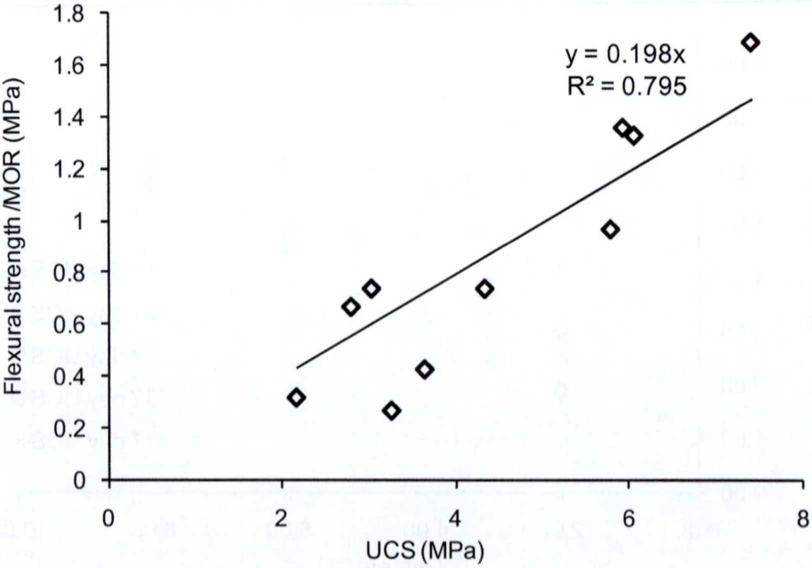

Fig. 5. MOR versus UCS

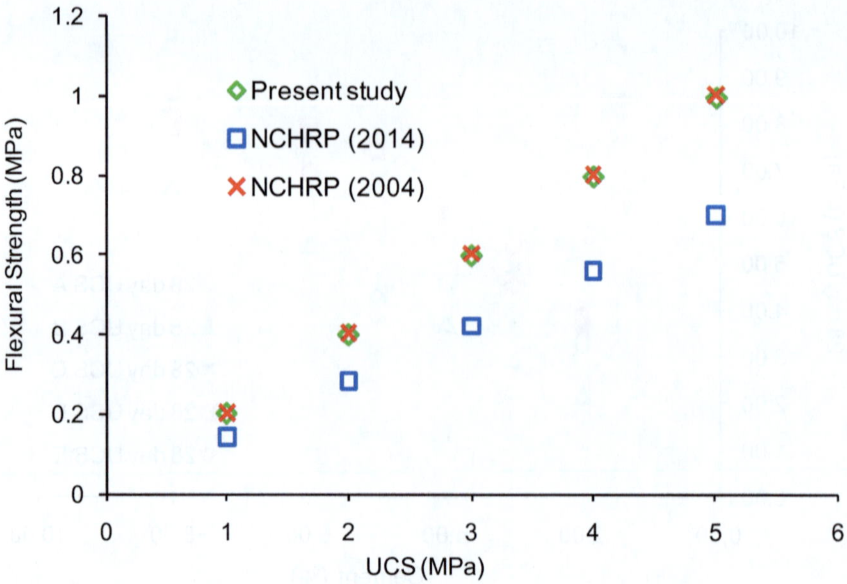

Fig. 6. MOR predicted from UCS using different correlations

respectively. The default value of MOR of soil cement adopted by IRC 37: 2012 and MEPDG for fatigue analysis is 0.7 MPa for UCS value of 5.17 MPa is very conservative for CLS as the predicted MOR obtained from Eq. 2 is 1.02 MPa. However, minimum value of MOR of CLS with 5% cement which is 0.74 MPa can be used as the design value of MOR for CLS.

$$MOR = 0.1985 \times UCS \quad R^2 = 0.79 \tag{5}$$

$$MOR = 0.14 \times UCS \tag{6}$$

$$MOR = 0.2 \times UCS \tag{7}$$

3.2 Flexural Modulus of CLS

Flexural moduli of 28 day cured lateritic soil samples with 2% cement and 5% cement at stress ratios of 0.2, 0.3 and 0.4 are illustrated in Fig. 7. It may be observed that, the modulus increases with stress ratio. This behavior of CLS is in line with gravel-cement, silt-cement and clay-lime (Wen et al. 2014). As already mentioned, flexural modulus at a stress ratio of 0.4 have been taken as the modulus of that sample as per the study of Austroads (2008).

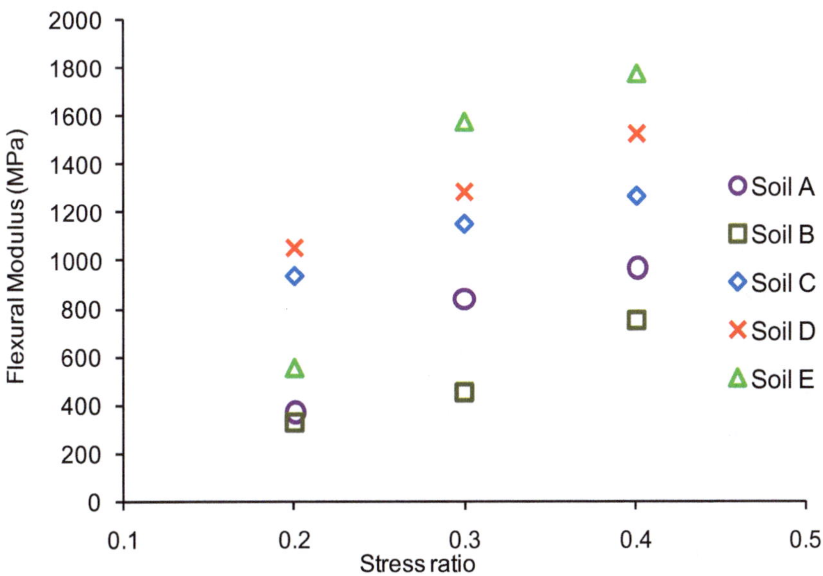

Fig. 7. Flexural modulus at various stress ratios

The relationship between flexural modulus and UCS is shown in Fig. 8 and presented in Eq. 8. However, more samples having wide a range of stiffness or strength values needs to be studied to develop a good correlation between flexural modulus and UCS. Similarly, the relationship between flexural modulus and UCS is shown in Fig. 9

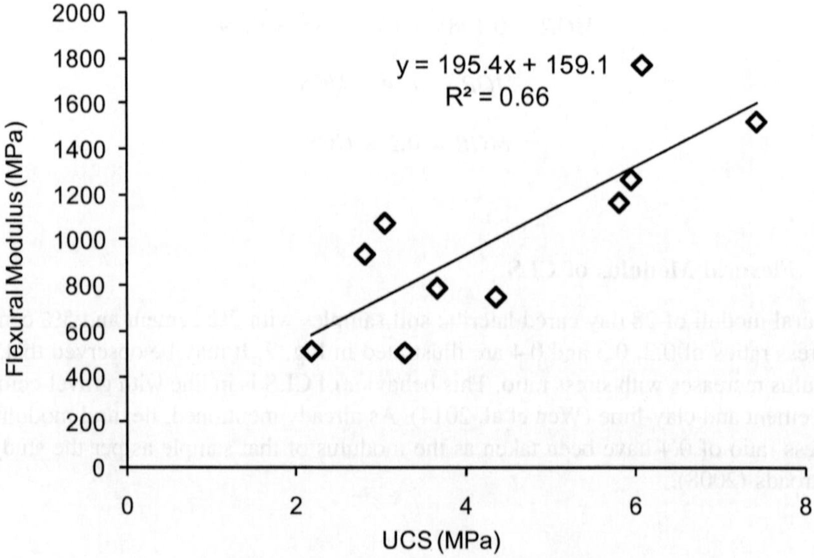

Fig. 8. Flexural modulus versus UCS

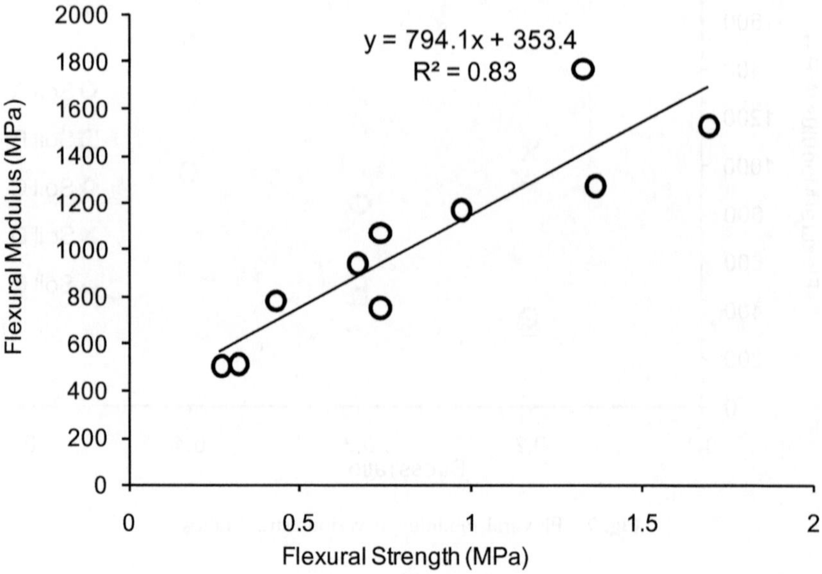

Fig. 9. Flexural modulus versus MOR

and presented in Eq. 9. It may be observed that the flexural modulus and MOR increases with increase in UCS which is in consistent with the earlier studies (Wen et al. 2014; Arulrajah et al. 2015; Austroads 2008).

Figure 10 shows a comparison between predicted flexural modulus and measured flexural modulus of CLS after 28 days curing periods. Figure 11 illustrates a comparison of predicted flexural modulus determined from UCS using various models such as Eq. 8 (present study), Eq. 1 (IRC 2012), Eq. 2 (ARA 2004) and Eq. 10 (Wen et al. 2014). It may be observed from Fig. 11 that the elastic modulus predicted from the relationship developed in present study is very close to the elastic modulus predicted from Eq. 10 (Wen et al. 2014). However, the design elastic modulus predicted from Eq. 1 (IRC 2012) is four times more than the elastic modulus predicted from the relationship developed in present study. Similarly, the design elastic modulus predicted from Eq. 2 (ARA 2004) are five times more than the elastic modulus predicted from the relationship developed in present study. Hence, it is very important to determine the elastic modulus value of cement stabilized soils by using appropriate laboratory of filed test which simulated the behaviour of stabilized layer in pavement

$$E_f = 195.43 \times UCS + 159.17 \quad \mathrm{R}^2 = 0.66 \tag{8}$$

$$E_f = 794.1 \times MOR + 353.4 \quad \mathrm{R}^2 = 0.83 \tag{9}$$

Where,

E_f = flexural modulus in MPa
UCS = Unconfined Compressive strength in MPa
MOR = Modulus of Rupture in MPa

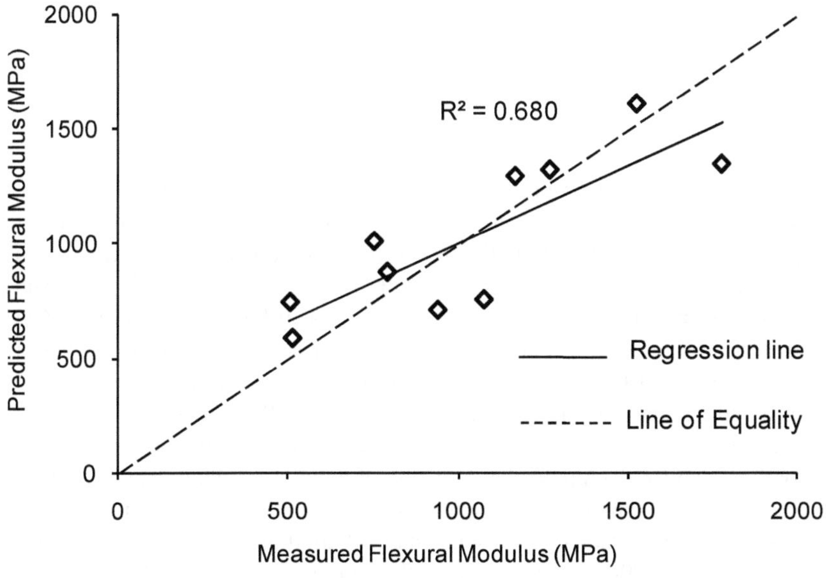

Fig. 10. Predicted verses measured flexural modulus values determined from UCS

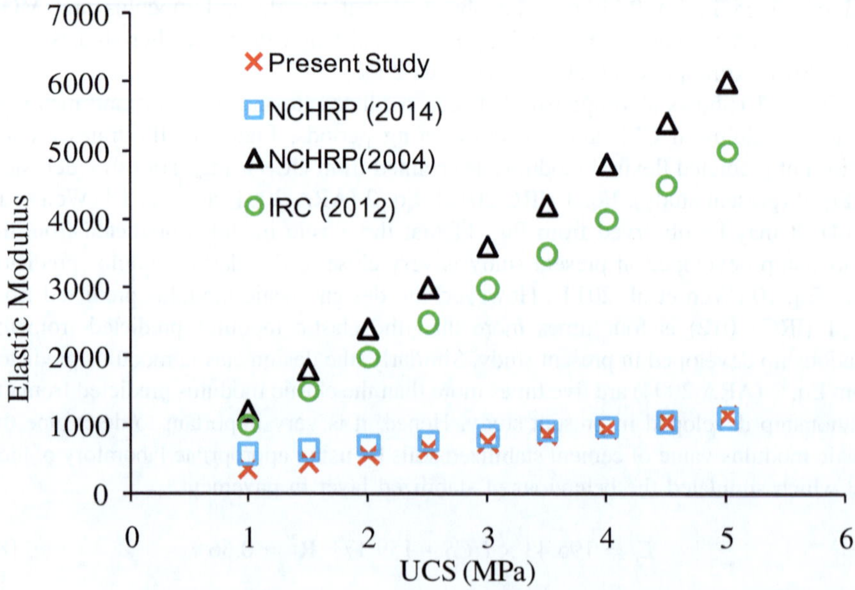

Fig. 11. Comparison of design modulus predicted from UCS

$$E_F^* = 131.08 \times UCS^* + 62382 \tag{10}$$

E_f^* = flexural modulus in psi
UCS^* = Unconfined Compressive strength in psi

4 Conclusions

A detailed laboratory testing programme have been conducted on cement stabilized granular lateritic soils collected from five different locations of eastern part of India to investigate the compressive strength, modulus of rupture and flexural modulus or design modulus. Cement stabilized granular lateritic soil at 2% cement is found suitable for sub-base layer from compressive strength criteria in both low. However, Cement stabilized granular lateritic soil at 5% cement content is found adequate for base layer of rural roads as per IRC:SP 7-2015. The relationship between modulus of rupture and UCS observed in this study is consistent with ARA (2004), Austroads (2012). Comparison has been made between predicted modulus of rupture and modulus determined from UCS of stabilized granular lateritic soil. Relationships have been developed to determine modulus of rupture of cement stabilized granular lateritic soil samples from UCS of CLS, for practical use. Design value of modulus of rupture, mentioned in IRC (2012), ARA (2004) is very close to the minimum modulus of rupture of CLS determined from this study. A relationship has also been established to determine the modulus of CLS from UCS. However, further study is required with more number of samples having a range of stiffness and/or strength values to develop a good correlation between flexural modulus and UCS. However, further studies of some important aspects like, shrinkage, durability and fatigue shall be conducted to assess the performance of stabilized granular lateritic soils, which are not within the scope of this paper.

References

Arulrajah, A., Disfani, M.M., Haghighi, H., Mohammadinia, A., Horpibulsuk, S.: Modulus of rupture evaluation of cement stabilized recycled glass/recycled concrete aggregate blends. Constr. Build. Mater. **84**, 146–155 (2015)

ASTM D1635: Standard Test Method for Flexural Strength of Soil-Cement Using Simple Beam Using Third-Point Loading

Austroads: Guide to Pavement Technology Part 2: Pavement Structural Design. Sydney, NSW: AUSTROADS Incorporated (2012)

Austroads: Review of definitions of modified Granular Materials and Bound Materials. AP-R434-13, Sydney (2013)

Austroads: The development and evaluation of protocols for the laboratory characterisation of cemented materials, by R Yeo, AP-T101/08, Sydney, NSW (2008)

Gass, B.G., Ventura, D.F.C., De Beer, M.: Erodibility of cemented materials. National Department of Transport, South Africa (1993)

Fall, M., Sawangsuriya, A., Benson, C.H., Edil, T.B., Bosscher, P.J.: On the investigations of resilient modulus of residual tropical gravel lateritic soils from Senegal (West Africa). Geotech. Geol. Eng. J. **26**(1), 109–111 (2008)

Gidigasu, M.D.: Laterite Soil Engineering: Pedogenesis and Engineering Principles. Elsevier, Amsterdam (1976)

IRC: 37-2012: Guidelines for the design of flexible pavements. The Indian Roads Congress, New Delhi

IRC: SP 89-2010: Guidelines for soil and granular material stabilization using cement, lime and flyash. The Indian Roads Congress, New Delhi

IRC: SP 72-2015: Guidelines for soil the design of flexible pavements for low volume roads. The Indian Roads Congress, New Delhi

Joel, M., Agbede, I.O.: Mechanical-cement stabilization of laterite for use as flexible pavement material. J. Mater. Civil Eng. **23**(2), 146–152 (2010)

ARA: Part 2. Design Inputs; Chapter 2. Material Characterisation. NCHRP, TRB (2004)

NAASRA 1986: Guide to stabilization in road works, National Association of Australian State Road Authorities, Sydney, NSW

Wen, H., Muhunthan, B., Wang, J., Li, X., Edil, T., Tinjum, J.M.: Characterization of cementitiously stabilized layers for use in pavement design and analysis. Report No-789. TRB (2014)

Paige-Green, P.: Recent developments in soil stabilization. In: ARRB Transport Research Ltd Conference, 19th 1998, Sydney, New South Wales Australia (1998)

Portelinha, F.H.M., Lima, D.C., Fontes, M.P.F., Carvalho, C.A.B.: Modification of a lateritic soil with lime and cement: an economical alternative for flexible pavement layers. J. Soil Rock **35**(1), 51–63 (2012)

Portland Cement Association: Soil-Cement Laboratory Handbook (1992)
Ravi, S.U.A., Suresha, S.N., Kashinath, B.: Characterisation of lateritic soil modified with pond ash and cement. Indian Highways **36**(6), 21–27 (2008)
Syed, I.M., Scullion, T.: Performance evaluation of recycled and stabilized bases in Texas. Transp. Res. Rec. **1757**, 14–21 (2001)
Vorobieff, G.: Stabilization practices in Australia. In: Proceedings of New Zealand Institute of Highway Technology (NZIHT) Stabilization of Road Pavements Seminar, pp. 1–14 (2004)
Williams, R.I.T.: Properties of cement stabilised materials. J. Inst. Highway Eng. **19**(2), 5–19 (1972)
Yeo, Y.S., Jitsangiam, P., Nikraz, H.: Mix design of cementitious base course. In: International Conference on Advances in Geotechnical Engineering, pp. 379–385 (2011)

An Experimental Study on Partial Replacement of Clayey Soil with an Industrial Effluent: Stabilization of Soil Subgrade

Godavarthi V.L.N. Murthy(✉), Atkuri Venkata Krishna, and Vedula V.N. Prabhakara Rao

Department of Civil Engineering, Sri Vasavi Engineering College, Tadepalligudem, Andhra Pradesh, India

Abstract. Rapid growth of infrastructure viz., construction of highways, embankments limits the construction on varied soil conditions. The construction sites at times pass through weak/expansive soils. In order to overcome weak and problematic sub grade world over, soil stabilization becomes inevitable. In this connection, mechanical stabilization of clayey soils is explored with partial replacement by Vitrified Tiles Sludge (VTS), an industrial waste. In this experimental study, expansive soils samples are collected from Palakol town in the West Godavari District, Andhra Pradesh, India. The experiments showed significant volume changes in the soil with a change in water content, corresponding decrease in strength and significant increase in compressibility. The current experimental investigation studies the variation in the geotechnical characteristics of expansive soil modified with sludge formed by crushing vitrified tiles (VTS) in aqueous medium. Our study reveals that there is a significant decrease in compressibility characteristics of the clay with an increase in the proportion of vitrified tile sludge. The composite soil proves to be good construction material for construction of various civil engineering structures such as embankment, earthen dams, and runways wherever the decrease in swelling and plasticity are the real targets.

Keywords: Expansive soil · Atterberg limits · Soil stabilization · Vitrified tiles sludge

1 Introduction

The study area comprising part of Godavari delta extends over Andhra coast from Rajahmundry to Narsapur. Geographically, the area lies between, 81.8040° E to 81.6966° E Long. and 16.4330° N to 17.0005° N Lat. and is covered by Survey of India topographical map Nos. 65 D & Hand 66 A & E. The north-western boundary touches the Madras- Calcutta grand trunk road passing through Vijayawada, Eluru and Rajahmundry. Administratively, the area falls in the jurisdiction of East and West Godavari districts of Andhra Pradesh. The delta is nearly flat and highly irrigated. The geological succession of the area comprises alluvial deposits (Recent period), Deccan trap (Eocene), Gondawana (Carboniferous), Cuddapah-Kurnool and Dharwars (Cambrian) and granite-gneisses (Archaean). The area receives rainfall both in south-west and

W. Frikha et al. (eds.), *Soil Testing, Soil Stability and Ground Improvement*,
Sustainable Civil Infrastructures, DOI 10.1007/978-3-319-61902-6_26

north-east monsoon periods. The actual rain fall in the East Godavari District is 449.7 mm and the actual rain fall in the West Godavari district is 442.0 mm. The main soil groups are Black cotton soils with the subgroup Typical Pellusterts and Chromusterts with the essential clay mineral as fine montmorillonitic. The area under study has many tile factories where in untreated effluent comprising sludge made up of grounded tiles and water is left into the cultivable land rendering them nearly useless.

Vitrified Tiles are the latest and largest growing industry alternate for many tiling requirements across the globe with far superior properties compared to natural stones and other manmade tiles. India and China are the largest regions to contribute to the 6900 million square meters of production every year. With an annual growth rate of 20% worldwide and 25% in India, Vitrified tile is the fastest growing segment in the tile industry. Vitrified tiles own 12% share of the overall tile production across the world. With the increase in production of vitrified tiles in India, there is growing concern about the huge generation of tile polishing dust.

The raw material composition of Vitrified tiles is:

- Quartz of 99% Silica,
- Potash Feldspar of 12% to 14% Alkalis,
- Soda Feldspar of 12% to 14% Alkalis,
- Strengthening agent, China clay, body stains for producing in various colors.

VTS is collected from a tile factory situated at Narayanapuram Village, Ungutur Mandal, West Godavari, Tadepalligudem - 534407, Andhra Pradesh, India. The physical properties of Vitrified Tile Sludge (VTS) are given in the Tables 1 and 2 below.

Table 1. Chemical properties of Vitrified Tile Sludge (VTS)

Materials	VTS (%)
SiO_2	63.29
Al_2O_3	18.29
Fe_2O_3	4.32
CaO	4.46
MgO	0.72
P_2O_5	0.16
K_2O	2.18
Na_2O	0.75
SO_3	0.10
CL^-	0.005
TiO_2	0.61
SrO_2	0.02
Mn_2O_3	0.05
L.O.I	1.61

The current experimental study is concerned with the selection of approximate type of soil to achieve a very high degree of compaction and to expose the compaction properties of clay. The results of the study can provide thoughts for applying clay soil in various applications of soil stabilization process.

Table 2. Engineering properties of pure vitrified tile sludge

% Medium sand	1.43
% Fine sand	95.7
% Silt and Clay	1.07
Specific gravity	2.46
MDD (g/cc)	1.58
OMC (%)	19.4

1.1 Objectives of the Study

The following are the main objectives for the present study:

(a) Evaluate the limiting VTS content to reduce the plasticity i.e., Atterberg limits viz., Liquid limit by 20% and Plastic limit by 10% off the native clayey soil
(b) To increase the sub grade soil stiffness of the soil by at least 25%.

2 Literature Review

Al-Rawasa et al. (2005) has tried to stabilize expansive soil with lime, cement, combinations of lime and cement, sarooj (Artificial pozzolana), heat treatment and have found that with the addition of 6% lime, both the swell percent and swell pressure reduced to zero and heat treatment reduced swelling potential to zero. Al-Rawas et al. (2002) investigated the effectiveness of using cement by-pass dust, copper slag, granulated blast furnace slag, and slag-cement in reducing the swelling potential and plasticity of expansive soils. Bell (1989) through his experiments on Clayey soil showed that addition of a small percentage of lime enhanced many of the engineering properties of the soil. Boardman et al. (2001) with the help of two clay minerals assessed to assess the time-dependent effects of mineral structural chemistry on the lime–clay reaction. Briptive (2004) emphasized the need of a quick or simple test to show the amount of lime required to react chemically with a soil to bring about these physical changes to an optimum degree. Buhler and Cerato (2007) and Modak et al. (2012) mixed highly expansive soil with lime and class C fly ash to study the plasticity reduction in highly expansive natural clays from Idabel, Oklahoma. Çokça (2001) has used, high-calcium and low-calcium class C fly ashes from the Soma and Tuncbilek thermal power plants, respectively, in Turkey, for stabilization of an expansive soil. An evaluation of the expansive soil-lime, expansive soil-cement, and expansive soil-fly ash systems is presented. Cristelo et al. (2013) have studied the effects of sodium-based alkaline activators and class F fly ash on soil stabilization. Eades (1966) tried to show through simple test, the amount of lime required to react chemically with a soil to bring about these physical changes to an optimum degree. Edil et al. (2006) have evaluated the effectiveness of self-cementing fly ashes derived from combustion of sub-bituminous coal for stabilization of soft fine-grained soils. Addition of fly ash resulted in appreciable increases in the California bearing ratio (CBR). Ferguson (1993) had reported that Ash treatment can effectively reduce the swell potential of fat clay soils and increases the sub grade support capacity of pavement sub grades. Greeves (1996) used expansive black cotton soil to assess the influence of ambient temperature in the progress of lime-soil reactions. Higgins (2005) reported the incorporation of ground granulated blast furnace slag to combat the soil expansion. Hunter (1988) studied the expansive reactions between lime and sulfate-bearing clay soils and reported that the long-term pozzolanic chemistry of normal lime-soil reactions found to be disrupted while Cation exchange, agglomeration, and carbonation are unaffected. Jones and Holtz (1973) studied the relationship between the mineralogical composition and water chemistry and their relationship to geotechnical properties and engineering structures failures. Kamon and Nontananandh (1991) prepared a product by burning industrial wastes along with lime to produce a by-product having cementing characteristics and showed that the resulting by-product is of promise for use in stabilizing a loam soil. Khattab et al. (2007) have evaluated the long-term stability characteristics of lime-treated expansive soils to improve the properties of expansive soils. Matsuo and Kamon (1981) have investigated a new stabilization method of soft clay soils using multivalent cations, trivalent cations, e.g. Fe^{3+} and $A1^{3+}$. The beneficial effects of this treatment are clarified not only for engineering properties but also for microscopic mechanisms (ionization process and aggregating phenomenon). McCarthy et al. (2014), investigated the engineering and durability properties of lime-stabilized soils treated with low-lime fly ash to limit swelling (heave) due to sulphate, associated with the ground improvement process. Turkoz et al. (2014) have evaluated the effect of magnesium chloride ($MgCl_2$) solution on the engineering properties of clay soils. The results show that dispersive and expansive clay soils can be effectively improved using an additive $MgCl_2$ solution. Puppala (2003) investigated with class F fly ash, sulphate-resistant cement, ground granulated blast furnace slag (GGBFS) and lime mixed with fibers as potential stabilizers in enhancing the strength and volume change properties of soft, expansive and sulphate-rich soils. Petry and Armstrong (1989) had developed performance-based testing of chemical stabilizers that simulate field conditions. The situations included a physical erosion test to determine dispersion of clays that have been treated, two swell test preparation sequences simulating injection of chemicals into clays, a wet-dry test sample preparation using field gradation specifications, and a three-dimensional swell test for stability of treated clays when subjected to wetting. Petry (1997) used Chemical stabilizing agents to improve the behavior of earth materials. Rao and Subba Rao (1994) have discussed the cause of ground heave of an inherently non-swelling, kaolinite-rich red soil (from Bangalore, India) due to prolonged spillage of concentrated (40%, weight/weight solution) caustic soda (sodium hydroxide) solution into the sub-soil through cracked drains in an industrial establishment. Treatment of the contaminated ground with 5% ferric chloride solution and simultaneously measures to minimize caustic soda solution spillage mitigate the soil against heave from caustic soda attack. Reddy et al. (2015) have employed three types of additives viz., (a) Cementitious: lime and fly ash (b) Non-cementitious: stone dust, and (c) Chemical additives: $CaCl_2$ and Na_2SiO_3 to evaluate the performance of stabilized expansive soils. They investigated the influence of valence of cations (viz., monovalent, divalent and trivalent) and mean particle diameter of additive(s) on percentage reduction of swelling characteristics. Their results revealed that chemical additives exhibit superior performance over cementitious and non-cementitious additives in reducing the swelling characteristics. Brooks (2009) studied the suitability of expansive soil as a construction material using rice husk ash (RHA) and flyash. Saha and Saha (1991) have upgraded expansive soil as a construction material using rice husk ash (RHA) and flyash, a cost comparison was made for the preparation of the sub-base of a highway project with and without the admixture stabilizations. When the RHA content of 12% and a flyash content of 25%, Unconfined Compressive Stress increased by 97% while CBR improved by 47%. Sivanna (1976) had studied the problem of damage to structures in expansive soils world wide and have proposed that sodium carbonate and calcium carbonate is stabilizing a black cotton soil. Sivapullaiah studied the effect of electrolytes on the shear strength of clayey soils. Abood et al. (2007) had investigated the effect of adding different chloride compounds including (NaCl, $MgCl_2$, $CaCl_2$) on the engineering properties of silty clay soil. The increase in the percentage of each of the chloride compounds increased the maximum dry unit weight and decreases the optimum moisture content. While the liquid limit, plastic limit and plasticity index decreased with the increase in salt content, the unconfined compressive strength increased with the increase in salt content. Nalbantoglu (2001) has shown that Fly ash treatment has tremendous potential as an economical method for the stabilization of the soil. Significant reduction in the swell potential and an increase in the hydraulic conductivity values were reported.

3 Need for Study

From the literature review it is observed that: Almost all of the researchers have studied the effect of (1) Industrial Byproducts (Fly ash, Rice Husk Ash), (2) Chemical additives (Lime, Caustic Soda, Calcium Carbonate, Sodium Carbonate, Magnesium Chloride, Mono-valence, Divalent and trivalent compounds).

Effect of treatment on soil properties like specific gravity, liquid limit, plastic limit, plasticity index, grain size analysis, shear strength of soil and bearing resistance have been studied. Long-term behavior of soil like consolidation properties, hydraulic conductivity and resilient modulus has not been studied by most researchers.

- There are many methods of laboratory experimentation to inhibit soil expansion but very few field methods to forecast soil expansion are available.
- Stress – strain behavior of the modified soil needs to be studied.
- Very minimal or no research has been conducted on industrial waste sludge despite the fact that they constitute environmental hazards and could have their usefulness in soil stabilization.

4 Experimental Studies

The methods adopted for the present study of investigation are presented below: Natural moisture content of the soil samples is determined as per IS: 2720 (Part II) "Methods of Test for Soils" (Water Content), BIS, New Delhi. The specific Gravity of the soil is determined as per 38. IS 2720 Part-III/Section 2 (1997), "Methods of Tests for soils" (Specific Gravity), BIS, New Delhi. The grain size analysis is conducted as per IS 2720 Part-IV (1995), "Methods of Tests for soils" (Grain Size Analysis), BIS, New Delhi.

The liquid, plastic limit (Atterberg's limits) are determined as per IS 2720 Part-V (1995), "Methods of Tests for soils" (Atterberg Limits), BIS, New Delhi. THE maximum unit weight and optimum moisture content for a given effort of compaction are obtained as per IS 2720 Part-VIII (1997), "Methods of Tests for soils" (Proctor's Compaction Test), BIS, New Delhi. The bearing resistance of the soil in soaked and un-soaked conditions is determined as per IS: 2720 (Part-16) (1979), "Methods of Tests for soils" (California Bearing Ratio), BIS, New Delhi. In CBR (Fig. 1) testing the rate of penetration of the plunger was kept at 1.25 mm/min and Proving ring capacity is 10 KN. Soaked CBR values at 2.5 mm and 5 mm penetration were calculated for all the samples.

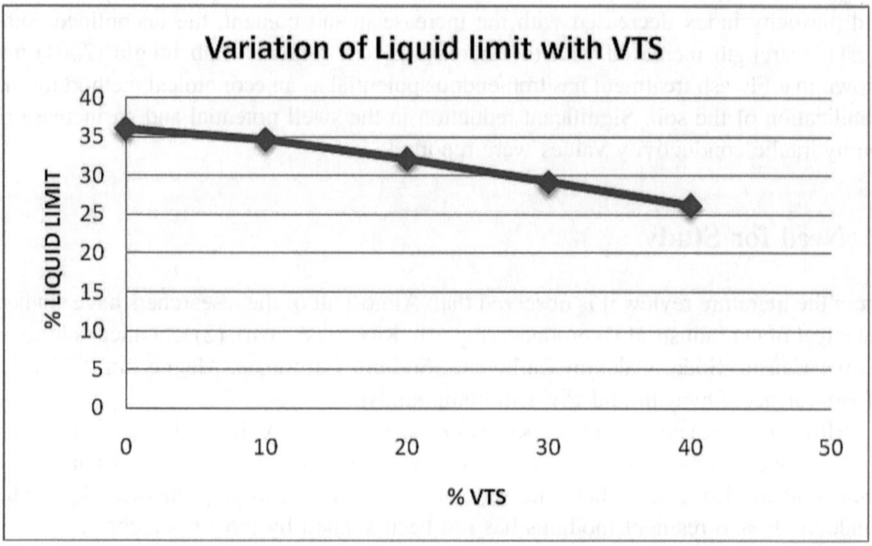

Fig. 1. Liquid limit of clayey soil admixed with different proportions of VTS

While the Unconfined Compressive strength of the soil is determined as per IS 2720 (Part-10) (1991), "Methods of Test for Soils", (Unconfined Compressive Strength), BIS, New Delhi. UCS testing is based on the fact that the minor principal stress is zero and the angle of internal friction of the soil is assumed to be zero. Strain controlled testing was followed and a strain rate of 1% per minute is used. At the time of testing loading is continued until the load on the specimen begins to decrease or until at least 20% of strain has been reached. In this case for the sample of 76 mm height 20% strain amounts to $0.2 \times 76 = 15.2$ mm.

Specimens for CBR and UCS testing are prepared at predetermined water content and unit weight. The free swell Index is determined as per IS: 2720 Part XL (1977), "Methods of Test for Soils", Determination of Free Swell Index of Soils, BIS, New Delhi.

5 Results and Discussion on Results

The native clayey soil is admixed with VTS by varying its proportion and the optimum content of the VTS is determined (Table 3). The admixed proportions of VTS considered in this study are 0%, 10%, 20%, 30% and 40%.

Table 3. Summary of the laboratory test results

Sample	Palakol
% Sand	6
% Silt	32
% Clay	62
Moisture content, n.m.c (%)	15.38
Specific gravity	2.71
Liquid limit	66.4
Plastic limit	32
OMC, %	22
Max dry unit weight $(\gamma_d)_{max}$ (g/cc)	1.56
Soaked CBR for 2.5 mm penetration	1.8%
Unsoaked CBR	3.2%
DFS	150

Table 4. Summary of the test results obtained with different proportions of the VTS with Clayey soil

VTS (%)	0	10	20	30	40
Liquid limit (%)	36.02	34.67	32.12	29.17	26.14
Plastic limit (%)	33.33	28.57	25	23.07	20
O.M.C (%)	18.2	17	15	14.4	10
Max dry unit weight (g/cc)	1.61	1.62	1.635	1.64	1.66
CBR for 2.5 mm penetration	2.50%	2.60%	5.20%	3.80%	3.72%
UCS	6.68	7.06	8.5	8.24	8.18

The following paragraphs describe the laboratory results (Table 4).

The above values are plotted against the different proportions of VTS admixed with the Clayey soil. The following are the observations found from these tests:

(a) As the admixed proportion of VTS increases from 10% to 40%, the liquid limit of the clayey soil is decreasing gradually (Fig. 1). The curve observed as falling trend. The similar trend is also observed in case of plastic limit of the soil (Fig. 2).

(b) In contrast, the dry unit weight of the soil is gradually increasing with slight reduction with OMC (Fig. 3).

(c) The soaked CBR values are gradually increased from 2.34% to 5.04% with 0% and 20% replacement and there onwards the CBR values are found to be decreasing with further increasing of the VTS content (Fig. 4).

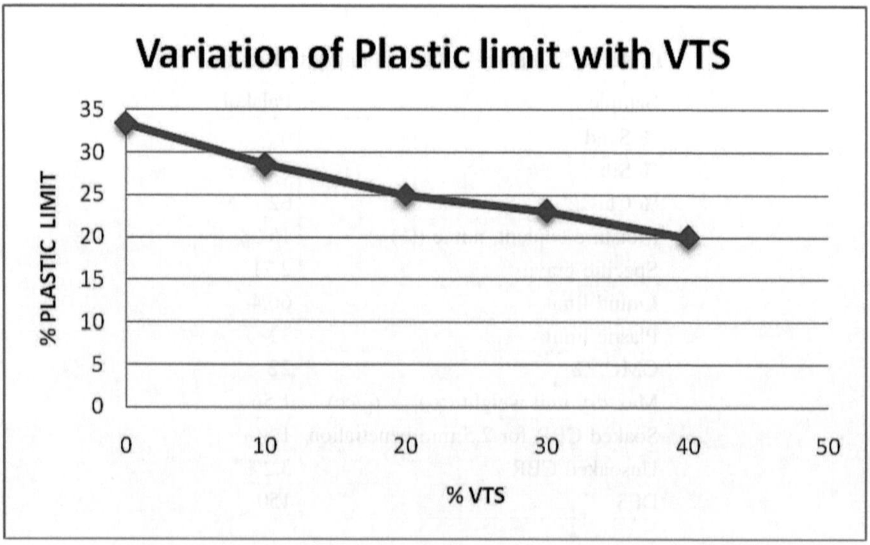

Fig. 2. Plastic limit of the soil obtained with different proportions with VTS

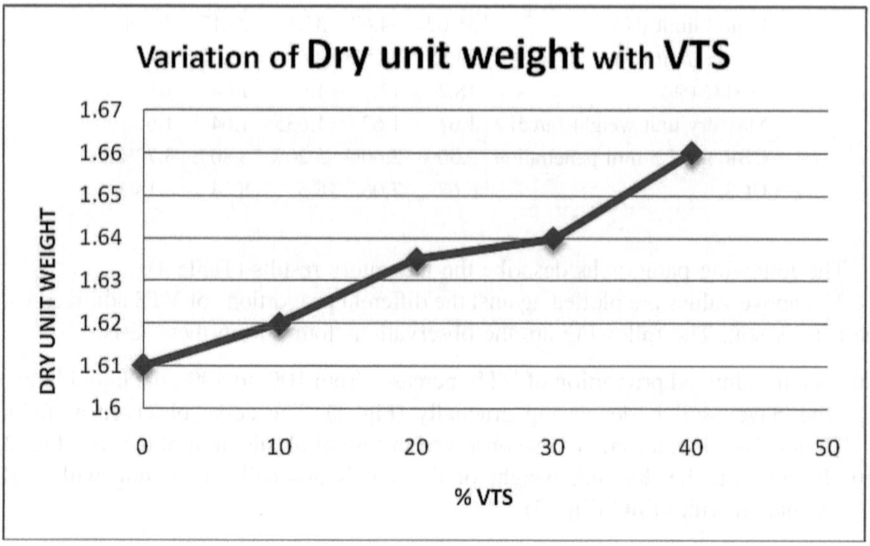

Fig. 3. Dry unit weight of the soil obtained with different proportions with VTS

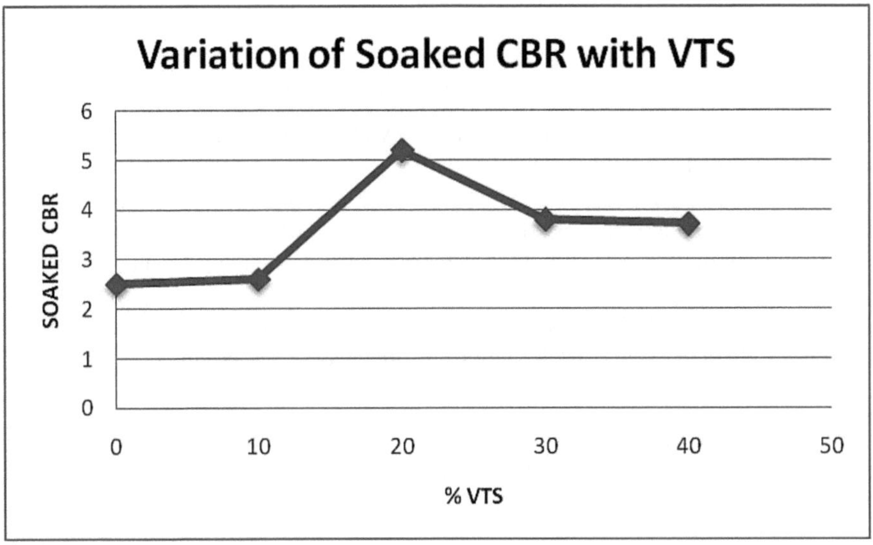

Fig. 4. Soaked CBR values of soil obtained with different proportions of VTS

(d) The UCS values are gradually increased from 6.68% to 8.5% with 0% and 20% replacement and thereafter, the UCS values are found to be decreasing with further increasing of the VTS content (Fig. 5).

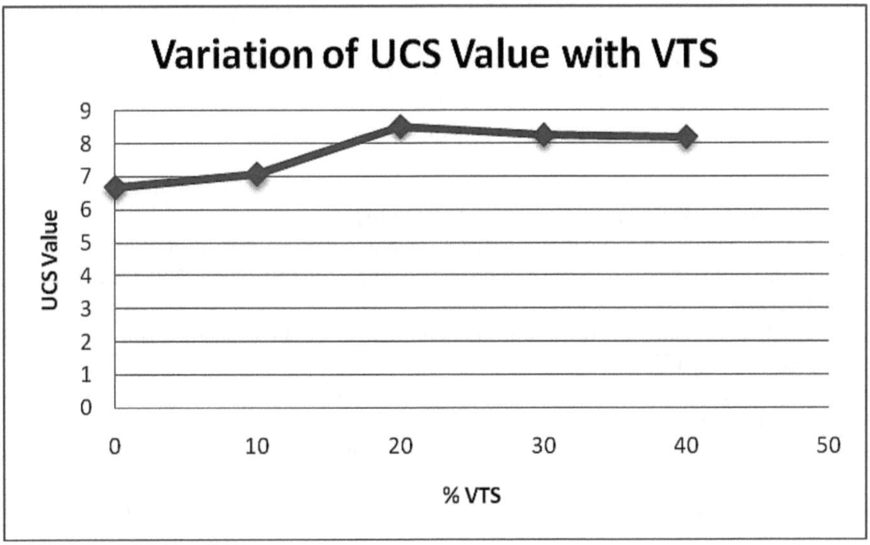

Fig. 5. UCS values of soil obtained with different proportions of VTS

6 Conclusions

The following are the conclusions drawn from the present study:

(a) Addition of varied Vitrified Tile Sludge (VTS) content has improved the properties of the clay sample.
(b) Replacement of Vitrified Tile Sludge by 20% is found to be optimum.
(c) The flow characteristics, plasticity characteristics and shrinkage characteristics of the soil sample are gradually decreasing with increase in the percentage of VTS in the clay sample.
(d) At the optimum VTS content of 20% replacement, the maximum dry density, soaked CBR value and UCS values are obtained as 1.635 gm/cc, 5.2% and 85, respectively.

Hence, it is concluded that the VTS as a waste product can also be used for soil stabilization purpose and the obtained results are based on this single study, further investigation is required to confirm the findings.

Acknowledgments. We sincerely acknowledge **Ms. Ch. Tanuja,** Assistant Professor at English in the Department of Basics Science & Humanities, Sri Vasavi Engineering College, Tadepalligudem, West-Godavari District, Andhra Pradesh, India, for rigorous round of proof reading at English.

References

Al-Rawas, A., et al.: A comparative evaluation of various additives used in the stabilization of expansive soils. Geotech. Test. J. **25**(2), 199–209 (2002)

Al-Rawasa, A.A., et al.: Build. Environ. **40**(5), 681–687 (2005). Elsevier

Bell, F.G.: Lime stabilization of clay soils. Bull. Int. Assoc. Eng. Geol. **39**, 67 (1989)

Boardman, D.I., et al.: Development of stabilization and solidification in lime-clay mixes. Geotechnique **51**, 533–543 (2001)

Britpave: Stabilized soils as sub-base or base for roads and other pavements. The British In-situ Concrete Paving Association, Camberley BP/08 (2004)

Buhler, R.L., Cerato, A.B.: Stabilization of Oklahoma expansive soils using lime and class c flyash. In: Problematic Soils and Rocks and In-situ Characterization, GSP 162, pp. 1–10. ASCE (2007)

Çokça, E.: Use of class C flyash for the stabilization of an expansive soil. J. Geotech. Geoenviron. Eng. **127**(7), 568–573 (2001). ASCE

Cristelo, N. et al.: Effects of alkaline-activated fly ash and portland cement on soft soil stabilization. Acta Geotech. (2013). doi:10.1007/s11440-012-0200-9

Eades, J.L., Grim, R.E.: A quick test to determine requirements for lime stabilization.Highw. Res. Board Bull. **139**(1966), 61–72 (1966)

Edil, T., et al.: Stabilizing soft fine-grained soils with fly ash. J. Mater. Civ. Eng. **18**(2), 283–294 (2006)

Ferguson, G.: Use of self commenting flyash as a soil stabilizing agent. Geotechnical Special Publication, vol. 36. ASCE, New York, pp. 1–15 (1993)

Greeves, H.M.: An introduction to lime stabilization. In: Proceedings, Seminar on Lime Stabilization, Loughborough University, pp. 5–12. Thomas Telford, London (1996)

Higgins, D.D.: Soil Stabilisation with ground granulated blastfurnace slag. Cementitious Slag Makers Association (CSMA) (2005)

Hunter, D.: Lime-induced heave in sulfate-bearing clay soils. J. Geotech. Eng. **114**(2), 150–167 (1988). ASCE

IS: 2720 Part XL: Methods of test for soils, Determination of Free Swelkl Index of Soils, BIS, New Delhi (1977)

IS: 2720 Part-16: Methods of tests for soils (California Bearing Ratio), BIS, New Delhi (1979)

IS 2720 Part-10: Methods of test for soils, (Unconfined Compressive Strength), BIS, New Delhi (1991)

IS 2720 Part-IV: Methods of tests for soils (Grain Size Analysis), BIS, New Delhi (1995)

IS 2720 Part-V: Methods of tests for soil (Atterberg Limits), BIS, New Delhi (1995)

IS 2720 Part-III: Methods of tests for soils (Specific Gravity), BIS, New Delhi (1997)

IS 2720 Part-VIII: Methods of tests for soils (Proctor's Compaction Test), BIS, New Delhi (1997)

Jones, D.E., Holtz, W.G.: Expansive soils- the hidden disaster. Civil Engg. **43**(8) (1973)

Kamon, M., Nontananandh, S.: Combining Industrial Wastes with Lime for Soil Stabilization. J. Geotech. Eng. **117**(1), 1–17 (1991). ASCE

Khattab, S.A.A., et al.: Long term stability characteristics of a lime-treated plastic soil. J. Mater. Civil Eng. **19**(4), 358–366 (2007)

Matsuo, S., Kamon, M.: Soil stabilisation by multivalent cations. In: Proceedings of the 10th ICSMFE, Stockholm, vol. 3, pp. 735–738 (1981)

McCarthy, M.J., Csetenyi, L.J., Dhir, R.K.: Engineering and durability properties of fly ash treated lime-stabilised sulphate-bearing soils. Eng. Geol. **174**, 139–148 (2014). Elsevier

Turkoz, M., et al.: The effect of magnesium chloride solution on the engineering properties of clay soil with expansive and dispersive characteristics. Appl. Clay Sci. **101**, 1–9 (2014)

Modak, P.R., et al.: Stabilization of black cotton soil using admixtures. Int. J. Eng. Innovative Technol. (IJEIT) **1**(5), 1–3 (2012)

Pappula, J., Wattanasanticharoen, E., Punthutaecha, K.: Experimental evaluations of stabilisation methods for sulphate-rich expansive soils. Proc. Inst. Civil Eng. Ground Improv. **7**(1), 25–35 (2003). ISSN 1755-0750

Petry, T.M., Armstrong J.C.: Stabilization of expansive clay soils, RR-1219, TRB, pp. 103–112 (1989)

Petry, T.M.: Performance based testing of chemical stabilizers. Transp. Res. Rec. Transp. Res. Board Nat. Acad. **1589**, 36–41 (1997)

Rao, S.M., Subba Rao, K.S.: Ground heave from caustic soda solution spillage - a case study. J. Soils Found. **34**(2), 13–18 (1994)

Reddy, N.G., et al.: Evaluating the influence of additives on swelling characteristics of expansive soil. Int. J. Geosynthetics Ground Eng. **1**(1), 7 (2015). Springer. doi:10.1007/s40891-015-0010-x. ISSN 2199-9260

Brooks, R.M.: Soil stabilization with flyash and rice husk ash. Int. J. Res. Rev. Appl. Sci. **1**(3), 209–217 (2009)

Saha, S., Saha, P.: Improvement of soils by use of chemicals. In: IGC - 91, Surat, vol. 1, pp. 429–432 (1991)

Sivanna, G.S.: Strength and Consolidation Characteristics of Black Cotton Soils with Chemical Additives- CaCl2 & KOH, Report Prepared by Karnataka Engg. Research Station, Krishnarajasagar, India, pp. 21–27 (1976)
Abood, T.T., et al.: Stabilization of silty clay soil using chloride compounds. J. Eng. Sci. Technol. **2**(1), 102–109 (2007)
Nalbantoglu, Z., Gucbilmez, E.: Improvement of calcareous expansive soils in semi-arid environments. J. Arid Environ. **47**(4), 453–463 (2001). Elsevier B.V

Ground Improvement of Titanium Dioxide Waste Spoils and Compressible Organics with In-Situ Mixing with Portland Cement and Surcharging

James J. Serpico(✉)

Geotechnical Services, Maser Consulting P.A., 331 Newman Springs Road, Suite 203, Red Bank, NJ 07701, USA
jserpico@maserconsulting.com

Abstract. Part of an overall $2.4 billion dollar, multi-use, brownfield redevelopment project along a major river and highway system in the Borough of Sayreville, New Jersey, USA, was a need to conduct ground improvement activities for approximately 40 acres of the 440-acre project for a large retail structure and portions of the roadway system. During the course of subsurface explorations it was determined that a 40-acre portion of the project contained 4 to 14 ft of titanium dioxide overlying highly compressible organics. The titanium dioxide materials were the result of abandoned plant processing activities in the production of white paint pigments.

The ground improvement program consisted of mixing the titanium dioxide byproduct, in-place, with Portland cement. The mixing was conducted using large excavators mixing set dimensioned cells to required depths. Over 375,000 cubic yards of materials were mixed in-place using this method. Many lessons were learned regarding design and construction considerations to overcome air quality, Portland cement delivery, mixing thoroughness, mix strength testing and quality control.

Once in-place cement mixing was completed a full-scale, 28-acre surcharge program was implemented to consolidate the underlying compressible organics to support the proposed development. The surcharge program was accomplished using prefabricated wick drains (PWDs) and surcharge fills consisting of Processed Dredge Materials (PDM). PDM typically consist of river and harbor dredge spoils mixed with at least 8% cement prior to reuse as fills. The surcharge program is being remotely monitored 24/7 through the use of digital piezometers, settlement plates and in-place inclinometers via a project website.

1 Introduction

The project area was formerly used in the production of white pigments by the National Lead Company. Beginning in the 1930s and continuing until the late 1970s the site produced white pigment (titanium dioxide) by processing raw ore materials barged to the site. Initially a sulfuric acid process was used until waste acid disposal costs moved the plant to consider a chloride process. The sulfuric acid process was halted in the mid-1970s. During titanium dioxide processing, process waste was discharged into

W. Frikha et al. (eds.), *Soil Testing, Soil Stability and Ground Improvement*,
Sustainable Civil Infrastructures, DOI 10.1007/978-3-319-61902-6_27

several areas at the site. The largest waste containment area was approximately 40 acres and was commonly known as the "acid lagoon" by locals and site workers. The pH of the groundwater in the lagoon area was generally found to range between 2.5 to 3.5 with some areas going as low as a pH of 1.8 prior to any material stabilization work. The strong acid nature of the groundwater resulted in a natural chemical process that produced high concentrations of heavy metals due to them leaching out of the soil. Additionally, over the years, transformers for pumping and electrical needs resulted in PCB (polychlorinated biphenyl) contamination in the lagoon itself. A site photo taken during mixing operations in May of 2014 (Fig. 1), presents a project scale and outlines areas to be discussed further in this case study.

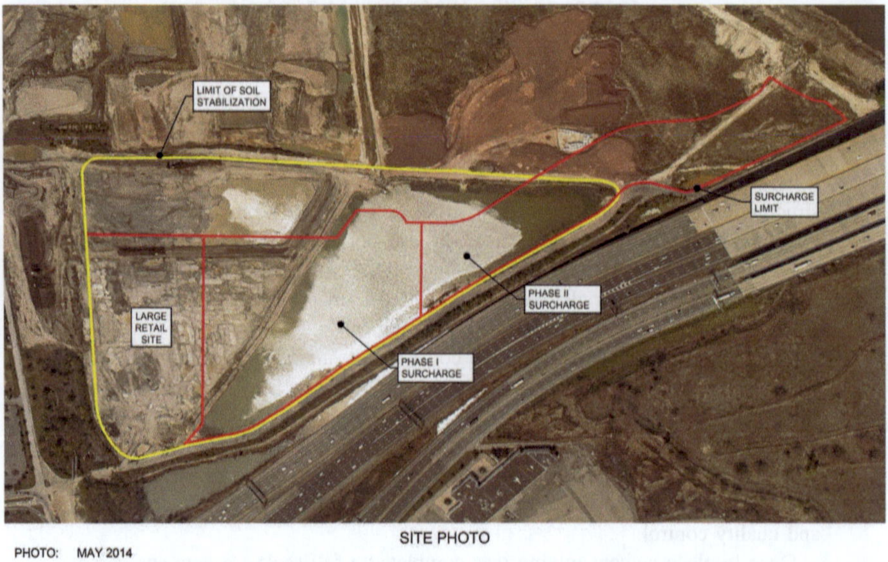

Fig. 1. Site photo taken May 2014 during mixing operations

2 Subsurface Conditions

Test boring access was limited in the lagoon area due to existing water levels at the time and the PCB contamination. The subsurface conditions in the lagoon area generally consisted of titanium dioxide waste materials directly overlying highly compressible organic silt and clay deposits. Directly underlying the highly compressible soils there are relatively free-draining sands and gravels. The titanium dioxide waste materials ranged in thickness from 4 to 14 ft with an average thickness over the area on the order of 5 ft thick. The highly compressible organic soils ranged in thickness from 8 to 35 ft with depth increasing as one moved toward the river. Groundwater was acidic as expected and ranged in depth from 3 to 5 ft below existing lagoon bottom grades.

Laboratory testing conducted on the compressible organics indicated that a modified compression index of 0.25 should be used in settlement evaluation based upon multiple consolidation tests and that the materials were normally consolidated. Atterberg Limit testing resulted in an average Liquid Limit value of 85% and a Plasticity Index of 45%. Natural moisture contents ranged from 58% to 78%.

3 Ground Improvement Considerations

As part of the site development layout and construction phasing considerations it became quickly apparent that the lagoon area needed to be addressed from an environmental and geotechnical viewpoint early on. After site remediation of high PCB levels at a few isolated areas in the lagoon, test pits were excavated in portions of the lagoon that had drained to better characterize and evaluate options for the stabilization of the titanium dioxide waste materials. Samples were collected for treatability studies for various mix designs. At the same time various options were considered to address the large volume of titanium dioxide waste materials anticipated in to be in the lagoon. Consideration was given to complete over-excavation and replacement with structural fills, mixing the titanium dioxide waste materials with large coarse aggregate (4″ to 6″ crushed stone), utilizing alternate mixing methods such as deep soil mixing, rotating head mixing and mixing with large track excavator with Portland cement. After consideration of off-site disposal costs due to contaminated soils and site limits to the lagoon area only for working due to the PCB contamination, mixing in-place with Portland cement was selected as the design solution.

3.1 In-Situ Soil Stabilization (ISS)

During contractor bidding, mixing in-place with large track excavators was ultimately selected to be the most cost effective solution. Treatability studies indicated that the titanium dioxide waste would need a minimum of 7.5% Portland cement to stabilize. However, the area of the Large Retail store required full-depth mixing completely through the compressible organics. Treatability evaluations for full-depth mixing indicated that Portland cement percentages would need to be increased to at least 10% due to impacts of the organics on resulting mixture strength. The contractor proposed three (3) different size mixing areas based upon cement delivery volumes; Small Size (10′ × 50′), Medium Size (20′ × 60′) and Large Size (20′ × 90′). The Small Size cells were for the area with full depth mixing under the Large Retail store with mix depths up to 16 ft anticipated in the design. Acceptance criteria for each mixed cell was established, by contract, to be evaluated with unconfined compressive strength tests on 2″ and 3″ diameter samples taken at designated areas per cell location. The mix acceptance strength was set at 65 psi at 28 days. The design mix strength requirement was based upon limited finite element evaluations of anticipated loadings that the area would be subject to during construction with the understanding that the majority of the lagoon area would only receive four (4) ft of ISS which in essence would "float" on any remaining un-mixed titanium dioxide waste materials and the highly compressible soils.

Dry Portland cement was to be placed directly over a cell area by direct pumping from a tanker truck using pneumatic methods then mixed in.

The ISS mixing started in August 2013 with an aggressive mobilization and within the first four (4) months of ISS, 783 cells of the estimated 1,200 cells required for the entire lagoon were mixed. However, acceptance rates for the first three (3) months were less than 40%. Discussions with the contractor in the field did not yield any appreciable change in procedures until the middle of November 2013 with introduction of a new Site Superintendent for the contractor. At this time, mixing was essentially stopped until March of 2014 while the contractor evaluated why the ISS results were not more successful. Review of pH levels, individual excavator operator performance, water added, mix depth compared to cell size and percent cement was conducted. Ultimately, the contractor increased the Portland cement content to a little over 11% and agreed to increase mix time duration. Figure 2 presents the number of cells mixed each month and the percent passing on the first mix. As one can see the changes the contractor made increased the success rate to what was anticipated within the contractor original bid estimates which was approximately 80%. Unfortunately the mixing changes occurred too late and the original contractor was forced, in July 2014 due to all the cell remixes that were needed, to cease work and terminate the business. While a new contractor was selected to complete the remaining work, the data was reviewed and it appeared that cell size had a significant impact. For the Small, Medium and Large cell sizes the success rate was 46%, 37% and 85%, respectively. The new contractor decided to accept the findings and used a large cell size with 11% cement and the success rate was generally greater than 80%. The final ISS cell was mixed in April 2015.

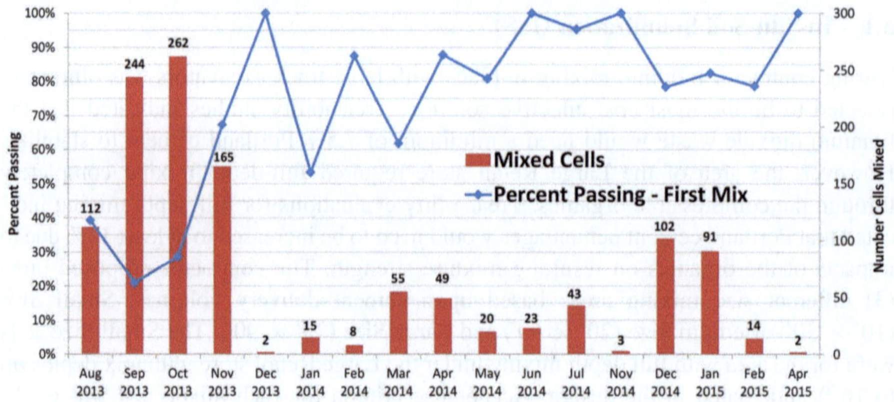

Fig. 2. Mixed cell acceptance rate per month

3.2 Surcharge Program

Once large areas of the lagoon were stabilized and approved, a site contractor started to install the necessary elements of the surcharge program. In April 2015 the installation of wick drains over a 28-acre area commenced. Due to time constraints for delivery of the large retail store and associated parking, the wick drains were spaced between 4 and 5 ft on a triangular grid. A total of 65,000 wick drains were installed. The material being used as structural base fills and surcharge fill materials for the project are Processed Dredge Materials (PDMs). Processed dredge materials typically consist of river and harbor dredge spoils mixed with at least 8% cement prior to reuse as fills which make them highly moisture sensitive and relatively impermeable. To account for the impermeable nature of the fills to be placed, the surcharge design included the use of strip drains to collect wick drain seepage and piping to manholes with 18-inches of clean fine gravels used as the drainage layer under the PDM materials. During wick drain installation, vibrating wire instrumentation consisting of (32) piezometers, eight (8) settlement sensors and one (1) in-place inclinometer (5 sensor locations with depth) were installed. All the instrumentation is wired to dataloggers which transmit real-time data to the project website for contractor use. The real time data allows monitoring so the site pore water pressures will not exceed established threshold limits. Additionally, (32) manual settlement plates were installed and observed on a bimonthly basis.

Base fills were placed over the entire surcharge area to elevations greater than final design grades to account for anticipated settlements. The PDM requires very thin lifts due to the relatively high moisture content of the material as delivered to the site which can range from 55% to 85% moisture. PDM use as a structural fill requires thin lifts over large areas which are aerated prior to compaction with a sheepsfoot roller. Based upon proctor tests (ASTM D-1557) conducted on the PDM materials, the optimum moisture content ranged from 25 to 35% with maximum dry densities ranging from 65 to 85 lb per cubic foot.

The surcharge fill placement was broken down into two (2) separate phases; Phase 1 was to cover the area associated with the parking, roadway and utility needs for the large retail store construction, and Phase II would utilize the surcharge fills from Phase I and prepare the ground for future roadway, parking and culvert construction. In computing the surcharge fill height for the areas the design had to account for the relatively low anticipated moist unit weight of the PDM materials placed in thick lifts and any impact of the ISS materials on surcharge load stress distribution. A final surcharge height of (19) ft was used in construction which included a 5-foot increase in height to account for any impacts of the ISS. Figure 3 presents the predicted versus actual settlement at settlement plate location M07 which is located in the middle of the Phase I surcharge.

The observed settlement is close to the predicted total (primary) settlement and it appears that the additional surcharge fill height added for the ISS concerns was necessary. Surcharge fills were cleared for removal 4.5 months after the start of surcharge placement which was estimated in the design of the wick drain spacing. Surcharge fills are currently being removed from Phase I and are being placed on Phase II.

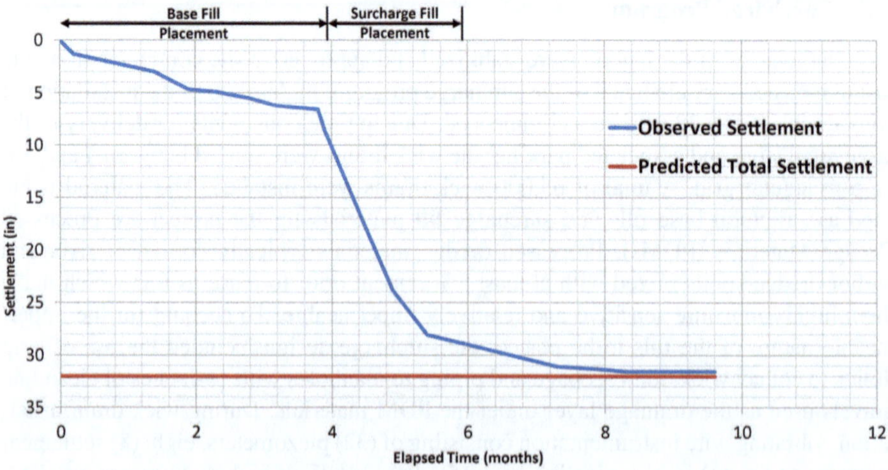

Fig. 3. Settlement observations-M07

4 Conclusions

While this is a single case study for the use of in-place soil mixing for structural support, there are many lessons to be learned from the field observations and testing.

1. Laboratory treatability studies, while important in establishing a baseline for design and pricing, should be followed-up with actual treatability evaluations in the field. Ideally, the contract should have a field treatability evaluation program as part of the mobilization. Construction documents and contracts should be structured that if the field observation success rate is not as anticipated, work should be stopped until further evaluation is done.
2. Placement of the dry Portland cement by pneumatic pumping created an issue with dust. The fine grained nature of the Portland cement meant that it would become airborne with very little wind. Air monitoring is recommended with contract provisions for dust remediation or work stoppage should air monitoring indicate unacceptable levels of dust transport.
3. The time of mixing each cell had a significant impact on the success rate.
4. Cell size had a significant impact on the success rate. Small, relatively narrow cells with deeper mixing depths had the lowest success rate. This is probably due to the use of large track excavator physical shape limitations on how it can operate in a confined, narrow trench. Larger cell sizes resulted in quicker mix times and higher success rates.
5. The real-time instrumentation was very helpful in confirming design assumptions as to stress distribution and saturated conditions. Piezometer data allowed for confirmation of pore-pressure distribution assumptions and construction rates. Settlement sensors were good in evaluating relative settlement but generally were not as accurate as the settlement values obtained from the manual settlement plates.

We were not able to evaluate if the settlement sensor readings were impacted by the installation process or heavy construction traffic impacting the sensor. Future surcharge programs at the site will focus on this issue.

Acknowledgements. The author appreciates the support and input from the redevelopment owner Brian O'Neill Jr., of the O'Neill Properties Group, the Civil/Site Project Manager for Maser Consulting P.A, Daniel Busch, P.E., and the Senior Geologist for Maser Consulting P.A., William Bolton.

Research and Application on the Prediction Method of Pearl Model of High Filling Subgrade Settlement

Xiangxing Kong(✉)

The First Highway Survey and Design Institute of China Communications Construction Company Ltd., Xi'an, China
103121153@qq.com

Abstract. According to the development law of high embankment settlement, the Pearl curve model is established. Combined with specific engineering example, the Pearl curve well reflects the change process of high embankment settlement, and the prediction result of the model with the measured settlement value is very accurate and reliable.

1 Introduction

The construction of expressway in the west, high filling and excavation of roadbed structure are generally occur [1, 2]. After completion construction of high fill embankment subgrade settlement often appear, even cause the destruction of pavement structure engineering accident [3]. In order to ensure the safety of the normal operation of highway, according to the law of development of high filling subgrade settlement observation data and combining site, Pearl model is applied to predict the settlement.

2 Mechanism of High Filling Subgrade Settlement

In the early period of construction, just after loading soil is in elastic status. The soil pore water do not exhaust [4, 5]. Due to the soil of instantaneous shear de-formation, with the increase of load the settlement has approximately linear increase. In the later period of construction, with increasing load, the load imposed in high embankment are also more and more big. The pore water in the foundation soil is gradually discharged, meanwhile excess pore water pressure gradually decreases. So compaction causes more de-formation, and soil in embankment will be in elastic-plastic state. With the continuous development of the plastic zone, settlement rate increases quickly, until the load no longer increases.

After construction, due to the pore water pressure close to disappear completely, the consolidation process is not complete, settlement will continue with the passage of time, but significantly decreased sedimentation rate [6, 7]. If the time of settlement is infinity, settlement will be at ultimate state, and settlement rate is zero, so the settlement at that time is really the final settlement. In fact, for the highway time takes for 15 years plus filling time.

W. Frikha et al. (eds.), *Soil Testing, Soil Stability and Ground Improvement*,
Sustainable Civil Infrastructures, DOI 10.1007/978-3-319-61902-6_28

3 The Forecast Model of Pearl Curve

Pearl curve is sometimes referred as logistic curve, which is based on the American biologists and demographer Raymond Pearl [8]. Because the curve can reflect the growth process of biological, so Pearl curve is widely applied in animal breeding, development of population statistics and analysis of product life cycle and so on. Pearl curve prediction model is expressed as below.

$$y(t) = \frac{L}{1 + ae^{-bt}} \tag{1}$$

Where: L, a, b are three parameters of the model, in which a > 0, b > 0. Figure 1 is the schematic diagram of typical Pearl curve.

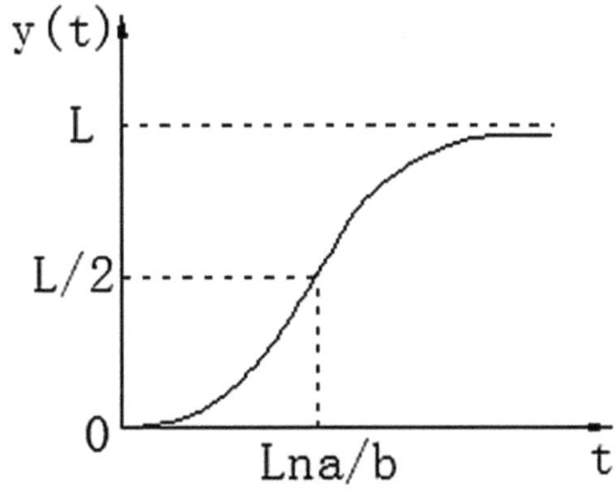

Fig. 1. The carve of Pearl model

Seen from Fig. 1, the change of Pearl curve occurs in the point by the concave. The point is the inflection point of the Pearl curve, which upper and lower halves are around the inflection point symmetry. Also Pearl curve graph is a long strip of S, so it is sometimes referred as S curve. It describes so regularly, and curve is relatively flat. The description of Pearl curve actually reflects the process of things happening, development, maturity and up to a certain limit. Because of the change of process and Pearl curve settlement of high embankment reflects accurately the process of things development, therefore Pearl curve model can reflect the law of development of high filling subgrade settlement, and is applied to predict the settlement.

4 Calculation of Parameters

Using 3 parameters of Pearl curve model method to solve [5] has the following two requirements:

(1) the settlement data items in the N time series is a multiple of 3, then the calculation time sequence can be divided into 3 segments, each segment containing items;
(2) before the argument the time interval is equal or the length of time equal to, and which is continuous and interval time series. The time of T are numbered sequentially starting from 1, t = 1, 2, 3,…, n. According to the requirement, then the number of items in time series is respectively (i = 1, 2,…, n). The time series is divided into 3 sections: in the first paragraph t = 1, 2, 3,…; in the second section, R, t = r + 1, r = 2, r + 3,…; in the third section t = 2r + 1, 2r + 2, 2r + 3, 2R,…, 3r.

Let S1, S2, S3 respectively for the 3 section numerical the sum, that is

$$S_1 = \sum_{t=1}^{r} \frac{1}{y(t)} \tag{2}$$

$$S_2 = \sum_{t=r+1}^{2r} \frac{1}{y(t)} \tag{3}$$

$$S_3 = \sum_{t=2r+1}^{3r} \frac{1}{y(t)} \tag{4}$$

$$\frac{1}{y(t)} = \frac{1}{L} + \frac{ae^{-bt}}{L} \tag{5}$$

$$b = \frac{\ln\frac{(s_1-s_2)}{(s_2-s_3)}}{r} \tag{6}$$

$$L = \frac{r}{s_1 - \frac{(s_1-s_2)^2}{(s_1-s_2)-(s_2-s_3)}} \tag{7}$$

$$a = \frac{(s_1 - s_2)^2(1 - e^{-b})L}{((s_1 - s_2) - (s_2 - s_3))e^{-b}(1 - e^{-rb})} \tag{8}$$

So far 3 parameters are calculated by the formula (1), which can be obtained from the Pearl prediction model. Based on the principle and the method of corresponding calculation program is compiled with Matlab.

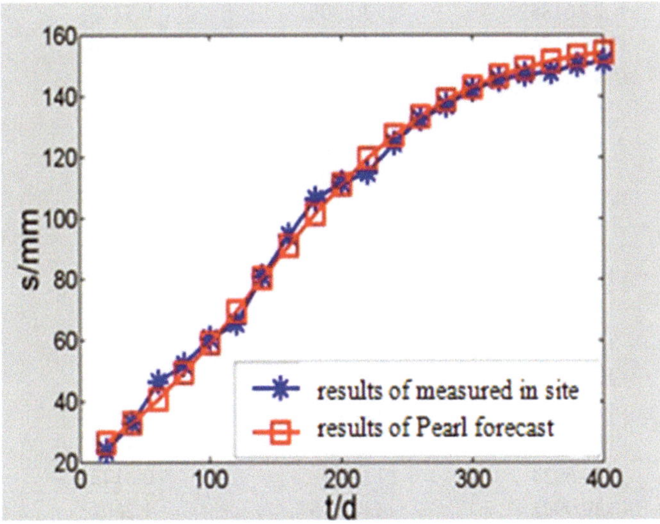

Fig. 2. Results of Pearl and forecasting and measured in K103+380

Table 1. Results of Pearl and forecasting and measured in K103+380

Time (d)	Pearl forecasting (mm)	Measured in site (mm)	Absolute error (mm)
20	26.3828	23.9	2.5
40	32.9468	32.9	0.0
60	40.6426	46.3	5.7
80	49.4299	52.1	2.6
100	59.1656	60.6	1.5
120	69.5986	65.7	3.9
140	80.3868	80.9	0.5
160	91.1385	94.4	3.3
180	101.4672	106.3	4.9
200	111.0450	111.6	0.6
220	119.6399	114.9	4.7
240	127.1288	124.6	2.5
260	133.4881	132.7	0.8
280	138.7712	137.0	1.8
300	143.0809	142.3	0.8
320	146.5447	145.2	1.3
340	149.2954	147.1	2.2
360	151.4591	148.0	3.5
380	153.1483	150.2	2.9
400	154.4594	151.4	3.1

Table 2. Results of Pearl and forecasting and measured in K123+560

Time (d)	Pearl forecasting (mm)	Measured in site (mm)	Absolute error (mm)
20	21.48777	21.755	0.2813
40	29.61882	30.305	0.7223
60	38.71763	43.035	4.5446
80	48.45608	48.545	0.0936
100	58.49492	56.62	1.9736
120	68.52094	61.465	7.4273
140	78.27012	75.905	2.4896
160	87.53908	88.73	1.2536
180	96.18579	100.035	4.0518
200	104.1238	105.07	0.996
220	111.3138	108.205	3.2724
240	117.7523	117.42	0.3498
260	123.4631	125.115	1.7388
280	128.4877	129.2	0.7498
300	132.8785	134.235	1.4279
320	136.6934	136.99	0.3122
340	139.9921	138.795	1.2601
360	142.8328	139.65	3.3503
380	145.2708	141.74	3.7166
400	147.3569	142.88	4.7125

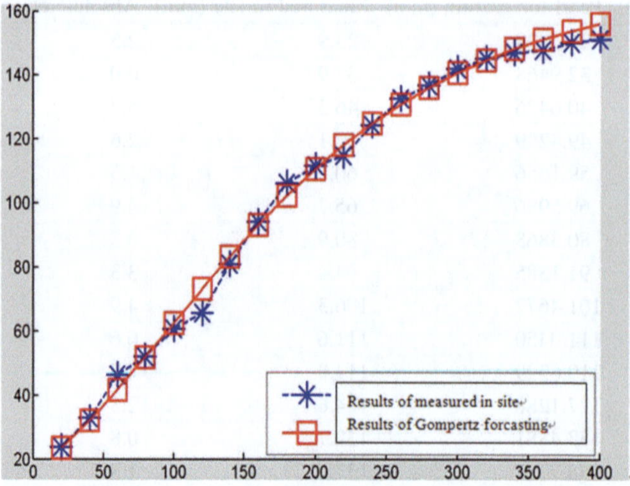

Fig. 3. Results of Pearl and forecasting and measured in K123+560

Table 3. Parameters of Pearl method

L	a	b
158.7895	6.5942	0.2730

5 Engineering Examples

Combined with the subgrade settlement observation data in Hunan province Changde-Zhangjiajie expressway, the K103+380, K123+560 are chosen as the case study. The high fill subgrade filler is typical, where is filled with low liquid limit clay and shale, an embankment height is about 13 m.
In Fig. 2, the time-scattered settlement is "S" shape. The first 12 points (20–240 days) data are selected as samples, Matlab6.5 is used to calculate related parameters of Pearl curve model (as shown in Table 1), then the results of Pearl curve model for settlement prediction of roadbed (as shown in Table 2) are obtained, as well as the sum squared error (prediction precision) 168.13 (Fig. 3 and Table 3).

6 Conclusion

In the forecasting of road foundation settlement, different methods can often provide valuable information.

(1) The settlement of high embankment has its own law of development, and Pearl curve is a good reflection of S shape relationship between settlement and time of the whole process of roadbed settlement, mean-while its settlement prediction is accurate and reliable.
(2) Using Pearl curve model in forecasting, in order to obtain good prediction sufficient measured data is basis.
(3) The aging situation also exists in the Pearl prediction model, and with time go recent settlement should be add as the original sample, so that the prediction effect is more better.

References

1. Xu, J., Cheng, K.K.: Road Engineering. Tongji University Press, Shanghai (1995)
2. Highway Traffic Division: The Highway Engineering Quality Problems Prevent Guide. China Communications Press, Beijing (2002)
3. Gu, X., Qian, H.J., et al.: The Foundation and Foundation. China Architecture Industry Press, Beijing (1993)
4. Zai, J., Mei, K.: Settlement of the whole process of prediction method. Rock Soil Mech. **21** (4), 1 322–325 (2000)
5. Zhao, M., Liu, Y., et al.: The soft soil roadbed settlement development law and its forecasting. J. Cent. South Univ. **35**(1), 157–161 (2004)

6. Pi, D.Y., Sun, Y.: Control and decision of multi model adaptive control. J. Algorithm **11**(1), 77–80 (1996)
7. Bates, J.M., Granger, C.W.J.: The combination of forecasts. J. Oper. Res. Soc. **20**(4), 451–468 (1969). doi:10.2307/3008764. http://www.jstor.org/stable/3008764
8. Tang, X., Wang, J., et al.: A new combination forecasting method of adaptive fuzzy variable weight. J. Univ. Electron. Sci. Technol. Algorithm **26**(3), 289–291 (1997)

Research on the Prediction Method of Gompertz Model for High Subgrade Settlement

Xiangxing Kong(✉)

The First Highway Survey and Design Institute of China Communications Construction Company Ltd., Xi'an, China
103121153@qq.com

Abstract. According to the development law of high embankment settlement, the Gompertz curve model is established. Combined with specific engineering example, the Gompertz curve well reflects the change process of high embankment settlement, and the prediction result of the model with the measured settlement value is very accurate and reliable.

1 Introduction

The construction of expressway in the west, high filling and excavation of roadbed structure are generally occur [1, 2]. After completion construction of high fill embankment subgrade settlement often appear, even cause the destruction of pavement structure engineering accident [3]. In order to ensure the safety of the normal operation of highway, according to the law of development of high filling subgrade settlement observation data and combining site, Gompertz curve model is applied to predict the settlement of the roadbed.

2 Mechanism of High Filling Subgrade Settlement

In the early period of construction, just after loading soil is in elastic status. The soil pore water do not exhaust [4, 5]. Due to the soil of instantaneous shear de-formation, with the increase of load the settlement has approximately linear increase. In the later period of construction, with increasing load, the load imposed in high embankment are also more and more big. The pore water in the foundation soil is gradually discharged, meanwhile excess pore water pressure gradually decreases. So compaction causes more de-formation, and soil in embankment will be in elastic-plastic state. With the continuous development of the plastic zone, settlement rate increases quickly, until the load no longer increases.

After construction, due to the pore water pressure close to disappear completely, the consolidation process is not complete, settlement will continue with the passage of time, but significantly decreased sedimentation rate. If the time of settlement is infinity, settlement will be at ultimate state, and settlement rate is zero, so the settlement at that time is really the final settlement. In fact, time generally is large enough, for the highway time takes for 15 years plus filling time.

W. Frikha et al. (eds.), *Soil Testing, Soil Stability and Ground Improvement*,
Sustainable Civil Infrastructures, DOI 10.1007/978-3-319-61902-6_29

3 The Forecast Model of Gompertz Curve

At present, there are two categories in calculation method of settlement of embankment. First, according to the constitutive model of consolidation theory combined with various soil, it is settlement calculation of the application of numerical methods, such as visco-elastoplastic finite element model method [6, 7]. The method of the calculation of the finite element method is very high, generally only used to calculate the key end of large engineering. And some constitutive model and engineering practice has a large gap, and needs a large number of soil tests to determine the soil parameters, so it is difficult to use in the general engineering design. Second, based on the measured data it is calculated settlement prediction method of quantity and time relations, such as hyperbola and exponential curve method. This method is concise and practical, and can satisfy the engineering precision requirement. Method for forecasting subgrade settlement of this selection is one of the Gompertz model.

Gompertz method is proposed by statisticians and mathematicians B. Gompertz, and it is also a growth curve, which are widely used in economics, management and statistics. Gompertz forecasting model is a growth curve model, and can be used to express and describe the process of growth and development. The practice indicated prediction model can be used as a reliable, effective method to forecast the settlement prediction of roadbed.

4 Calculation of Parameters

The three parameters in the model can be obtained using the three estimation. First of all, make it into a modified exponential form, then the data is divided into three sections (if not divisible by 3, can increase and decrease the individual data which is a multiple of 3), each containing a number of segments, i.e., seek and obtain.

$$y_t = e^{(k+ab^t)} \tag{1}$$

$$y' = k + ab^t \tag{2}$$

$$y_1 = k + ab^1 \tag{3}$$

$$y_2 = k + ab^2 \tag{4}$$

$$y_3 = k + ab^3 \tag{5}$$

$$y_t = k + ab^T \tag{6}$$

$$\sum\nolimits_1 y_t = \sum_{t=1}^{n} y_t = nk + ab(b^0 + b^1 + b^2 + \cdots + b^{n-1}) \tag{7}$$

$$\sum_2 y_t = \sum_{t=n+1}^{2n} y_t = nk + ab^{n+1}(b^0 + b^1 + b^2 + \cdots + b^{n-1}) \tag{8}$$

$$\sum_3 y_t = \sum_{t=2n+1}^{3n} y_t = nk + ab^{2n+1}(b^0 + b^1 + b^2 + \cdots + b^{n-1}) \tag{9}$$

$$(b^0 + b^1 + b^2 + \cdots + b^{n-1}) = \frac{b^n - 1}{b - 1} \tag{10}$$

$$\sum_1 y_t = nk + ab\frac{b^n - 1}{b - 1} \tag{11}$$

$$\sum_2 y_t = nk + ab^{n+1}\frac{b^n - 1}{b - 1} \tag{12}$$

$$\sum_3 y_t = nK + ab^{2n+1}\frac{b^n - 1}{b - 1} \tag{13}$$

$$b = \sqrt[n]{\frac{\sum_3 y_t - \sum_2 y_t}{\sum_2 y_t - \sum_1 y_t}} \tag{14}$$

$$a = \frac{b - 1}{(b^n - 1)^2 b}\left(\sum_2 y_t - \sum_1 y_t\right) \tag{15}$$

$$k = \frac{1}{n}\left(\sum_1 y_t - ab\frac{b^n - 1}{b - 1}\right) \tag{16}$$

$$k = \frac{1}{n}\left[\frac{\sum_1 y_t \sum_3 y_t - (\sum_2 y_t)^2}{\sum_1 y_t + \sum_3 y_t - 2\sum_2 y_t}\right] \tag{17}$$

By using the method of analogy, Gompertz prediction parameters can be calculated as below:

$$a = \frac{b - 1}{(b^n - 1)^2 b}\left(\sum_2 \ln y_t - \sum_1 \ln y_t\right) \tag{18}$$

$$b = \sqrt[n]{\frac{\sum_3 \ln y_t - \sum_2 \ln y_t}{\sum_2 \ln y_t - \sum_1 \ln y_t}} \tag{19}$$

$$k = \frac{1}{n}\left[\frac{\sum_1 \ln y_t \sum_3 \ln y_t - (\sum_2 \ln y_t)^2}{\sum_1 \ln y_t + \sum_3 \ln y_t - 2\sum_2 \ln y_t}\right] \tag{20}$$

5 Engineering Examples

Combined with the subgrade settlement observation data in Hunan province Changde-Zhangjiajie expressway, the K103+380, K123+560 are chosen as the case study. The high fill subgrade filler is typical, where is filled with low liquid limit clay and shale, an embankment height is about 13 m (Fig. 1).

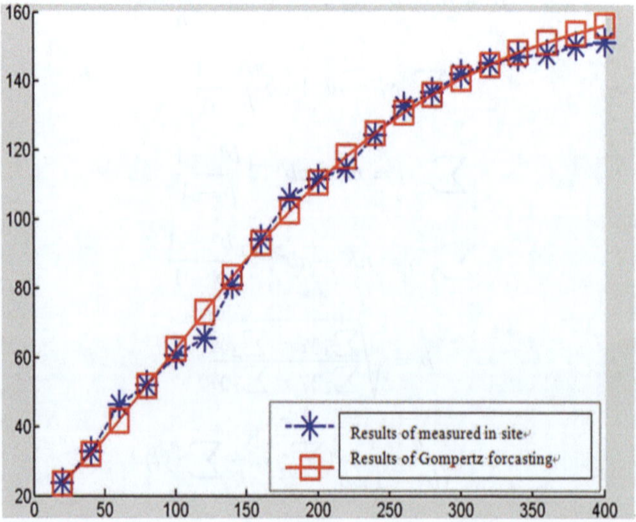

Fig. 1. Results of Gompertz and forecasting and measured in K123+560

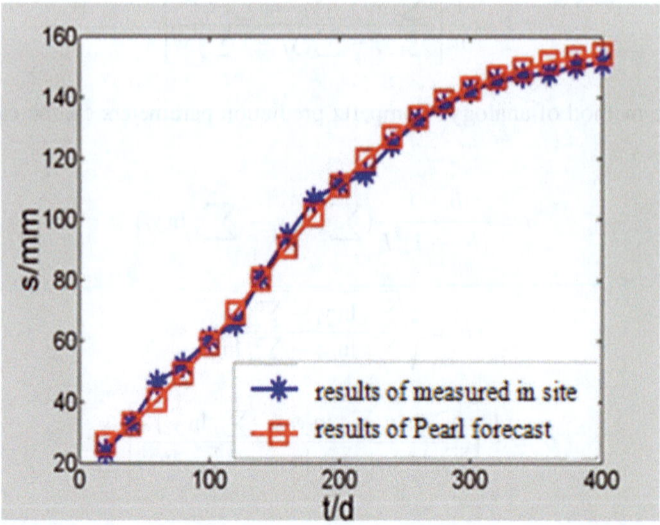

Fig. 2. Results of Pearl and forecasting and measured in K103+380

In Fig. 2, the time-scattered settlement is "S" shape. The first 12 points (20–240 days) data are selected as samples, Matlab6.5 is used to calculate related parameters of Gompertz curve model (as shown in Table 1), then the results of Gompertz curve model for settlement prediction of roadbed (as shown in Table 2) are obtained, as well as the sum squared error (prediction precision) 168.13 (Table 3).

Table 1. Results of Gompertz and forecasting and measured in K123+560

Time (d)	Gompertz forecasting (mm)	Measured in site (mm)	Absolute error (mm)
20	23.6187	23.9	0.3
40	32.1777	32.9	0.7
60	41.7554	46.3	4.6
80	52.0064	52.1	0.1
100	62.5736	60.6	2.0
120	73.1273	65.7	7.5
140	83.3896	80.9	2.5
160	93.1464	94.4	1.3
180	102.2482	106.3	4.1
200	110.6040	111.6	1.0
220	118.1724	114.9	3.2
240	124.9498	124.6	0.3
260	130.9612	132.7	1.7
280	136.2502	137.0	0.7
300	140.8721	142.3	1.4
320	144.8878	145.2	0.3
340	148.3601	147.1	1.3
360	151.3503	148.0	3.3
380	153.9166	150.2	3.7
400	156.1125	151.4	4.8

Table 2. Results of Pearl and forecasting and measured in K103+380

Time (d)	Pearl forecasting (mm)	Measured in site (mm)	Absolute error (mm)
20	25.32749	22.944	2.383488
40	31.62893	31.584	0.044928
60	39.0169	44.448	5.4311
80	47.4527	50.016	2.5633
100	56.79898	58.176	1.37702
120	66.81466	63.072	3.742656
140	77.17133	77.664	0.49267
160	87.49296	90.624	3.13104

(*continued*)

Table 2. (continued)

Time (d)	Pearl forecasting (mm)	Measured in site (mm)	Absolute error (mm)
180	97.40851	102.048	4.63949
200	106.6032	107.136	0.5328
220	114.8543	110.304	4.550304
240	122.0436	119.616	2.427648
260	128.1486	127.392	0.756576
280	133.2204	131.52	1.700352
300	137.3577	136.608	0.749664
320	140.6829	139.392	1.290912
340	143.3236	141.216	2.107584
360	145.4007	142.08	3.320736
380	147.0224	144.192	2.830368
400	148.281	145.344	2.937024

Table 3. Parameters of Gompertz method

K	a	b
5.1264	−2.3314	0.8426

6 Conclusion

In the forecasting of road foundation settlement, different methods can often provide valuable information.

(1) The settlement of high embankment has its own law of development, Gompertz model can express S shape relationship between settlement and time of the whole process of subgrade. The settlement reflects very good, but also can be accurately applied in prediction of high embankment.
(2) Using the Gompertz model to forecast the settlement of roadbed, in order to obtain good prediction effect enough measured data is basis. The measured settlement value should be checked, and delete obvious error.
(3) In order to improve the prediction accuracy, the selection of measured sample points should be increased, and with the passage of time, the settlement of recent is added as the original sample, so that the prediction effect is better.

References

1. Highway Traffic Division: The Highway Engineering Quality Problems Prevent Guide. China Communications Press, Beijing (2002)
2. Xu, J., Cheng, K.K.: Road Engineering. Tongji University Press, Shanghai (1995)

3. Zai, J.M., Mei, K.: Settlement of the whole process of prediction method. Rock Soil Mech. **21**(4), 322–325 (2000)
4. Gu, X., Qian, H.J., et al.: The Foundation and Foundation. China Architecture Industry Press, Beijing (1993)
5. Pi, D.Y., Sun, Y.: Control and decision of multi model adaptive control. J. Algorithm **11**(1), 77–80 (1996)
6. Zhao, M., Liu, Y., et al.: The soft soil roadbed settlement development law and its forecasting. J. Cent. South Univ. **35**(1), 157–161 (2004)
7. Bates, J.M., Granger, C.W.J.: The combination of forecasts. J. Oper. Res. Soc. **20**(4), 451–468 (1969). doi:10.2307/3008764. http://www.jstor.org/stable/3008764

Study on Methods for Predicting the Settlements of Soft Clay Roadbed

Baotong Shi and Xiangxing Kong(✉)

The First Highway Survey and Design Institute of China Communications Construction Company Ltd., Xi'an, China
{117021654,103121153}@qq.com

Abstract. The soft clay has the characteristics of high compressibility, high water content, low strength and poor permeability, which will lead to large settlements under long-term loads, therefore, the prediction of settlements of soft clay roadbed has been one of the important research topics in Geotechnical Engineering. In this paper, the commonly-used methods in practice to predict the settlements of soft clay roadbed have been summarized, and the advantages and disadvantages have been evaluated.

1 Introduction

The soft clay has the characteristics of high compressibility, high water content, low strength and poor permeability, which will lead to large settlements under long-term loads. Therefore, the settlement of soft clay roadbed is one of the prominent problems during the construction of railway and highway in soft clay area, and many engineering accidents are caused by the large settlement of subgrade. Thus, the prediction of settlements of soft clay roadbed has been one of the important research topics in Geotechnical Engineering.

As the settlement of soft clay usually develops with time, the settlement vs. time curve is usually used to describe the settlement properties of soft clay [1–3]. The most common method to predict the long term settlement is fitting the settlement data measured in the field. Many traditional fitting curves have been used to predict the settlement of soft clay, such as exponential curve, hyperbolic curve, Poisson curve and Asaoka method [4, 5]. Furthermore, in recent years, more and more new methods based on new theories have been introduced to improve the prediction of settlement, such as Grey forecasting method [6], Logistic method [7], and Gompertz model [8]. In this paper, the commonly used methods mentioned above to predict the settlement of soft clay have been evaluated and compared with each other.

2 Mechanism of the Settlement of Soft Clay Roadbed

The total settlement of soft clay roadbed contains three components: immediate settlement, consolidation settlement and secondary consolidation settlement. The development of settlement can be divided into the following four stages (as shown in Fig. 1):

W. Frikha et al. (eds.), *Soil Testing, Soil Stability and Ground Improvement*,
Sustainable Civil Infrastructures, DOI 10.1007/978-3-319-61902-6_30

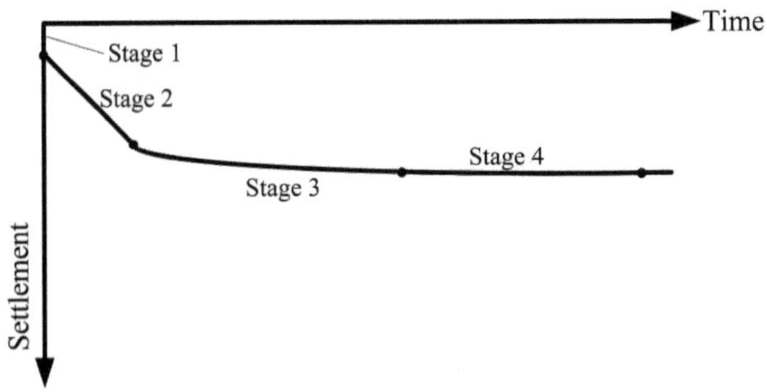

Fig. 1. Sketch view of settlement vs. time

(1) at the initial status, the soil is elastic and the change of settlement with load is linear;
(2) then with the increase of load, the settlement of soft clay and the corresponding rate increase and the nonlinear is obvious;
(3) although the load keeps constant, the settlement continues to increase due to consolidation and rheological property of clay, but the settlement rate declines;
(4) when the time is long enough, the settlement of clay will reach a limit state, the corresponding settlement rate is zero and the settlement arrives at its final value.

The purpose to study the properties of soft clay is to develop theoretical or empirical models to predict the settlement of soft clay at any time and any load. In recent years, many fitting curve methods based on measured field settlement data have been proposed, we will introduce these models detailedly in the next section.

3 Fitting Curve Methods

3.1 Exponential Curve

According to the consolidation theory proposed by Terzaghi, the relation between pore water pressure with time can be described by exponential expression. For soil with linear elastic property, the consolidation degree of soil defined by using stress equals to that defined by using strain. the consolidation degree can be expressed as follows:

$$U = 1 - \alpha \exp(-\beta t) \tag{1}$$

Where α and β are consolidation parameters, of which the expressions relate to drainage condition and are shown in Table 1.

Then according to the measured settlement data, the average consolidation degree can be expressed as,

Table 1. Theoretical values of α and β

Drainage condition	α	β
Consolidation with vertical drainage	$\frac{8}{\pi^2}$	$\frac{\pi^2 c_v}{4H^2}$
Drainage wells run through the soil layer	$\frac{8}{\pi^2}$	$\frac{8c_h}{F_n d_e^2} + \frac{\pi^2 c_v}{4H^2}$
Drainage wells do not run through the soil layer	$\frac{8}{\pi^2 \rho}$	$\frac{8c_h}{F_n d_e^2}$

$$U = \frac{s_t - s_d}{s_f - s_d} \tag{2a}$$

Where s_t is the measured settlement at time t, s_d and s_f are immediate consolidation settlement and final consolidation settlement.

Combined Eq. (1) with Eq. (2a), the following expression can be obtained,

$$s_t = (s_f - s_d)[1 - \alpha \exp(-\beta t)] + s_d \tag{2b}$$

In exponential curve fitting method, three data points on the settlement vs. time curve are selected and the time interval between the points must be the same, that is,

$$\Delta t = t_3 - t_2 = t_2 - t_1 \tag{3}$$

Then put the three points into Eq. (2b), three equations can be obtained and the expressions of β, s_f, and s_d are shown as follows,

$$\beta = \frac{1}{\Delta t} \ln\left(\frac{s_2 - s_1}{s_3 - s_2}\right) \tag{4a}$$

$$s_f = \frac{s_3(s_2 - s_1) - s_2(s_3 - s_1)}{(s_2 - s_1) - (s_3 - s_2)} \tag{4b}$$

$$s_d = \frac{s_t - s_f[1 - \alpha \exp(-\beta t)]}{\alpha \exp(-\beta t)} \tag{4c}$$

The above equations are the controlling equations in exponential curve.

3.2 Hyperbolic Curve

For soft clay with high compressibility, the stress vs. strain curve is non-linear, so the consolidation process of soft clay does not necessarily conform exponential curve.

When 0.6 < U < 0.9, the relation between U and time factor T can be expressed by equilateral hyperbola. The theory background of hyperbolic curve is given as follows:

The settlement of soft clay at time t can be calculated by the following equation,

$$s = s_0^{'} + \frac{t - t_0^{'}}{\alpha_0^{'} + \beta^{'}(t - -t_0^{'})} \tag{5}$$

Rewriting Eq. (5), then,

$$\frac{t - t_0^{'}}{s - s_0^{'}} = \alpha_0^{'} + \beta^{'}(t - -t_0^{'}) \tag{6}$$

Where and are the intercept and slope of vs. curve.

3.3 Asaoka Method

The one dimensional consolidation equation described by using volumetric strain can be expressed as follows,

$$c_v \frac{\partial^2 \varepsilon_v}{\partial z^2} = \frac{\partial \varepsilon_v}{\partial t} \tag{7}$$

On the basis of Eq. (7), Asaoka employed a series to approximate the expression,

$$s + \alpha_1 \frac{ds}{dt} + \alpha_2 \frac{d^2 s}{dt^2} + \ldots + \alpha_n \frac{d^n s}{dt^n} = b \tag{8}$$

Where s is consolidation settlement, αi(i = 1, 2, 3, …, n) is consolidation coefficient, b is a constant accounting for boundary condition.

According to Asaoka' suggestion, the parameters in Eq. (8) can be determined by using field measured data. Furthermore, Eq. (7) can also be simplified as recursive relation as follows,

$$s_j = \beta_0 + \sum_{i=1}^{n} \beta_i s_{j-1} \tag{9}$$

Taking the fixed boundary condition, the solution for the equation is given by,

$$s(t) = s_\infty - (s_\infty - s_0)\exp(-\frac{t}{\alpha_1}) \tag{10}$$

where s0 and are the initial and final settlement of soft clay, respectively, α1 = 5h2/(12cv).

3.4 Poisson Curve

Poisson curve is the logistic curve, which is suitable for incremental or decayed "S" shaped curve. The expression of the commonly-used Poisson curve is given by,

$$s_t = \frac{k}{1 + ae^{-bt}} \tag{11}$$

where a, b and k are parameters that should be determined by measure data.

The parameters in the expression of Poisson curve can be fitted by using the measured s_t curve, the main procedures are shown as follows:

Taking a time interval of and the total number of time series is 3n, then divide the time series into three sub-series and each sub-series has a number of n data points, that is,

$$\begin{cases} \text{sub} - \text{series } 1 & t = 1,\ 2,\ 3,\ \ldots, n \\ \text{sub} - \text{series } 2 & t = n+1,\ \text{n}+2,\ \text{n}+3,\ \ldots, 2n \\ \text{sub} - \text{series } 3 & t = 2n+1,\ 2\text{n}+2,\ 2\text{n}+3,\ \ldots, 3n \end{cases} \tag{12}$$

Then assuming that S1, S2 and S3 are the sum of the reciprocal of settlements corresponding to each time sub-series, the expressions to determine the parameters a, b and k are given as follows,

$$\begin{cases} a = \frac{(S_1 - S_2)^2 (1 - e^{-b}) k}{[(S_1 - S_2) - (S_2 - S_3)] e^{-b} (1 - e^{-nb})} \\ b = \frac{\ln(S_1 - S_2)/(S_2 - S_3)}{n} \\ k = \frac{n}{S_1 - \frac{(S_1 - S_2)^2}{(S_1 - S_2) - (S_2 - S_3)}} \end{cases} \tag{13}$$

Combining Eq. (13) with Eq. (12), the settlement of clay at any time st can be calculated, and the final settlement equals to the ultimate value of s_t, when the time t trends to infinity, that is,

$$s_\infty = \lim_{t \to \infty} s_t = k \tag{14}$$

3.5 Grey Forecasting Method

The Grey forecasting method is always referred to as GM model, and the most common GM Model has isometric time sequence, that is, GM(1, 1) model, the expression of which is given by the following differential equation,

$$\frac{dx^{(1)}}{dt} + cx^{(1)} = u \tag{15}$$

The corresponding time response equation is,

$$\hat{x}^{(1)}(t) = [x^{(1)}(0) - \frac{u}{c}]e^{-ct} + \frac{u}{c} \tag{16}$$

The discrete response equation is,

$$\hat{x}^{(1)}(k) = [x^{(1)}(0) - \frac{u}{c}]e^{-c(k-1)} + \frac{u}{c} \tag{17}$$

3.6 Logistic Model

Logistic model is a kind of growth model that is widely used in the fields of ecology and demography. The differential form of logistic model is as follows,

$$\frac{ds}{dt} = rs(1 - \frac{s}{K}) \tag{18}$$

According to Eq. (18), Logistic curve contains three parameters that should be determined. If the measured settlement data can be obtained, there are three methods used to determine these three parameters: three section method, grey model and non-linear regression method. The main processes of the three methods can be found in other literatures.

3.7 Gompertz Model

Gompertz model is also a kind of growth curve, the mathematical expression of which shows as follows,

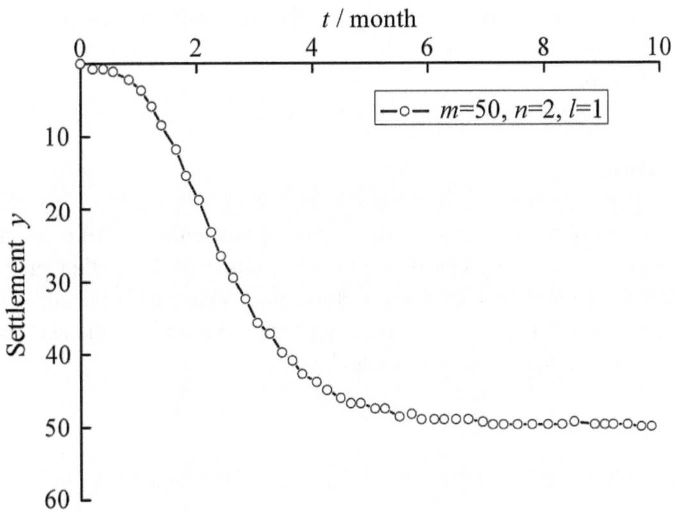

Fig. 2. Typical Gompertz curve

$$y = m\ \exp[-e^{(n-lt)}] \tag{19}$$

where m, n and l are constants, t is time series.

The typical Gompertz curve is shown in Fig. 2. From the figure, it is found that the theoretical curve agrees well with measured settlement curve. Therefore, it is feasible to use Gompertz model to predict the settlement of soft clay.

4 Discussions

In Sect. 3, seven methods to predict the settlement of soft clay have been reviewed detailedly. Although all of them have been applied to predict the settlements of some highway subgrade successfully, each method has its own application scope.

The exponential curve conforms to the one-dimensional consolidation theory proposed by Terzaghi and the parameter β has definite physical meaning. However, the method to determine the parameters α and β based on three data points is subjective, which will influence the accuracy.

According to Terzaghi's one dimensional consolidation theory, the hyperbolic method will make the relation between consolidation degree U and time factor T simpler than the exponential curve. Therefore, the hyperbolic method is more suitable for engineers in the field. However, the disadvantage of this method is that the consolidation parameters can not be reflected.

Compared with the consolidation of Terzaghi, it is found that Asaoka method predicts the settlement trends once the degree of consolidation exceeds 0.6. Furthermore, the final settlement of the soft clay can also be predicted well.

The Poisson curve can describe well the S-shaped relationship between settlement and time. The applicability of Poisson curve for other settlement vs. time curves is questionable and should be given more attentions.

The four methods mentioned above are traditional methods used for predicting the settlement of soft clay, while the Grey forecasting method, Logistic model and Gompertz model are new methods suitable for special kind of soft clay or different boundary conditions.

5 Conclusion

In this paper, seven different methods for predicting the settlement of soft clay roadbed have been reviewed detailedly based on the presentation of the settlement mechanism of soft clay. The applicability of each method has been analyzed. Although all the methods introduced in this paper have been applied into practices, the accuracy of each method should be studied in further researches.

References

1. Zhao, M., Liu, Y., Cao, W.: The developing regularity and forecasting of settlement in soft clay roadbed. J. Cent. South University (Nat. Sci.) **35**(1), 157–161 (2004)
2. Chen, S., Wang, X., Xu, X., Yu, F., Qin, S.: Three-point modified exponential curve method for predicting subgrade settlements. Rock Soil Mech. **32**(11), 3355–3360 (2011)
3. Zai, J., Mei, G.: Forecast method of settlement during the complete process of construction and operation. Rock Soil Mech. **21**(4), 322–325 (2000)
4. Tan, S.A.: Hyperbolic method for settlement in clays with vertical drains. Can. Geotech. J. **31**(1), 125–131 (1994)
5. Asaoka, A.: Observational procedure of settlement prediction. Soil Found. **18**(4), 87–101 (1978)
6. Lei, X., Bai, S., Meng, Q.: The application of Grey forecasting to analyzing soft foundation settlements. Rock Soil Mech. **21**(2), 145–147 (2000)
7. Xu, H., Shi, B., Li, X.: Logistic growth model and its applicability for predicting settlement during the whole process. Rock Soil Mech. **26**(3), 387–391 (2005)
8. Yu, C., Liu, S.: A study on prediction of embankment settlement with the Gompertz model. Rock Soil Mech. **26**(4), 82–86 (2005)

Geotechnical Investigations on Highway BR 135 for Earthworks, Asphalt Paving and Special Art Works in Subsection Municipalities Manga-Itacarambi, North of Minas Gerais

Rideci Farias[1(✉)], Haroldo Paranhos[1], José Jailson Nogueira[2], Marco Aurélio Diana Costa[3], and Roberto de Pimentel Sousa Junior[4]

[1] UCB/Reforsolo Engenharia/UniCEUB/IesPlan, Brasília, Brazil
rideci.reforsolo@gmail.com, reforsolo@gmail.com
[2] VALEC, Brasília, Brazil
jailson.np@gmail.com
[3] UniCEUB, Brasília, Brazil
marco_aurelio_d@hotmail.com
[4] UCB/UNB, Brasília, Brazil
eng.robertopimentel@gmail.com

Abstract. This article presents part of the results and discussions of the geotechnical studies carried out in the area and in materials collected in the BR 135 in the section between the counties of Manga and Itacarambi in the state of Minas Gerais. The study consisted in the execution of field work and laboratory work. In the field investigations were performed (auger, SPT and rotary), collecting material (soil and rock), densities "in situ" and identification/tactile-visual classification of materials. In the laboratory, with soils and rocks collected, were performed additional tests necessary to the design and implementation of the road in question, in that stretch. The studies focused along the existing highway axis and the track side, and side guidance in the predicted contours, concentrated deposits of soil, deposits concentrated of gravel, sands deposits and deposits of rocky material. With implementation of Auger investigations, it was found that the soils tactile-visual characteristics are similar to the predominance of yellow/red sandy soils. It was found that along the highway there are competent rocky material and also friable, mixed, which are not local, but come from deposits in the region and which were used in the primary coating of the road. In the carrying out field work, was done a broad recognition of the BR 135 insertion region that recognition was found that the region is covered by a wide layer of typical sandy soil (Red and Yellow Latosol). In the case of rocky material, it was found that there is a high occurrence of carbonate rocks in the region. This rock is exploited and used as building material, especially on pavements and as aggregates for concrete (Gravel 1, 2 and Gravel Gravel powder). Being a characteristic rock materials in the region, it was used for granulometric tests of sand equivalency, organic matter content, form of content, durability, abrasion "Los Angeles" and adhesiveness. The check of sands deposits, local people reported that most of the sand used in the region comes from dredging the San Francisco River in the

W. Frikha et al. (eds.), *Soil Testing, Soil Stability and Ground Improvement*,
Sustainable Civil Infrastructures, DOI 10.1007/978-3-319-61902-6_31

municipality of Itacarambi. two locations sands were also collected in the municipality of Mango, also the banks of the São Francisco River.

Keywords: Geotechnical investigation · BR 135 · Manga · Itacarambi

1 Introduction

This article aims to present some of the results of the geotechnical studies carried out in the area and collected materials on the BR 135 in the stretch between the towns of Manga and Itacarambi in the State of Minas Gerais. The studies consisted in performing field work and laboratory. On the field were executed investigations (Auger, SPT and rotating), collecting material (soil and rock), in situ density and identification/classification of materials. In the lab were carried out further testing required for the design and implementation of the road between the municipalities Manga and Itacarambi, State of Minas Gerais.

The studies were concentrated along the existing axis and in the sides of the highway to be paved in the guideline and sides of the contours to be executed, concentrated deposits of soil, concentrated deposits of gravel, sand deposits and deposits of stonelike material.

The field work was conducted in the months of June, July and November, and the laboratory work were executed in the months of July, August, November and December 2010.

The field studies were reconciled with the Office so that the information they were consistent with the topographic study provided.

With the implementation of the Auger investigation, it was found that the tactile-visual characteristics soils similar to the predominance of yellow/red sandy soils.

2 Location Studies of Execution Area

The area of execution of study is located between the municipalities of Manga and Itacarambi in the state of Minas Gerais. In terms of organization, the area was divided into four sections, and was classified as follows:

Stretch 1: Segment between the municipalities of Manga and São João das Missões;
Stretch 2: Segment between the City of São João das Missões to the start of paved stretch, passing for the small city called Rancharia;
Stretch 3: Manga municipality contour;
Stretch 4: Outline of the City of São João das Missões;

Figures 1 and 2, from Google Earth, show the insertion area of the enterprise in question.

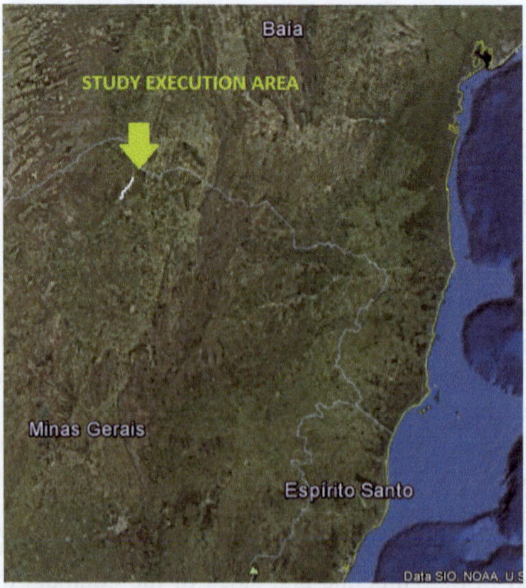

Fig. 1. Macro view of the studies area

Fig. 2. Macro view of studies running area (Excerpt: Manga - Itacarambi).

Figures 3 and 4, also from Google Earth, present, respectively, the outlines provided in the municipalities of Manga, and São João da Missões.

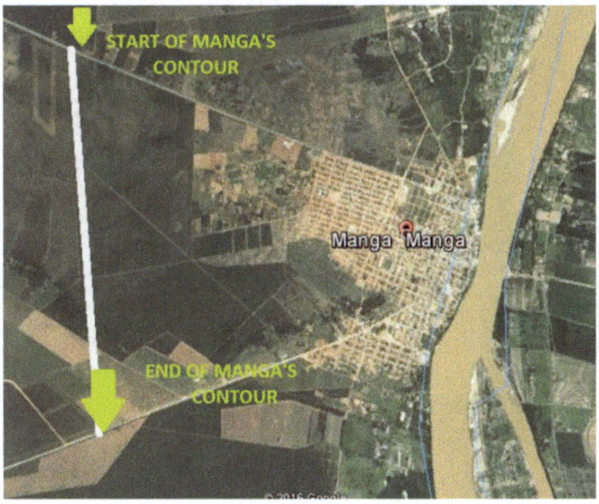

Fig. 3. View contouring Manga/MG. Home (approximate coordinates: Start 14° 44′04.6″ S, 043° 58′ 36.9″ W, and end 14 46′25.8″ S, 043° 58′ 19.1″ W).

Fig. 4. View the outline of St. John of the Missions. Home (approximate coordinates: Home 14° 52′15,875″ S, 044° 04′ 44,751″ W, and end 14 53′20.6″ S, 044° 04′ 39.3″ W).

3 General on the Region of Materials

In carrying out of field work was done a detailed recognition of the region of insertion of BR 135. In recognition was found that the region is covered by a wide layer of typical sandy soil (Red Latosol and Yellow Latosol), Figs. 5 and 6. Along the track

Fig. 5. Material of the primer coating and typical sandy soil found in the region.

Fig. 6. Start of Manga-São João das missões. Also observe the implementation of the survey auger.

axis and contours, has the typical profile shown in Figs. 7, 8, 9 and 10. Samples of the surface layers of these profiles were collected but not used in the implementation of laboratory tests. For soil concentrated study, we used an area near the municipality of São João da Missões.

In the case of rocky material, it was found that there is a high occurrence of carbonate rocks in the region. This material is exploited and used as building material, especially on pavements and as aggregates for concrete (Gravel 1, 2 and Gravel powder).

Fig. 7. Typical via banks. Manga-São João das Missões. (GPS 14º 50′28.6″ S, 44º 03′ 12″ W)

Fig. 8. Excerpt from Manga-São João das Missões. Observe typical rocky material in the surface layer.

Note: It is verified that along the way there is the existence of competent rocky material and also friable, mixed, which are not local, but come from deposits in the region and which were used in the primary coating of the road.

Aiming the possible deployment of a bridge on the outline provided in São João das

Fig. 9. Excerpt from Manga-São João das Missões. Detail of the typical rocky material in the surface layer.

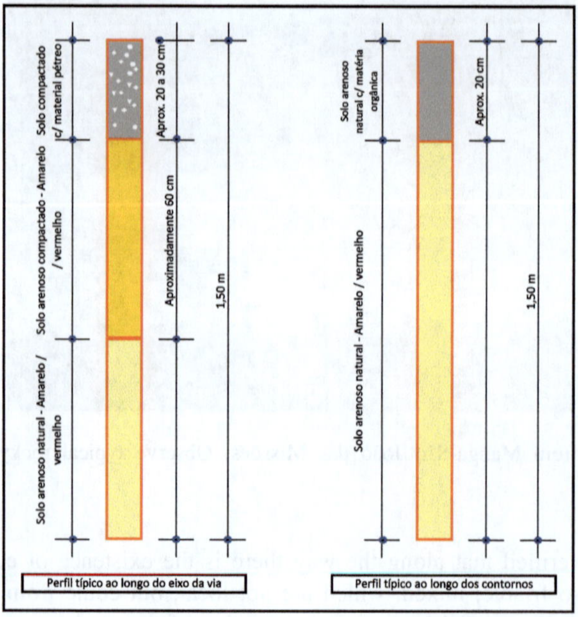

Fig. 10. Typical profiles prevailing in the local collection of soil samples

Missions, mixed investigations were performed (SPT + Rotating), Figs. 11 and 12.

4 Results of Studies Performed

Fig. 11. Implementation of the SPT survey on the river Itacarambi possible to bridge the implementation outline of São Joao Mission.

Fig. 12. Execution of the rotary drilling on the river Itacarambi possible to bridge the implementation outline of São Joao Mission.

As already mentioned, the study consisted in the execution of field work and laboratory. In the field site investigations were performed (auger and rotary SPT), collecting material (soil and rock), densities "in situ" and identification/material classification. In the laboratory, with soils and rocks collected, the following tests were performed: (a) moisture content; (b) Liquidity Limits and Plasticity; (c) Granulometri by sieving and sedimentation, using as a dispersant sodium hexametaphosphate; (d) real density of soil grains; (e) compression; (f) California Support Index (ISC or CBR); (g) content of organic matter; (h) Sand Equivalency; (i) Abrasion "Los Angeles"; (j) Adhesion; (k) Durability; (l) index of form; (m) Material Classification; (n) Analysis of SPT; (o) Analysis of Rotating SPT. The following part summaries are presented the results of studies performed.

4.1 Soil Collected in Axis of Projected via

Figure 13 shows a summary of the average of the physical and mechanical characteristics of the soil along the Via to be paved. Through this figure, has the follow results: in granulometric terms the predominant soil is sandy; not liquid and not plastic; Sand equivalency equal to 33%; Group Index equal to zero; The predominant soil-2-4

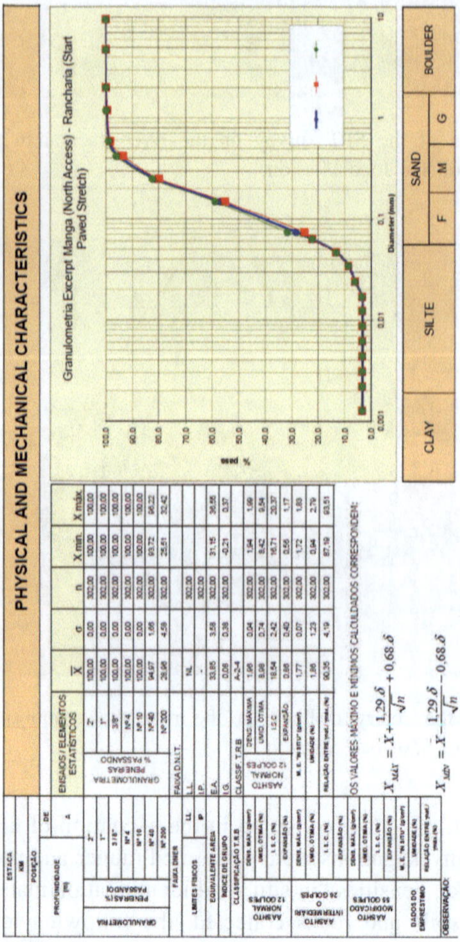

Fig. 13. Summary and physical characteristics and soil mechanics for kids stretch (North)

(gravel or silty or clayey sands) the classification TRB (HRB); optimum moisture content in the normal energy equal to 8%; ISC (CBR) 18%; Expansion of the compacted material after 96 h of immersion of 0.86%; apparent density of 1,77 g/cm^3; natural moisture of 1.86%.

4.2 Deposits of Soil and Gravel

For areas of lending materials, we studied a concentrated deposit of soil and three deposits of gravel. The field soil has similar characteristics to that shown in Fig. 13. The soil collected in this field was also used to study soil-gravel mixture. Two Soil Gravel mixtures were studied. A mixture of 20% and another 30% Gravel, being respectively 80% and 70% of soil. Figures 14 and 15 show the results obtained for these mixtures (soil and gravel). In the case of compression it is found that the normal

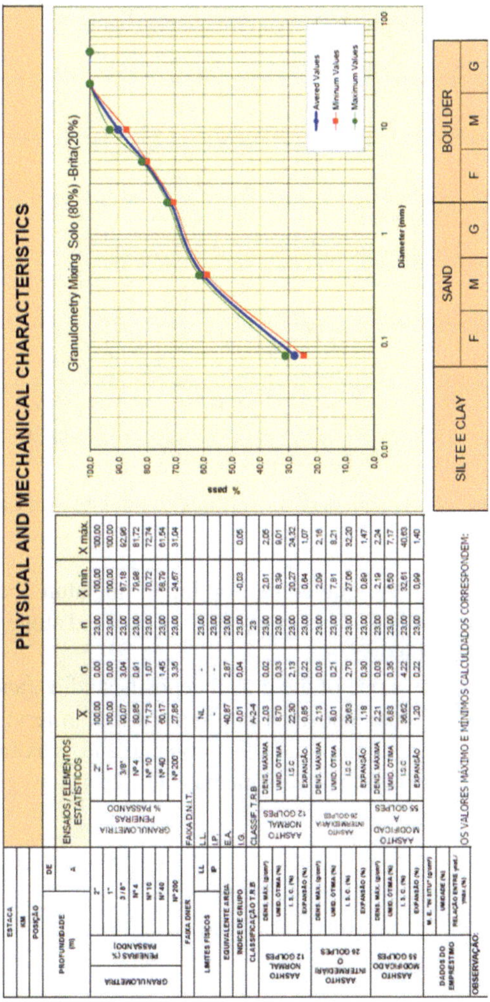

Fig. 14. Study summary for soil mix (80%) - Gravel (20%).

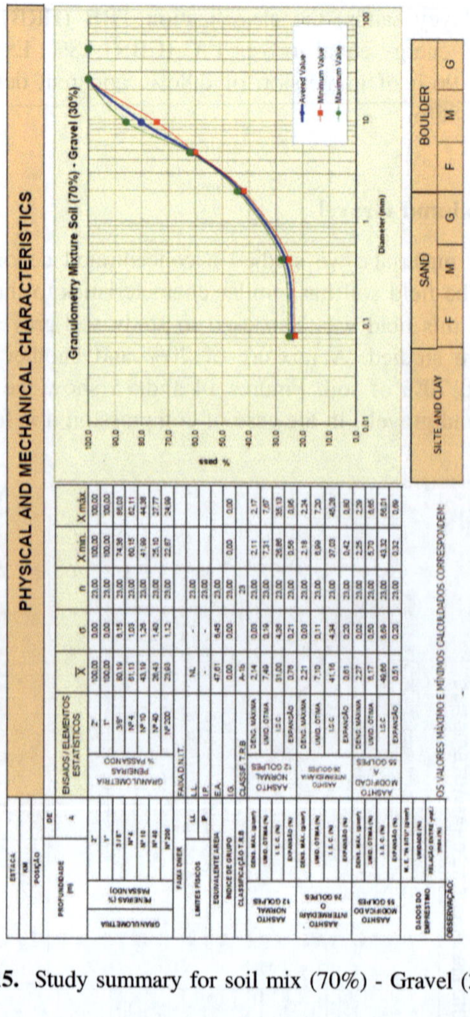

Fig. 15. Study summary for soil mix (70%) - Gravel (30%).

power sample with 20% dry gravel has optimum moisture content and density, respectively 2,03 g/cm^3 and 8.70%. Already 30% of crushed stone features 2.14 g/cm^3 and 7.49% of optimum moisture.

Next, in Fig. 16, is shows one of the gravel deposits studied, and Fig. 17 presents a summary of this study for this deposit. It is observed that in terms of compacting the material has medium for optimum moisture content and maximum dry density in normal power (13.2%; 1.90 g/cm^3), intermediate (12.7%; 1.96 g/cm^3) and modified (12.0%; 2.01 g/cm^3).

Fig. 16. View of the gravel deposits studied.

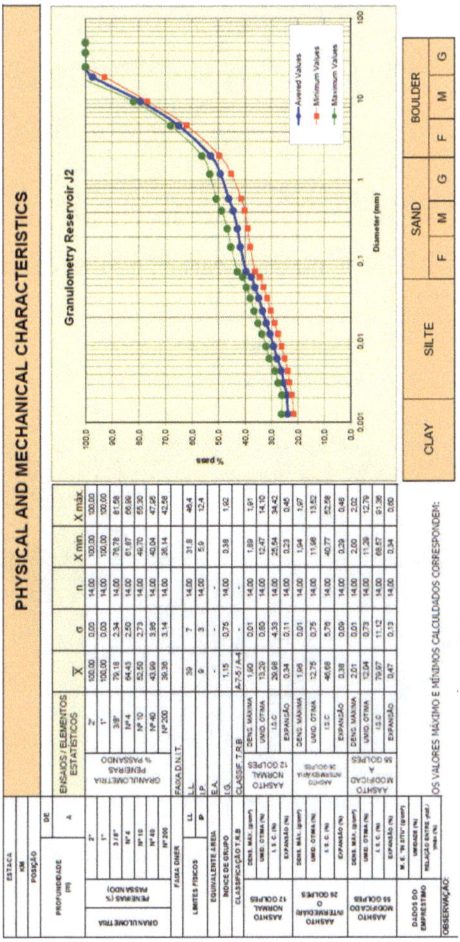

Fig. 17. Study summary for the gravel mine.

4.3 Sands Deposits

During the check of sands deposits, local people reported that most of the sand used in the region comes from the dredging of the river San Francisco in the municipality of Itacarambi. On the way to the site (Fig. 18) and in conversations, it was found that are marketed three kinds of sand called by traders "fine sand", "pebble" and "sand with pebble", as shown in Fig. 19. In addition to these in two locations, sands were collected in Manga municipality, also in the margins of the São Francisco river.

Fig. 18. Sand dredgers area view in Itacarambi/MG on the banks of the São Francisco River (GPS 15° 06′18.7″ S, 44° 05′ 41.1″ W).

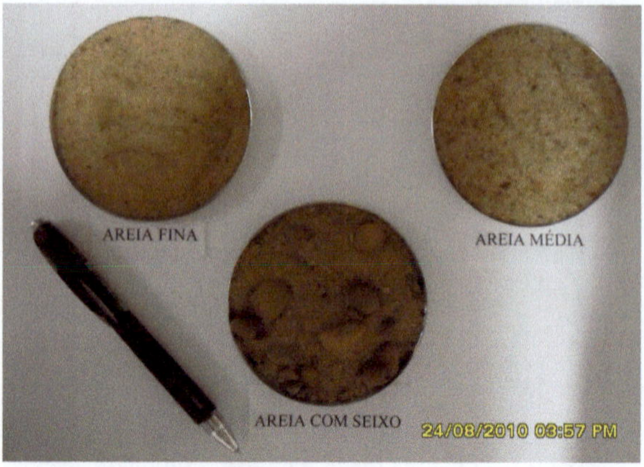

Fig. 19. Sands collected on the banks of the São Francisco River in Itacarambi/MG.

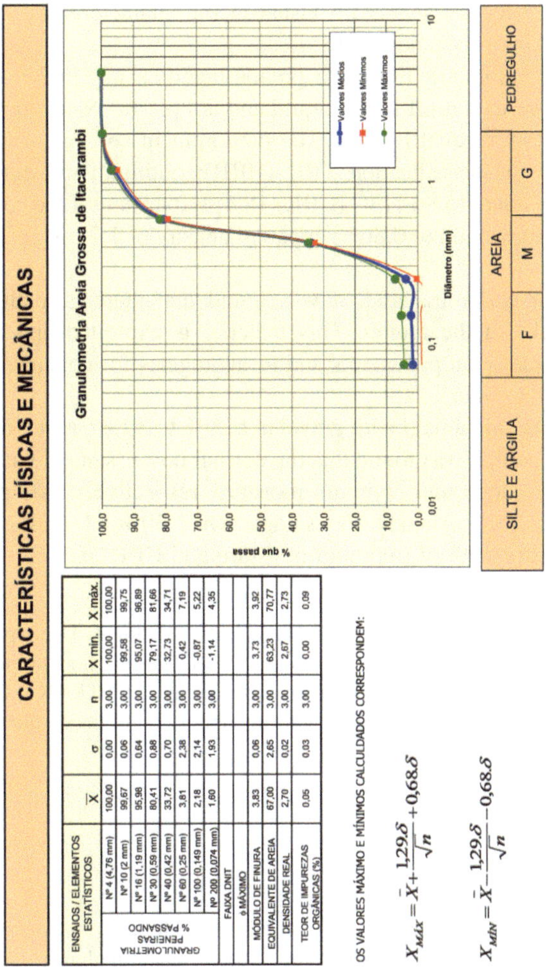

Fig. 20. Study summary for the dredged coarse sand of the São Francisco River in the municipality of Itacarambi.

Figure 20 shows the summary of the study to the coarse sand dredged from the São Francisco River in the municipality of Itacarambi. It is observed that the sand in terms of the equivalent material has 67% with low organic impurities.

5 Analysis and Conclusions

This article aims to present some of the results of geotechnical studies carried out in the area and materials collected in the BR 135 in the section between the municipalities of Manga and Itacarambi in the state of Minas Gerais. With support for the implementation of field work and laboratory and analyzes, we conclude that:

(a) the area is covered by soils with visual tactile characteristics similar to the predominance of latosols sandy yellow/red;
(b) in terms of granulometric the soil is predominantly sandy in the center of the track to be paved, presents itself as not liquid and not plastic; Sand equivalency equal to 33%; Group Index equal to zero; The predominant soil A-2-4 (gravel or silty or clayey sands) the classification TRB (HRB); optimum moisture content in the normal energy equal to 8%; ISC (CBR) 18%; Expansion of the compacted material after 96 h of immersion of 0.86%; apparent density of 1,77 g/cm^3; natural moisture of 1.86%;
(c) In the case of rocky material, it was found that there is a high occurrence of carbonate rocks in the region. This material is exploited and used as building material, especially on pavements and as aggregates for concrete (Gravel 1, 2 and Gravel Gravel powder);
(d) by mixing the soil in a field with gravel in a 20% to 30% Gravel, being respectively 80% and 70% soil, it was found that the normal power sample with 20% gravel has a specific mass dried and optimum moisture, respectively 2,03 g/cm^3 and 8.70% with 22.30% CBR. Already the sample with 30% of crushed stone features 2.14 g/cm^3 and 7.49% of optimum moisture and CBR of 30%.

Acknowledgments. The Reforsolo Engenharia Ltda, National Department of Transport Infrastructure (DNIT), Department of Engineering and Construction of the Brazilian Army (DEC), Centre of Excellence in Transportation Engineering (CENTRAN.); Catholic University of Brasilia (UCB), IesPlan and UniCEUB with important contributions that made possible the realization of this work.

References

DNIT (DNER-ES 301/97): Pavimentação - sub-base estabilizada granulometricamente
DNIT (DNER-ES 303/97): Pavimentação - base estabilizada granulometricamente
MEDRADO, Wallen Alexandre. Caracterização Geotécnica de Solo da Região Norte de Minas Gerais para Aplicação em Obras Rodoviárias. Dissertação de Mestrado. UFOP (2009)
NORMAS Diversas ABNT-NBR/DNIT para ensaios em laboratório
NORMAS Diversas ABNT-NBR/DNIT para trabalhos de campo

Geotechnical Investigations and Proposal for Stabilization and Recovery Slope on Highway BR-010 Stretch in Aparecida do Rio Negro to Goiatins, Tocantins, Brazil

Rideci Farias[1], Haroldo Paranhos[1], Jovino Rachid Araújo[2], Leonardo Ramanho Sales[3], and Roberto Pimentel[4](✉)

[1] UCB, Reforsolo Engenharia, UniCEUB, IesPlan, Brasília, Brazil
rideci.reforsolo@gmail.com, reforsolo@gmail.com
[2] CMT Engehharia, Brasília, Brazil
jovino@cmtengenharia.com.br
[3] UniCEUB, Brasília, Brazil
leonardoramalhosales@gmail.com
[4] UnB, Brasília, Brazil
eng.robertopimentel@gmail.com

Abstract. This article presents the results of geotechnical studies and proposals to stabilize embankments referring to cuts 16, 23, 25, 30 and 31 of the BR-010 referring to Aparecida do Rio Negro a Goiatins, State of Tocantins. The study consisted in the execution of a field incursions program for verification of geotechnical conditions along the indicated stretch, running lab tests, and stability analysis of embankments in order to indicate solutions to cuts made necessary for the completion of the highway. In terms of characterization of the problem, it presents part of the photographic survey conducted in November 2007 with the conditions of the most critical slopes. There is in the pictures that the side slopes to the highway were in the process of degradation and (or) progressive destabilization requiring urgent interventions aimed at mitigating the ongoing problems. In terms of geometry, the slope had heights up to approximately 35 m. To determine the parameters necessary to verify the stability of embankments using three undeformed blocks corresponding soil to cuts 25 and 31. Samples of identifications are presented and the summary of the results of geotechnical tests performed and certain parameters. In laboratory tests, it was found that the strength parameters (cohesion and angle of friction) diverged somewhat from the usual laboratory parameters obtained in soils. This behavior was due to the fact that the heterogeneity of the samples for the presence mainly of discontinuities in the collected blocks. Before the run stability analysis and field surveys, it was found, the need for actions aimed at solving problems for marginal slope to the highway. Thus, due to the proximity of the intense rainy season, it was proposed the reconstruction of slope with surface protection of them aimed at stabilization and protection against erosion, as well as the execution of superficial and deep drainage. Aiming to further verify the possible degradation of the slopes by rain, was sought in INMET's website (National Institute of Meteorology) precipitation to the region. It was found that for the time in January, February and March rainfall is considerably

W. Frikha et al. (eds.), *Soil Testing, Soil Stability and Ground Improvement*,
Sustainable Civil Infrastructures, DOI 10.1007/978-3-319-61902-6_32

high, which could further compromise the integrity of the executed cuts. Faced with this possibility, he indicated consider the emergency actions that could prevent this degradation. Actions such as the reduction and protection of embankments shown in the study.

1 Introduction

This article presents the results of geotechnical studies and proposals to stabilize embankments referring to cuts 16, 23, 25, 30 and 31 of the BR-010 referring to the Black River Aparecida excerpt Goiatins, State of Tocantins, made in the year 2007. The study consisted in the execution of a field incursions program for verification of geotechnical conditions along the indicated section, as well as the execution of laboratory tests, and stability analysis of embankments in order to indicate the solutions made cuts necessary to completion of the highway.

2 Problem Characteristics

In terms of characterization of the problem proceeded to the assessment and field analysis accompanied by an extensive survey photographic held in November 2007 with the conditions of the slopes 16, 23, 25, 30 and 31. It can be seen in Figs. 1, 2, 3, 4, 5, 6, 7, 8, 9, 10, 11, 12 and 13 which the side slope to the highway, in some cases, were in the process of degradation and (or) progressive destabilization requiring urgent interventions aimed at mitigating the ongoing problems. In terms of geometry, the slope presented heights up to about 35 m, as evidenced in the analyzed sections and shown in this article.

Fig. 1. General view of part of slope 16.

Fig. 2. General view of part of cut 16 with superficial erosion.

Fig. 3. General view of part of the cut 16 with superficial erosion.

Fig. 4. View of the slope of cut 23.

Fig. 5. Cut 25, still intact (September 2007).

Fig. 6. Slope of cut 25 starting degradation process of slope.

Fig. 7. Material of lower horizon of cut 25.

Fig. 8. Traction cracks in the berm slope of cut 25.

Fig. 9. Traction cracks on berm slope of cut 25.

Fig. 10. Saturation from the slope basis in cut 25.

Fig. 11. Upwelling water in cut 25 slope.

Fig. 12. Failure stretch in cut 25 slope.

Fig. 13. Cut 31 (slope basis saturation).

3 Geotechnical Testes Performed

To determine the parameters necessary to verify the stability of slopes, we used three undisturbed samples of soil corresponding to Sections 25 and 31. Table 1 shows the sample to the Court 25, but also the summary of the results of the conducted geotechnical testing and certain parameters. The three undisturbed soil samples were used for executing the natural moisture of assays, bulk density and direct shear to define the strength parameters. The tests were performed in accordance with relevant standards and (or) the usual procedures used in geotechnical laboratory. For direct shear tests (ASTM D3080/D3080M: 2011), the bodies of the test piece were tested in the flooded condition in order to verify possible critical mechanical behavior of soil.

Table 1. Summary of geotechnical tests realized and determined parameters for sample of cut 25.

CUT 25 - Stake 1410 (slope right side) - Level 310,394 m		
Specific weight natural	γ_{nat} (kN/m^3)	22,17
Natural humidity	w (%)	9,38
Effective cohesion (Top)	C′ (kPa)	78 (Condição Inundada)
Effective angle of friction (Pico)	ϕ′ (graus)	19 (Condição Inundada)
Effective cohesion (Residual)	C (kPa)	27 (Condição Inundada)
Effective friction angle (Residual)	ϕ′ (graus)	15 (Condição Inundada)
Specific gravity apparent dry (kN/m^3)	γ_d (kN/m^3)	20,27
Specific weight of solids (or grains)	γ_s (g/cm^3)	2,81
Specific weight saturated (kN/m^3)	γ_{sat} (kN/m^3)	22,86
Void index	_	0,36
Porosity	n (%)	26,43
Saturation level (%)	Sr (%)	73,38

4 Comments About Soil Samples Collected

It appears from Table 1, the example of other tests performed for other samples not shown in this article, the strength parameters (cohesion and friction angle) diverge somewhat from the usual parameters obtained for laboratory soil. This behavior was due to the fact that the heterogeneity of the sample by the presence mainly of discontinuities in the collected blocks.

Such discontinuities were also found in the field when the three surveys conducted in the months of September, November and December 2007, to evaluate the cuts run along the highway.

These discontinuities contributed directly to the technical difficulties encountered by the Geotechnical Laboratory when the molding of the samples for the appropriate tests. Such difficulties caused repetitions of several tests in which there was no convergence results.

5 Slope Stability Analysis Performed

Due to the difficulties, both in the collection of the blocks as the molding of the test bodies, obtaining a few parameters for ordinary soils of the North Central region of Brazil, it was performed several analyzes of stability of embankments through computer program SLOPE/W, version 2007, marketed by Geo-slope International. These analyzes understood the use of the parameters obtained in the laboratory, as well as simulations with more realistic parameters that represented stability with safety factor close to 1 (one) to the slope of the Court 25, in compliance with consistent conditions of embankments.

After several tests in order to find more realistic parameters for soil that made up the slopes of the cuts 25 and 31, corrected to the upper and lower horizon soil obtaining the values shown in Tables 2 and 3 (horizon) and 6 (lower horizon).

Table 2. Strength parameters fixed for upper horizont soil.

Parametrosde Resistencia para o Horizonte Superior		
Coesão Efetiva	C (kPa)	14
Ângulo de atrito efetivo	ϕ′ (graus)	26

Table 3. Strength parameters fixed for below horizont of soil.

Parametros de Resistencia para o Horizonte Superior		
Coesão Efetiva	C (kPa)	44
Ângulo de atrito efetivo	ϕ′ (graus)	26

For the cohesion values, remained the value of 14 kPa found for the upper horizon, and it is a high cohesion value for characteristic soils of the region, decreased to 78 kPa to 44 kPa related to the lower horizon. The value of 44 kPa due to the fact that several stability analyzes with a view to the compatibility of what was happening on the field. It is surprising that value because the soil had low cohesion when desconfinado, but when confined cohesion tended to be higher. For the case in question on which the slope face exposure the soil tended to expand with the consequent breakdown necessitating thus of at least surface protection of the slope.

It should be noted that the field located various types of breaks were observed, highlighting the planar, circular and small spalling. Within this complexity, it broke up in most cases the reproduction of circular breaks that are quite common in soils. Simplification adopted in part of the study to circulate break comes the already mentioned complexity in terms of heterogeneity and the difficulties encountered, both in the selection of collecting the material and the molding of the samples in order to obtain parameters that could represent the field of similarity. Within the subject matter, for the analysis and conclusions as to the positioning in the appropriate interventions, aiming stable geometries, one should consider - in addition to computational studies with different characteristics, but also the importance of surveys in situ by professionals who are able to position solutions address the various problems that may be identified.

In terms of stability analyzes were done for several cuts 16, 23, 25, 30 and 31 represent, in general, we found the slopes along the portions studied. Below, we present the studies to the slope of the cut 25 because it is one of the most critical checked in the field. Initially shows the stabilities of the cut 25 for the parameters obtained in the

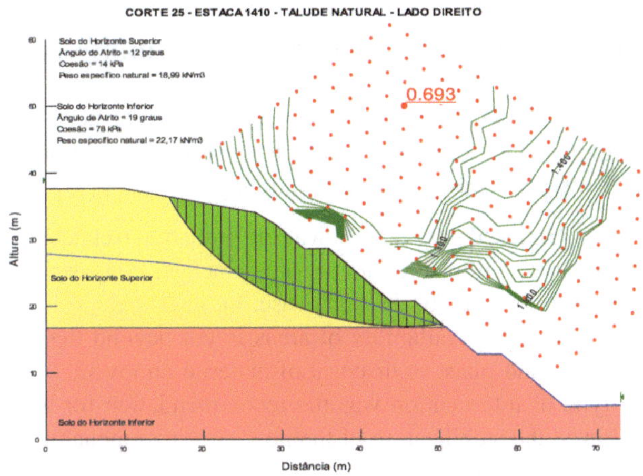

Fig. 14. Cutting 25 stability with lab parameters (FS = 0.693) – Stake 1410 (natural slope – right side).

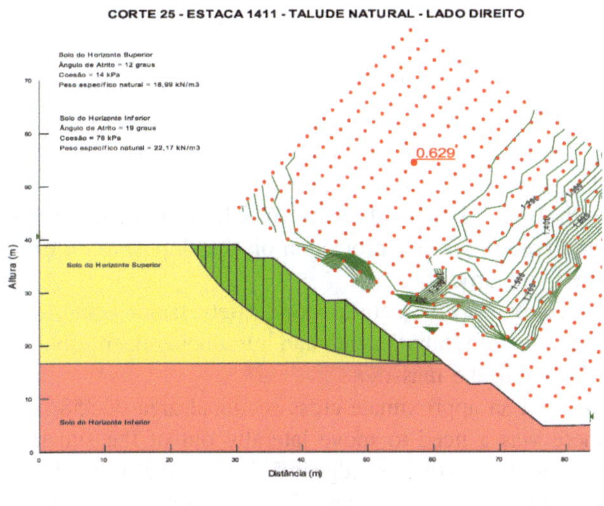

Fig. 15. Cutting 25 stability with lab parameters (FS = 0.629) – Stake 1411 (natural slope – right side).

laboratory just to check, since the obtaining of strength parameters has been compromised, as mentioned above. Subsequently, we present the analysis with more realistic parameters in compliance with those found soils and conditions prevailing in the field.

5.1 Stability with Parameters Obtained in Lab

Next, in Figs. 14 and 15, the stability analyzes are presented, with the strength parameters obtained in our laboratory for the cuts 25 on cuttings 1410 and 1411.

6 Considerations About Performed Slope Stability Analysis

Before the run stability analysis and field surveys, there was the need for actions aimed at solving the problems that the marginal slope to the highway had presented. For this, it was indicated two types of interventions following:

(a) The first type of proposed intervention was the reduction (retaludamento) of slopes to stable geometries within the engineering practice, considering the economy, environment, insertion area of the enterprise, in other factors. In this case, it recommended to check the availability of areas to cut beyond the highway right of way, if necessary, and place of disposal of material removed;
(b) The second type of intervention was the recommendation for adoption of structures that allowed the stabilization of the now existing embankments. Within this practice there are several solutions such as soil nailing, bolting, root cuttings, etc.

For the case in question, due to the proximity to the most intense rainy season, there was the need to intervene as soon as possible. Within the time that was available, then recommended him retaludamento sections with the execution of deep drainage and surface. For deep drainage usually use the DHP - Drain subhorizontal Deep and surface drainage is commonly contemplated by the vestment of drains and channels.

For simulations of retaludamento proceeded to several stability analysis in the studied slope. By way of example, is presented in Figs. 16 and 17 the analysis of stability for the cut 25 in the most critical Cutting (Cutting 1411). For these simulations, broke up the natural sections with the execution of cuts for defining stable geometries within the practice of engineering. After several simulations came to embankments with 1 V: 1.5 H for the cuts 16, 23, 30 and 31. This primarily by the need to install any type of surface protection on the slopes, as an example geocell in order to prevent erosive processes.

For cutting 25, broke up the natural slope safety factor equal to 1.007 (Fig. 16), with the implementation of retaludamento and introducing deep subhorizontal drains to relegation in N.A. within the massif.

Initially, simulated an approximate cross-sectional area of 358 m^2 Stake 1411. In this situation, there was a need to move laterally out of the situation of the road - counting from the bottom up - the first edge in 5 m, second at 10 m, the third in 15 m, the fourth at 18.5 m and fifth in 20 m. Figure 17 shows such a situation where we got the 1,118 Safety factor that represents a low value for the stability of the slope in question.

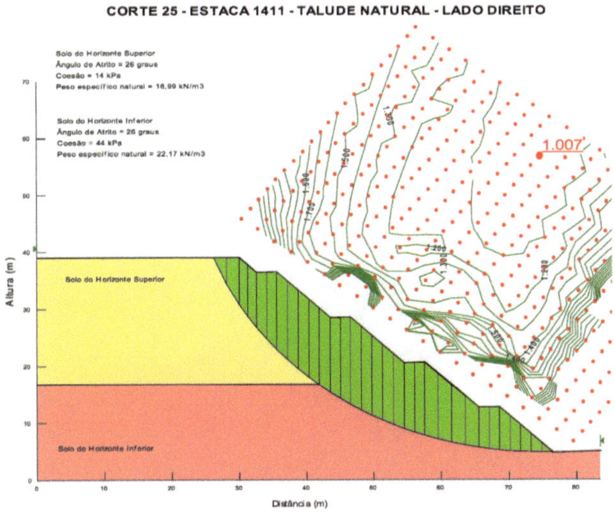

Fig. 16. Cutting 25 Stability with fixed parameters – Estaca 1411 (natural slope – right side).

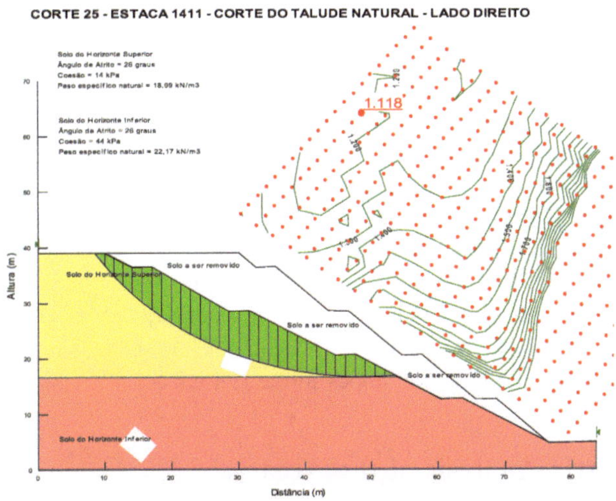

Fig. 17. Cutting 25 stability with proposal remotion of soil – Stake 1411 (natural slope – right side).

In addition to retaludamento, simulated to the lowering of N.A. inside slope which represents the introduction of the sub horizontal drains for the removal of excess water. In this situation it obtained the Safety Factor of 1.488, considered satisfactory for the type of work in question. Figure 18 shows the situation analyzed.

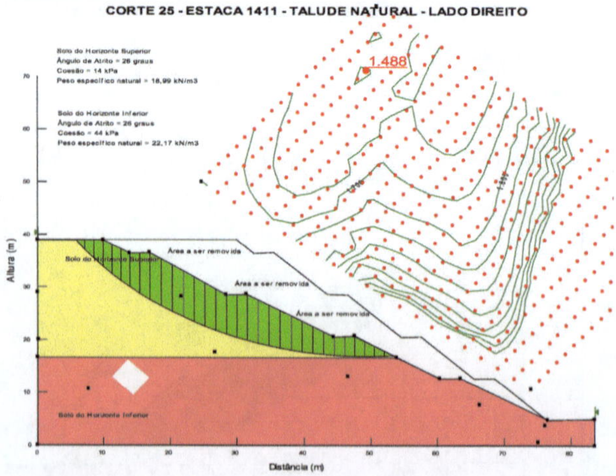

Fig. 18. Cutting 25 Stability – Stake 1411 (natural slopes – right side) – Low NA (introdution of deep subhorizontal drain).

7 Recommendations for Stabilization and Recovery of the Studied Slopes

Before the run stability analysis and field surveys, it was found, as already mentioned, the need for actions aimed at solving the problems that the marginal slope to the highway had presented. Thus, due to the proximity of the intense rainy season approaching, it proposes the retaludamento Slashes 16, 23, 25, 30 and 31, with surface protection of slopes aimed at stabilization and protection against erosion that occurred and came to occur, as well as the execution of superficial and deep drainage.

Another recommendation made was the draining mattress running on the highway over the cuts implemented in order to prevent possible upwelling of water that could compromise the surface layers of the highway, as the sub-base, base and coating, that the saturation finding the bases of slopes in the run cuts.

8 Pluviometric Analysis of the Area

In order to verify the possible degradation of the slopes by rain, was sought in INMET's website (National Institute of Meteorology) precipitation for the region in previous years, and the period of the executed cuts. It was found in Figs. 19, 20, 21, 22, 23 and 24 in the months of January, February and March there is heavy rainfall, which could further compromise the integrity of the executed cuts. Faced with this possibility, it had to consider the urgency of actions that could prevent degradation. Actions such as the reduction and protection of embankments listed in this article and the complete work developed.

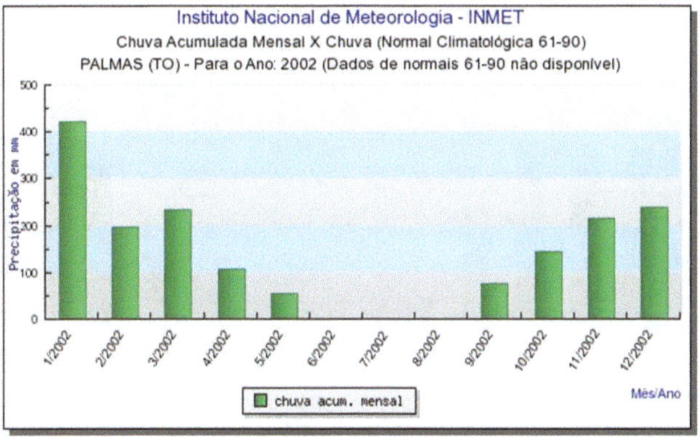

Fig. 19. Accumulated rainfall of 2002.

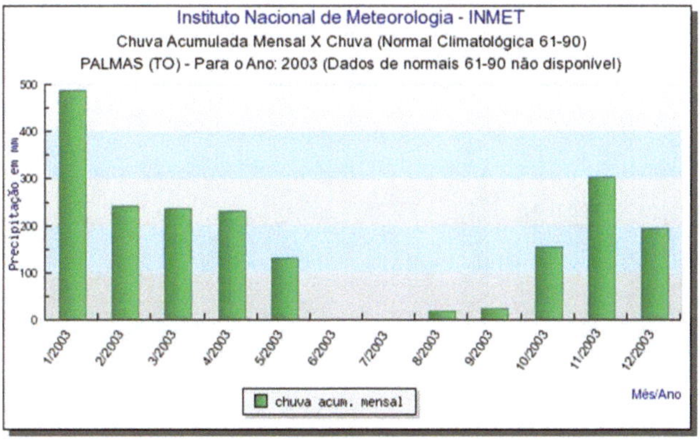

Fig. 20. Accumulated rainfall of 2003.

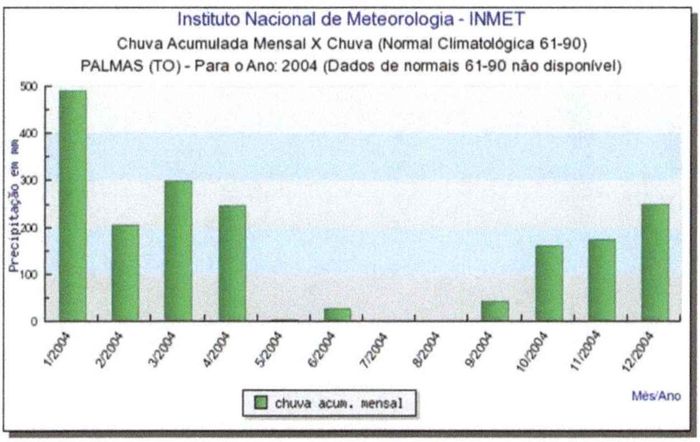

Fig. 21. Accumulated rainfall of 2004.

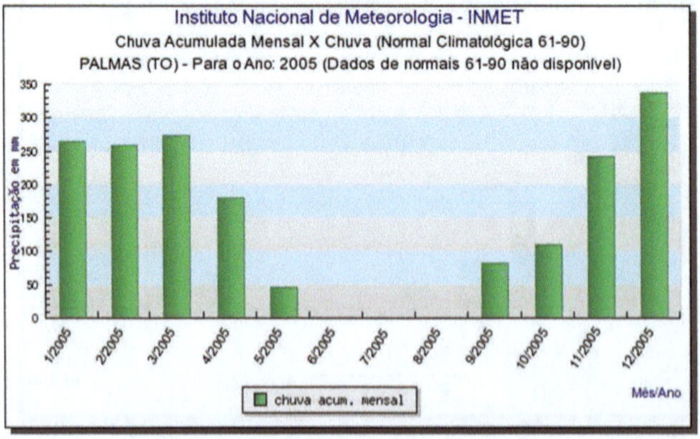

Fig. 22. Accumulated rainfall of 2005.

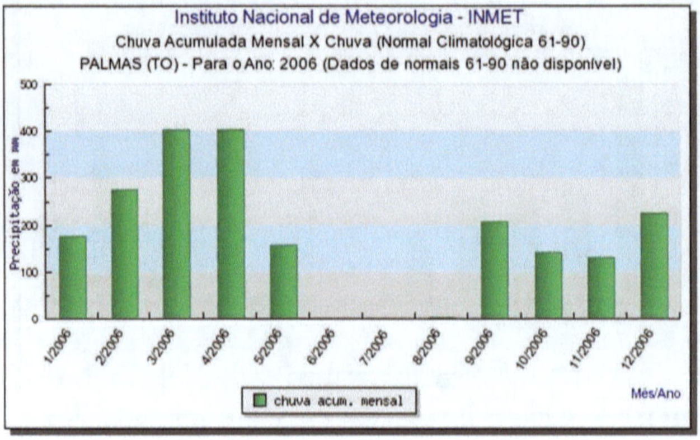

Fig. 23. Accumulated rainfall of 2006.

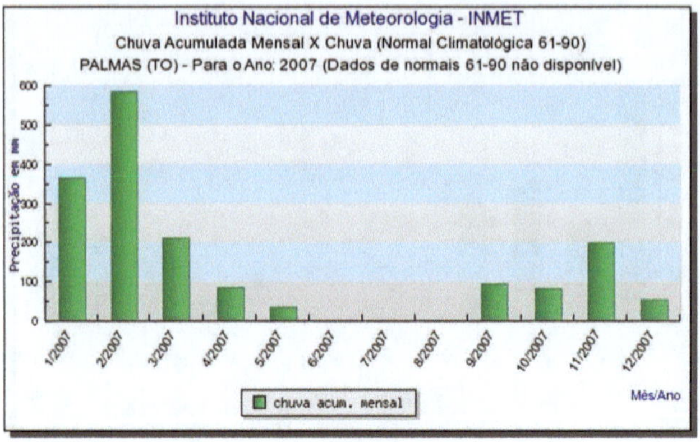

Fig. 24. Accumulated rainfall of 2007.

Acknowledgments. The Reforsolo Engenharia Ltd., CMT Engineering, Catholic University of Brasilia (UCB), University Center of Brasilia (UniCEUB) and the Institute of Higher Education Plateau with important contributions that made possible the realization of this work.

References

ABNT/ NBR 7181: Solo – Análise granulométrica

ABNT/NBR 6457: Solo – Amostras de solo – Preparação para ensaios de compactação e ensaios de caracterização

ASTM D3080/D3080M:2011 - Standard Test Method for Direct Shear Test of Soils Under Consolidated Drained Conditions

Geo-slope international: Stability modeling with SLOPE/W, an engineering methodology (2007a)

Geo-slope international: Stress deformation modeling with SIGMA/W, an engineering methodology (2007b)

INMET (Instituto Nacional de Meteorologia). http://www.inmet.gov.br/portal/

Author Index

W. Frikha et al. (eds.), *Soil Testing, Soil Stability and Ground Improvement*,
Sustainable Civil Infrastructures, DOI 10.1007/978-3-319-61902-6

MIX
Papier aus verantwortungsvollen Quellen
Paper from responsible sources
FSC® C105338

If you have any concerns about our products,
you can contact us on
ProductSafety@springernature.com

In case Publisher is established outside the EU,
the EU authorized representative is:
Springer Nature Customer Service Center GmbH
Europaplatz 3, 69115 Heidelberg, Germany

Printed by Libri Plureos GmbH
in Hamburg, Germany